# Condensed Matter Physics: Concepts and Applications

# Condensed Matter Physics: Concepts and Applications

Editor: Raymond Stevens

NY RESEARCH PRESS
New York

Published by NY Research Press
118-35 Queens Blvd., Suite 400,
Forest Hills, NY 11375, USA
www.nyresearchpress.com

Condensed Matter Physics: Concepts and Applications
Edited by Raymond Stevens



International Standard Book Number: 978-1-63238-661-8 (Hardback)



**Cataloging-in-Publication Data**

Condensed matter physics : concepts and applications / edited by Raymond Stevens.
p. cm.
Includes bibliographical references and index.
ISBN 978-1-63238-661-8
1. Condensed matter. 2. Physics. I. Stevens, Raymond.
QC173.454 .C66 2019
530.41--dc23

# Contents

# Preface

In my initial years as a student, I used to run to the library at every possible instance to grab a book and learn something new. Books were my primary source of knowledge and I would not have come such a long way without all that I learnt from them. Thus, when I was approached to edit this book; I became understandably nostalgic. It was an absolute honor to be considered worthy of guiding the current generation as well as those to come. I put all my knowledge and hard work into making this book most beneficial for its readers.

The branch of physics concerned with the microscopic and macroscopic properties of matter is known as condensed matter physics. It specifically studies condensed phases, which have a high population of constituents, and interact strongly with each other. It builds on the principles of quantum mechanics, electromagnetism and statistical mechanics to study all condensed phases of matter. Modern theoretical condensed matter physics involves the application of numerical computation and mathematical tools to understand high-temperature superconductivity, gauge symmetries and topological phases. The aim of this book is to present researches that have transformed this discipline and aided its advancement. It presents the complex subject of condensed matter physics in the most comprehensible and easy to understand language. The readers would gain knowledge that would broaden their perspective about this field.

I wish to thank my publisher for supporting me at every step. I would also like to thank all the authors who have contributed their researches in this book. I hope this book will be a valuable contribution to the progress of the field.

**Editor**

# 1

# Magnetocaloric effect in the spin-1/2 Ising-Heisenberg diamond chain with the four-spin interaction

L. Gálisová

Department of Applied Mathematics and Informatics, Faculty of Mechanical Engineering, Technical University in Košice, Letná 9, 042 00 Košice, Slovak Republic

The magnetocaloric effect in the symmetric spin-1/2 Ising–Heisenberg diamond chain with the Ising four-spin interaction is investigated using the generalized decoration-iteration mapping transformation and the transfer-matrix technique. The entropy and the Grüneisen parameter, which closely relate to the magnetocaloric effect, are exactly calculated to compare the capability of the system to cool in the vicinity of different field-induced ground-state phase transitions during the adiabatic demagnetization.

**Key words:** *Ising–Heisenberg diamond chain, four-spin interaction, phase diagram, magnetocaloric effect*

## 1. Introduction

The magnetocaloric effect (MCE), which is characterized by an adiabatic change in temperature (or an isothermal change in entropy) arising from the application of the external magnetic field, has been known for more than a hundred years [1]. This interesting phenomenon has also got a long history in the cooling applications at various temperature regimes. The first successful experiment of the adiabatic demagnetization, which was used to achieve the temperatures below 1K with the help of paramagnetic salts, was performed in 1933 [2]. Nowadays, the MCE is a standard technique for achieving the extremely low temperatures [3].

It should be noted that the MCE in quantum spin systems has again attracted much attention of researchers. Indeed, various one- and two-dimensional spin systems have recently been exactly numerically investigated in this context [4–19]. The main features of the MCE which have been observed during the examination of various spin models include: an enhancement of the MCE owing to the geometric frustration, an enhancement of the MCE in the vicinity of quantum critical points, the appearance of a sequence of cooling and heating stages during the adiabatic demagnetization in spin systems with several magnetically ordered ground states, as well as a possible application of the MCE data for the investigation of critical properties of the system at hand.

In this paper, we investigate the MCE in a symmetric spin-1/2 Ising–Heisenberg diamond chain with the Ising four-spin interaction, which is exactly solvable by combining the generalized decoration-iteration mapping transformation [20–22] and the transfer-matrix technique [23, 24]. As has been shown in our previous investigations [25], the considered diamond chain has a rather complex ground state, which predicts the appearance of a sequence of cooling and heating stages in the system during adiabatic demagnetization. The main aim of this work is to compare the adiabatic cooling rate of the system (an enhancement of the MCE) near different field-induced ground-state phase transitions. Bearing in mind this motivation, we investigate the entropy and the Grüneisen parameter during the adiabatic demagnetization process, as well as the isentropes in the $H-T$ plane.

The paper is organized as follows. In section 2, we first briefly present the basic steps of an exact analytical treatment of the symmetric spin-1/2 Ising–Heisenberg diamond chain with the Ising four-spin interaction. Exact calculations of the quantities related to the MCE, such as the entropy and the Grüneisen parameter, are also realized in this section. In section 3, we briefly recall the ground state of the system, and then the most interesting results for the entropy as a function of the external magnetic field, the isentropes in the $H-T$ plane and the adiabatic cooling rate of the system versus the applied magnetic field are also presented here. Finally, some concluding remarks are drawn in section 4.

## 2. Model and its exact solution

Let us consider a one-dimensional lattice of $N$ inter-connected diamonds in the external magnetic field, which is defined by the Hamiltonian (see figure 1)

$$\begin{aligned}\hat{\mathscr{H}} = \sum_{k=1}^{N} \Big[ & J_{\mathrm{H}}\Delta\left(\hat{S}^x_{3k-1}\hat{S}^x_{3k}+\hat{S}^y_{3k-1}\hat{S}^y_{3k}\right)+J_{\mathrm{H}}\hat{S}^z_{3k-1}\hat{S}^z_{3k}+J_{\mathrm{I}}\left(\hat{S}^z_{3k-1}+\hat{S}^z_{3k}\right)\left(\hat{\sigma}^z_{3k-2}+\hat{\sigma}^z_{3k+1}\right) \\ & +K\hat{S}^z_{3k-1}\hat{S}^z_{3k}\hat{\sigma}^z_{3k-2}\hat{\sigma}^z_{3k+1}-H\left(\hat{S}^z_{3k-1}+\hat{S}^z_{3k}\right)-H\left(\hat{\sigma}^z_{3k-2}+\hat{\sigma}^z_{3k+1}\right)/2\Big]. \end{aligned} \quad (2.1)$$

Here, the spin variables $\hat{S}^{\gamma}_k$ ($\gamma = x, y, z$) and $\hat{\sigma}^z_k$ denote spatial components of the spin-1/2 operators, the parameter $J_{\mathrm{H}}$ stands for the XXZ Heisenberg interaction between the nearest-neighbouring Heisenberg spins and $\Delta$ is an exchange anisotropy in this interaction. The parameter $J_{\mathrm{I}}$ denotes the Ising interaction between the Heisenberg spins and their nearest Ising neighbours, while the parameter $K$ describes the Ising four-spin interaction between both Heisenberg spins and two Ising spins of the diamond-shaped unit. Finally, the last two terms determine the magnetostatic Zeeman's energy of the Ising and Heisenberg spins placed in an external magnetic field $H$ oriented along the $z$-axis.

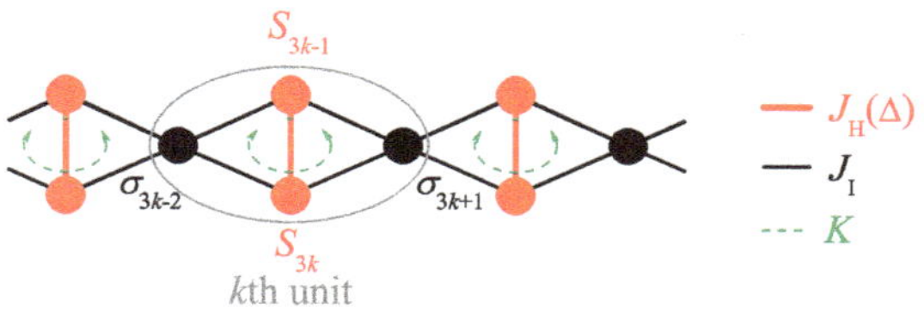


**Figure 1.** (Color online) A part of the symmetric Ising-Heisenberg diamond chain with the four-spin interaction. The black (red) circles denote lattice positions of the Ising (Heisenberg) spins. The ellipse demarcates spins belonging to the $k$th diamond unit.

It is worth mentioning that the considered quantum-classical model is exactly solvable within the framework of a generalized decoration-iteration mapping transformation [20–22] (for more computational details see our recent works [25] and [7]). As a result, one obtains a simple relation between the partition function $\mathcal{Z}$ of the investigated symmetric spin-1/2 Ising-Heisenberg diamond chain with the four-spin interaction and the partition function $\mathcal{Z}_{\mathrm{IC}}$ of the uniform spin-1/2 Ising linear chain with the nearest-neighbour coupling $R$ and the effective magnetic field $H_{\mathrm{IC}}$

$$\mathcal{Z}(T, J_{\mathrm{I}}, J_{\mathrm{H}}, K, \Delta, H, N) = A^N \mathcal{Z}_{\mathrm{IC}}(T, R, H_{\mathrm{IC}}, N). \quad (2.2)$$

The mapping parameters $A$, $R$ and $H_{\mathrm{IC}}$ emerging in (2.2) can be obtained from the "self-consistency" condition of the applied decoration-iteration transformation, and their explicit expressions are given by relations (4) in reference [7] with the modified $G$ function, which is given by equation (6) of reference [25]. It should be mention that the relationship (2.2) completes our exact calculation of the partition function because the partition function of the uniform spin-1/2 Ising chain is well known [23, 24].

At this stage, exact results for other thermodynamic quantities follow straightforwardly. Using the standard relations of thermodynamics and statistical physics, the Helmholtz free energy $\mathscr{F}$ of the symmetric spin-1/2 Ising-Heisenberg diamond chain with the four-spin interaction can be expressed through the Helmholtz free energy $\mathscr{F}_{\mathrm{IC}}$ of the uniform spin-1/2 Ising chain

$$\mathscr{F} = -T\ln\mathcal{Z} = \mathscr{F}_{\mathrm{IC}} - NT\ln A \quad (2.3)$$

(we set the Boltzmann's constant $k_B = 1$). Subsequently, the entropy of the investigated diamond chain can be calculated by differentiating the free energy (2.3) with respect to the temperature $T$. In our case, the resulting equation for the entropy behaves numerically better if the derivation is taken with respect to the inverse temperature $\beta = 1/T$

$$S = -\left(\frac{\partial \mathcal{F}}{\partial T}\right)_H = \beta^2 \left(\frac{\partial \mathcal{F}}{\partial \beta}\right)_H = \ln \mathcal{Z}_{\mathrm{IC}} + N \ln A - \beta \partial_\beta \ln \mathcal{Z}_{\mathrm{IC}} - N\beta \partial_\beta \ln A. \tag{2.4}$$

Here, the functions $\partial_\beta \ln \mathcal{Z}_{\mathrm{IC}}$ and $\partial_\beta \ln A$ satisfy in general the equations

$$\begin{aligned} \partial_x \ln \mathcal{Z}_{\mathrm{IC}} &= \frac{N}{2}\left[\frac{1}{2} + \frac{s^2 - Q^2}{Q(c+Q)} + \frac{s}{Q}\right]\partial_x \ln G_- + \frac{N}{2}\left[\frac{1}{2} + \frac{s^2 - Q^2}{Q(c+Q)} - \frac{s}{Q}\right]\partial_x \ln G_+ \\ &\quad - N\left[\frac{1}{2} + \frac{s^2 - Q^2}{Q(c+Q)}\right]\partial_x \ln G_0 + \frac{Ns}{2Q}\partial_x(\beta H), \qquad (2.5) \\ \partial_x \ln A &= \frac{1}{4}\left[\partial_x \ln G_- + \partial_x \ln G_+ + 2\partial_x \ln G_0\right] \qquad (2.6) \end{aligned}$$

with $s = \sinh(\beta H_{\mathrm{IC}}/2)$, $c = \cosh(\beta H_{\mathrm{IC}}/2)$ and $Q = \sqrt{\sinh^2(\beta H_{\mathrm{IC}}/2) + \exp(-\beta R)}$. For $x = \beta$ the partial derivatives $\partial_x \ln G_\mp$ and $\partial_x \ln G_0$ emerging in equations (2.5) and (2.6) read

$$\begin{aligned} \partial_\beta \ln G_\mp &= \frac{(J_{\mathrm{I}} \mp H)\sinh(\beta J_{\mathrm{I}} \mp \beta H) - \left(\frac{J_{\mathrm{H}}}{4} + \frac{K}{16}\right)\cosh(\beta J_{\mathrm{I}} \mp \beta H)}{\cosh(\beta J_{\mathrm{I}} \mp \beta H) + \exp\left(\frac{\beta J_{\mathrm{H}}}{2} + \frac{\beta K}{8}\right)\cosh\left(\frac{\beta J_{\mathrm{H}}\Delta}{2}\right)} \\ &\quad + \frac{\left(\frac{J_{\mathrm{H}}}{4} + \frac{K}{16}\right)\cosh\left(\frac{\beta J_{\mathrm{H}}\Delta}{2}\right) + \frac{J_{\mathrm{H}}\Delta}{2}\sinh\left(\frac{\beta J_{\mathrm{H}}\Delta}{2}\right)}{\exp\left(-\frac{\beta J_{\mathrm{H}}}{2} - \frac{\beta K}{8}\right)\cosh(\beta J_{\mathrm{I}} - \beta H) + \cosh\left(\frac{\beta J_{\mathrm{H}}\Delta}{2}\right)}, \qquad (2.7) \\ \partial_\beta \ln G_0 &= \frac{H\sinh(\beta H) - \left(\frac{J_{\mathrm{H}}}{4} - \frac{K}{16}\right)\cosh(\beta H)}{\cosh(\beta H) + \exp\left(\frac{\beta J_{\mathrm{H}}}{2} - \frac{\beta K}{8}\right)\cosh\left(\frac{\beta J_{\mathrm{H}}\Delta}{2}\right)} \\ &\quad + \frac{\left(\frac{J_{\mathrm{H}}}{4} - \frac{K}{16}\right)\cosh\left(\frac{\beta J_{\mathrm{H}}\Delta}{2}\right) + \frac{J_{\mathrm{H}}\Delta}{2}\sinh\left(\frac{\beta J_{\mathrm{H}}\Delta}{2}\right)}{\exp\left(-\frac{\beta J_{\mathrm{H}}}{2} - \frac{\beta K}{8}\right)\cosh(\beta H) + \cosh\left(\frac{\beta J_{\mathrm{H}}\Delta}{2}\right)}. \qquad (2.8) \end{aligned}$$

Next, let us calculate the quantity called Grüneisen parameter for the investigated model, which closely relates to the MCE. In general, the Grüneisen parameter $\Gamma_H$ can be coupled with the adiabatic cooling rate $(\partial T/\partial H)_S$ by using basic thermodynamic relations [26, 27]:

$$\Gamma_H = -\frac{(\partial M/\partial T)_H}{C_H} = -\frac{(\partial S/\partial H)_T}{T(\partial S/\partial T)_H} = \frac{1}{T}\left(\frac{\partial T}{\partial H}\right)_S, \tag{2.9}$$

where $M$ is the total magnetization of the system and $C_H$ is the specific heat at a constant magnetic field $H$. In our case, a direct substitution of the entropy (2.4) into expression (2.9) yields to the following comprehensive form of the Grüneisen parameter $\Gamma_H$ for the symmetric spin-1/2 Ising-Heisenberg diamond chain with a four-spin interaction (2.1):

$$\Gamma_H = -\frac{\partial_H \ln \mathcal{Z}_{\mathrm{IC}} + N\partial_H \ln A - \beta \partial^2_{\beta H} \ln \mathcal{Z}_{\mathrm{IC}} - N\beta \partial^2_{\beta H} \ln A}{\beta^2 \partial^2_{\beta\beta} \ln \mathcal{Z}_{\mathrm{IC}} + N\beta^2 \partial^2_{\beta\beta} \ln A}. \tag{2.10}$$

The first two functions $\partial_H \ln \mathcal{Z}_{\mathrm{IC}}$ and $\partial_H \ln A$ occurring in the numerator of the fraction (2.10) satisfy the general equations (2.5) and (2.6), respectively, where the derivatives $\partial_x \ln G_\mp$ and $\partial_x \ln G_0$ are given as follows for $x = H$:

$$_{H}\ln G_{\mp} = \frac{\mp\beta\exp\left(-\frac{\beta J_{\mathrm{H}}}{2} - \frac{\beta K}{8}\right)\sinh(\beta J_{\mathrm{I}} \mp \beta H)}{\exp\left(-\frac{\beta J_{\mathrm{H}}}{2} - \frac{\beta K}{8}\right)\cosh(\beta J_{\mathrm{I}} \mp \beta H) + \cosh\left(\frac{\beta J_{\mathrm{H}}\Delta}{2}\right)}, \tag{2.11}$$

$$\partial_{H}\ln G_{0} = \frac{\beta\exp\left(-\frac{\beta J_{\mathrm{H}}}{2} - \frac{\beta K}{8}\right)\sinh(\beta H)}{\exp\left(-\frac{\beta J_{\mathrm{H}}}{2} - \frac{\beta K}{8}\right)\cosh(\beta H) + \cosh\left(\frac{\beta J_{\mathrm{H}}\Delta}{2}\right)}. \tag{2.12}$$

Other functions $^{2}_{\beta H}\ln\mathcal{Z}_{\mathrm{IC}}$, $\partial^{2}_{\beta H}\ln A$, $\partial^{2}_{\beta\beta}\ln\mathcal{Z}_{\mathrm{IC}}$ and $\partial^{2}_{\beta\beta}\ln A$ that emerge in (2.10) can be obtained by differentiating (2.5) and (2.6) with respect to $H$ and $\beta$, respectively, provided that $x = \beta$. However, the resulting expressions for these functions are too cumbersome to be written down here explicitly.

## 3. Results and discussion

In this section, we present the results for the entropy as a function of the external magnetic field, isentropes in the $H - T$ plane and the cooling rate during the adiabatic demagnetization for the symmetric spin-1/2 Ising–Heisenberg diamond chain with the Ising four-spin interaction. We assume the Ising and Heisenberg pair interactions $J_{\mathrm{I}}$ and $J_{\mathrm{H}}$ to be antiferromagnetic ($J_{\mathrm{I}} > 0, J_{\mathrm{H}} > 0$), since it can be expected that the magnetic behaviour of the model with the antiferromagnetic interactions in the external longitudinal magnetic field should be more interesting compared to its ferromagnetic counterpart.

### 3.1. Ground state

In view of a further discussion, it is useful firstly to comment on possible spin arrangements of the investigated diamond chain at zero temperature. Typical ground-state phase diagrams constructed in the $\Delta - H/J_{\mathrm{I}}$ plane for the model in the external magnetic field, including all possible ground states, are displayed in figure 2. As can be seen from this figure, three different phases appear in the ground state regardless of the nature of the four-spin interaction $K$: the semi-classically ordered ferrimagnetic phase $\mathrm{FRI}_1$ with the perfect antiparallel alignment between the nearest-neighbouring Ising and Heisenberg spins, the quantum ferrimagnetic phase QFI, where all nodal Ising spins occupy the spin state $\sigma^z = 1/2$ and the pairs of Heisenberg spins reside at a quantum superposition of spin states described by the antisymmetric wave function $(|1/2,-1/2\rangle - |-1/2,1/2\rangle)/\sqrt{2}$, as well as the saturated paramagnetic phase SPP, where all Ising and Heisenberg spins are oriented towards the external-field direction. Furthermore,

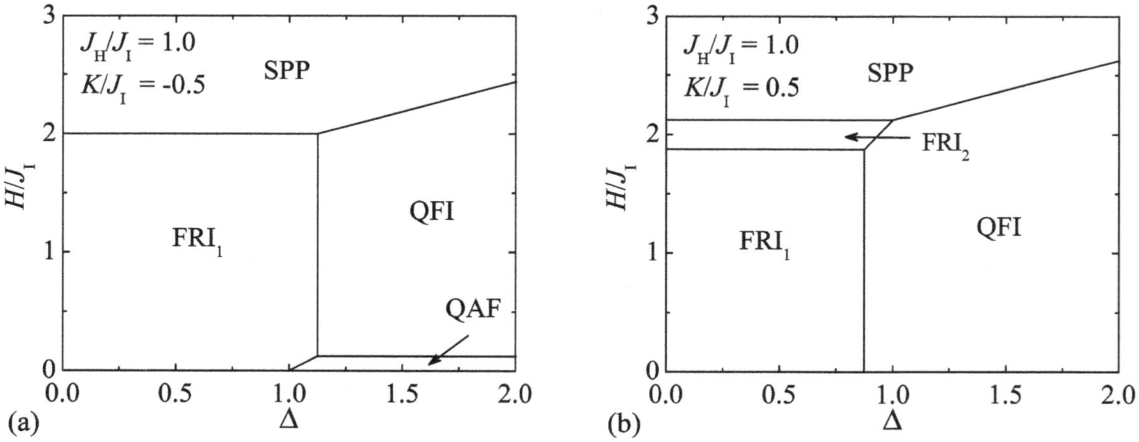


**Figure 2.** (Color online) Ground-state phase diagrams in the $\Delta - H/J_{\mathrm{I}}$ plane for the spin-1/2 Ising–Heisenberg diamond chain with the fixed interaction ratio $J_{\mathrm{H}}/J_{\mathrm{I}} = 1.0$ and the fixed four-spin interaction (a) $K/J_{\mathrm{I}} = -0.5$, (b) $K/J_{\mathrm{I}} = 0.5$.

two other interesting phases QAF and $FRI_2$ with a perfect antiferromagnetic order in the Ising sublattice can also be found in the ground state depending on whether the four-spin interaction $K$ is considered to be ferromagnetic ($K < 0$) or antiferromagnetic ($K > 0$), respectively. For more details on the magnetic order of relevant ground states see our recent work [25].

### 3.2. Entropy

Now, let us turn our attention to the entropy of the investigated diamond chain as a function of the external magnetic field. Figure 3 shows several isothermal dependencies of the entropy per one spin $S/3N$ (recall that the system is composed of $N$ Ising spins and $2N$ Heisenberg spins) versus the magnetic field $H/J_I$, corresponding to the spin-1/2 Ising–Heisenberg diamond chain with the fixed interaction ratio $J_H/J_I = 1.0$ and the fixed ferromagnetic (antiferromagnetic) four-spin interaction $K/J_I = -0.5$ ($K/J_I = 0.5$). It should be mention that the values of the exchange anisotropy parameter $\Delta$ are chosen so as to reflect all possible field-induced ground-state phase transitions. Evidently, the plotted entropy isotherms are almost unchanged down to temperature $T/J_I = 0.5$ for any choice of the parameters $\Delta$ and $K$. In the limit $T/J_I \to \infty$, the entropy per spin approaches its maximum value $S_{max}/3N = \ln 2 \approx 0.69315$ for any finite value of the applied magnetic field $H/J_I$, since the spin system is disordered at high temperatures, while it monotonously decreases upon an increase of $H/J_I$ when the temperature $T/J_I$ is finite. Below $T/J_I = 0.5$, the entropy as a function of the magnetic field exhibits irregular dependencies that develop into pronounced peaks located around the transition fields as the temperature is lowered. Finally, almost all these peaks split into isolated lines at critical fields when the temperature reaches the zero value. The only exception is the low-temperature peak observed around the critical field $H_c/J_I = 2.0$, corresponding to the field-induced phase transition between the phases $FRI_1$ and SPP, which completely vanishes at $T/J_I = 0$ [compare the lines for $T/J_I = 0.03$ and 0 in figure 3 (a)]. The residual entropy takes the finite

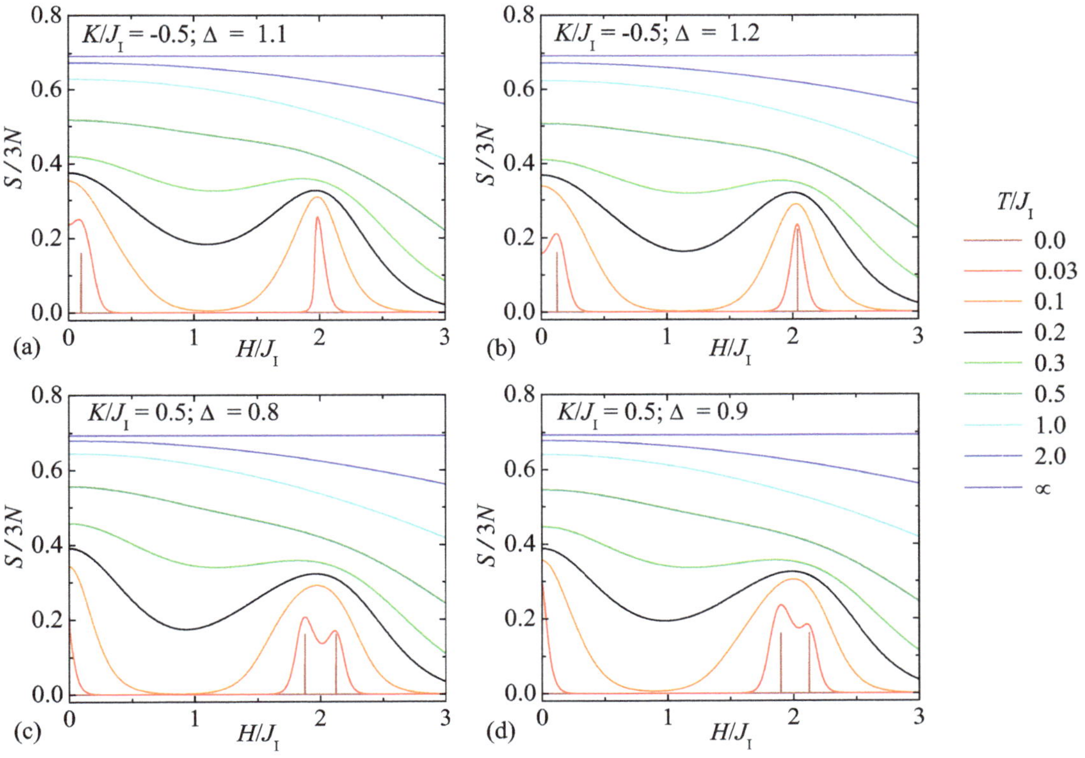


**Figure 3.** (Color online) Isothermal dependencies of the entropy versus the external magnetic field at various temperatures for the model with the interaction ratio $J_H/J_I = 1.0$ and the ferromagnetic four-spin interaction $K/J_I = -0.5$ when (a) $\Delta = 1.1$, (b) $\Delta = 1.2$, as well as the antiferromagnetic four-spin interaction $K/J_I = 0.5$ when (c) $\Delta = 0.8$, (d) $\Delta = 0.9$.

value $S_{res} = \ln 2$ at this critical point, because just one Ising spin is free to flip in the system and the spin arrangement of its nearest Ising neighbours (and consequently all others) is unambiguously given through the Ising four-spin interaction. Of course, this contribution vanishes in the thermodynamic limit $N \to \infty$, and the residual entropy normalized per spin is $S_{res}/3N = 0$, which implies that the mixed-spin system is not macroscopically degenerate at the phase transition $FRI_1$–SPP. However, the macroscopic non-degeneracy of the investigated diamond chain found at $H_c/J_I = 2.0$ can be observed merely if the four-spin interaction $K$ is ferromagnetic, since the ground-state phase transition $FRI_1$–SPP occurs only for $K < 0$ according to the ground-state analysis (see figure 2 as well as figure 2 in the reference [25]). By contrast, isolated lines appearing in zero-temperature entropy isotherms at other critical fields for $K > 0$ as well as $K < 0$, whose heights are given by the values of the residual entropy $S_{res}/3N = \ln 2^{1/3} \approx 0.23105$ and/or $\ln[(1+\sqrt{5})/2]^{1/3} \approx 0.16040$, clearly point to the macroscopic ground-state degeneracy of the system at these points. The former residual entropy $S_{res}/3N = \ln 2^{1/3}$ found at the ground-state phase transition QFI–SPP is the result of breaking up (forming) the antisymmetric quantum superposition of up-down states of the Heisenberg spins at each unit cell, whereas the latter one $S_{res}/3N = \ln[(1+\sqrt{5})/2]^{1/3}$ is closely associated with destroying (forming) a perfect antiferromagnetic order in the Ising sublattice at critical fields during the (de)magnetization process.

## 3.3. Isentropes and Grüneisen parameter

In the last part, let us proceed to the investigation of the MCE in its classical interpretation as an adiabatic change of the temperature of the considered model under field variation. For this purpose, the isentropes in the $H - T$ plane are plotted in figure 4. The values of the interaction parameters $J_H/J_I, K/J_I$ and $\Delta$ are chosen as in figure 3. Comparing figure 4 with ground-state phase diagrams shown in figure 2 one can note that the displayed sets of $T(H)$ curves exhibit a pronounced valley-peak structure, which perfectly reproduces the field-induced phase transitions of the ground state. The most obvious drop/grow of the temperature can be found in the vicinity of critical fields, where the system undergoes

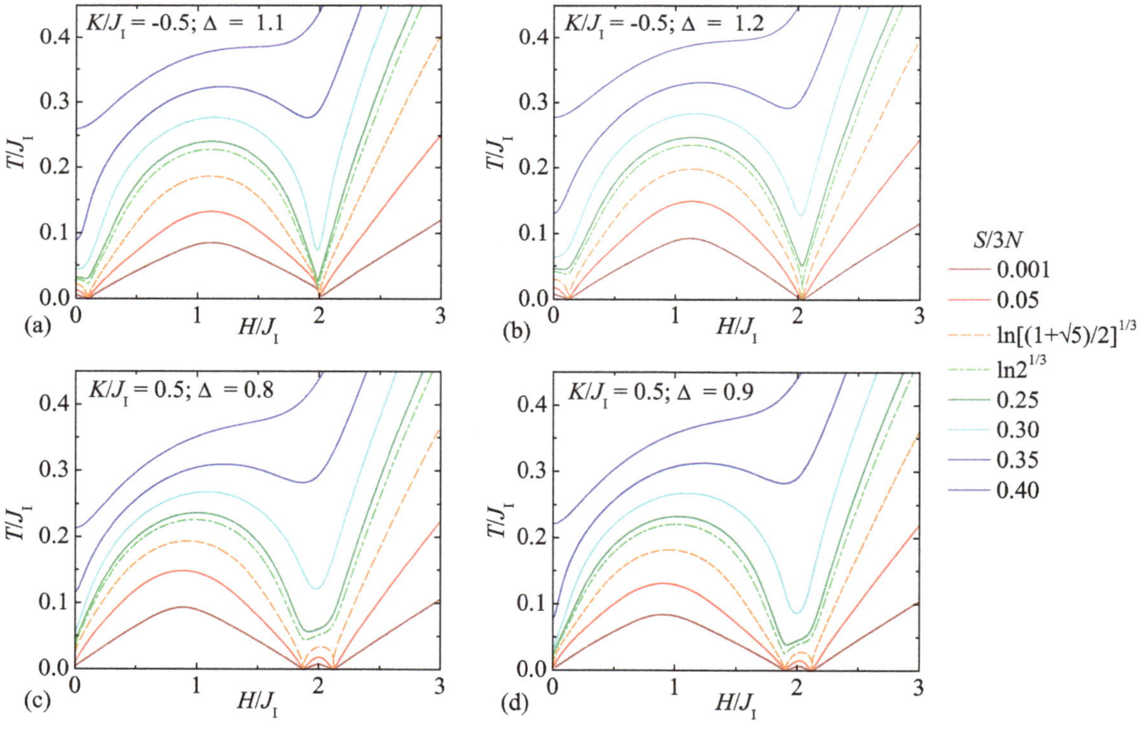


**Figure 4.** (Color online) The isentropes at the entropy per spin $S/3N = 0.001, 0.05, \ln[(1+\sqrt{5})/2]^{1/3}, \ln 2^{1/3}, 0.25, 0.3, 0.35$ and 0.4 in the $H - T$ plane. The values of the interaction parameters $J_H/J_I, K/J_I$ and $\Delta$ are chosen as in figure 3.

zero-temperature phase transitions. It should be pointed out that this relatively fast cooling/heating of the system near critical points clearly indicates the existence of a large MCE. As can be also found from figure 4, the temperature of the system reaches the zero value at critical fields if the entropy is less than or equal to its residual value at these points (see also figure 3 showing the isothermal dependencies of the entropy versus the external magnetic field at various temperatures for better clarity).

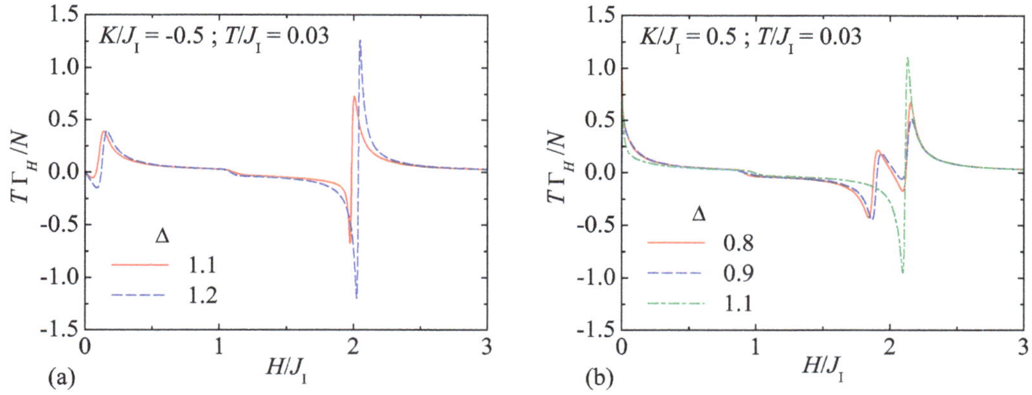


**Figure 5.** (Color online) The Grüneisen parameter multiplied by the temperature versus the external magnetic field at the fixed temperature $T/J_{\mathrm{I}} = 0.03$ for the model with the interaction ratio $J_{\mathrm{H}}/J_{\mathrm{I}} = 1.0$ and the set of parameters (a) $K/J_{\mathrm{I}} = -0.5$, $\Delta = 1.1, 1.2$; (b) $K/J_{\mathrm{I}} = 0.5$, $\Delta = 0.8, 0.9, 1.1$.

To discuss the adiabatic cooling rate of the system around the ground-state phase transitions in more detail, the Grüneisen parameter multiplied by the temperature $T\Gamma_H$ versus the external magnetic field $H/J_{\mathrm{I}}$ at the relatively low temperature $T/J_{\mathrm{I}} = 0.03$ is depicted in figure 5 by assuming $J_{\mathrm{H}}/J_{\mathrm{I}} = 1.0$, $K/J_{\mathrm{I}} = -0.5, 0.5$ and a few values of $\Delta$. Recall that the product $T\Gamma_H$ represents, in fact, the cooling rate $(\partial T/\partial H)_S$ during the adiabatic (de)magnetization [see equation (2.9)]. As one can see from figure 5, the displayed low-temperature $T\Gamma_H(H)$ curves pass through the minimum upon increasing the applied magnetic field, change sign from negative to positive values and adopt a maximum within a narrow interval of each critical field, where the system undergoes zero-temperature phase transitions. The sign change in $T\Gamma_H \propto (\partial S/\partial H)_T$ close to the critical fields clearly indicates the presence of a maximum in the corresponding isothermal dependencies of the entropy versus the field in the vicinity of the field-induced ground-state phase transitions (see the lines for $T/J_{\mathrm{I}} = 0.03$ in figure 3) and, therefore, we can say that it tracks the accumulation of the entropy due to the competition between neighbouring ground states. Moreover, it is also evident from figure 5 that high-field peaks of the $T\Gamma_H(H)$ curves plotted for the values of $\Delta = 1.2$ in figure 5 (a) and $\Delta = 1.1$ in figure 5 (b), emerging at the fields $H/J_{\mathrm{I}} \approx 2.049$ and 2.129, respectively, are significantly higher than the others. According to the ground-state phase diagrams shown in figure 2, these peaks, whose heights are $T\Gamma_H^{\mathrm{peak}} \approx 1.26311$ and 1.10248, appear somewhat above the critical fields associated with the ground-state phase transition QFI–SPP. Other peaks of the heights $T\Gamma_H^{\mathrm{peak}} \approx 0.73064$ (see the full red line in figure 5 (a) for $\Delta = 1.1$), 0.67272 and 0.51832 (see full red and dashed blue lines in figure 5 (b) for $\Delta = 0.8$ and 0.9, respectively), which can be observed in the field region $H/J_{\mathrm{I}} > 2.0$, occur just above the phase boundaries $\mathrm{FRI}_1$–SPP and $\mathrm{FRI}_2$–SPP, respectively. It is thus clear that the cooling effect observed during the adiabatic demagnetization around the ground-state phase transition QFI–SPP is approximately twice of the cooling effect, which can be detected around the phase transitions $\mathrm{FRI}_1$–SPP and $\mathrm{FRI}_2$–SPP in this $H-T$ range. From these observations one may conclude that the enhancement of the MCE found just around the phase transitions is extremely sensitive to the nature of the degeneracy of the model at these points. Actually, the MCE is the most pronounced around the ground-state boundary QFI–SPP, where strong thermal excitations of the decorated Heisenberg spins are present at low (but non-zero) temperatures due to breaking up the antisymmetric quantum superpositions of their up-down states at $T/J_{\mathrm{I}} = 0$. By contrast, vigorous low-temperature fluctuations of the Ising spins in the vicinity of other field-induced ground-state phase transitions cause a less pronounced or only a relatively weak cooling effect during the adiabatic demagnetization.

The effect of the Ising four-spin interaction on the enhanced MCE in the investigated model is depicted in figure 6. Figure 6 (a) demonstrates the situation around the field-induced phase transition QFI–SPP, while figure 6 (b) shows the situation around the phase boundaries $FRI_1$–SPP and $FRI_2$–SPP. As can be seen from figure 6 (a), the peak of the low-temperature $T\Gamma_H(H)$ dependence plotted for $K/J_I = -0.7$ is higher than those, which appear in $T\Gamma_H(H)$ curves plotted for $K/J_I = -0.5$ and $-0.1$. The similar behaviour can be found for $K > 0$: the stronger the antiferromagnetic Ising four-spin interaction $K$ is, the higher peaks can be observed in $T\Gamma_H(H)$ dependencies [see the curves plotted for $K/J_I = 0.1, 0.5$ and $0.7$ in figure 6 (a)]. Thus, one may conclude that the adiabatic cooling rate of the system increases with the strengthening of the Ising four-spin interaction $K$ just around the phase boundary QFI–SPP, regardless of its nature. It is clear from figure 6 (b) that the effect of the interaction $K$ on the adiabatic cooling rate of the system in the vicinity of other field-induced phase transitions $FRI_1$–SPP and $FRI_2$–SPP is the same (note that the peaks of the $T\Gamma_H(H)$ curves plotted for $K/J_I = -0.7, -0.5$ appear somewhat above the phase boundary $FRI_1$–SPP, while the peaks of the $T\Gamma_H(H)$ curves plotted for $K/J_I = 0.5, 0.7$ emerge just above the phase transition $FRI_2$–SPP, see also figure 2).

For completeness, let us briefly look at the effect of the temperature on the adiabatic cooling rate of the system. For this purpose, figure 7 illustrates the Grüneisen parameter multiplied by the temperature $T\Gamma_H$ versus the field $H/J_I$ for the set of parameters $K/J_I = -0.5, \Delta = 1.2$ [figure 7 (a)] and $K/J_I = 0.5, \Delta = 0.9$ [figure 7 (b)], by assuming three different temperatures. As expected, the adiabatic cooling rate $T\Gamma_H$ gradually diminishes as the temperature increases. Finally, for sufficiently high temperatures, e.g., for $T/J_I = 0.5$, the product $T\Gamma_H$ takes only positive values for all magnetic fields, which implies that the thermal fluctuations are already strong enough to drive the system to excited states where no quantum phase transition effects can be seen.

## 4. Conclusions

In the present paper, we have studied the MCE for the symmetric spin-1/2 Ising–Heisenberg diamond chain with the Ising four-spin interaction, which is exactly solvable by combining the generalized decoration-iteration transformation and the transfer-matrix technique. Within the framework of this approach, we have exactly derived the entropy and Grüneisen parameter, that closely relates to the MCE. We have also obtained the isentropes in the $H-T$ plane.

We have illustrated that the MCE in the low-entropy and/or low-temperature regimes indicate the field-induced phase transition lines seen in ground-state phase diagrams. More specifically, field-induced ground-state phase transitions perfectly manifest themselves in the form of maxima in low-temperature isothermal dependencies of the entropy versus the external magnetic field, or equivalently in the form

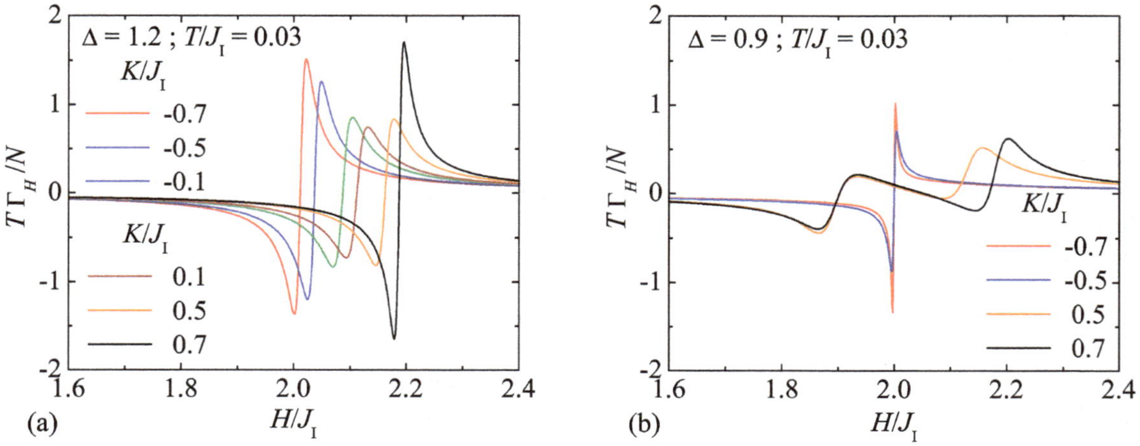


**Figure 6.** (Color online) The Grüneisen parameter multiplied by the temperature $T\Gamma_H$ versus the external magnetic field $H/J_I$ at the fixed temperature $T/J_I = 0.03$ for the model with the interaction ratio $J_H/J_I = 1.0$ and the exchange anisotropy (a) $\Delta = 1.2$; (b) $\Delta = 0.9$, by assuming a few different values of the Ising four-spin interaction $K/J_I$.

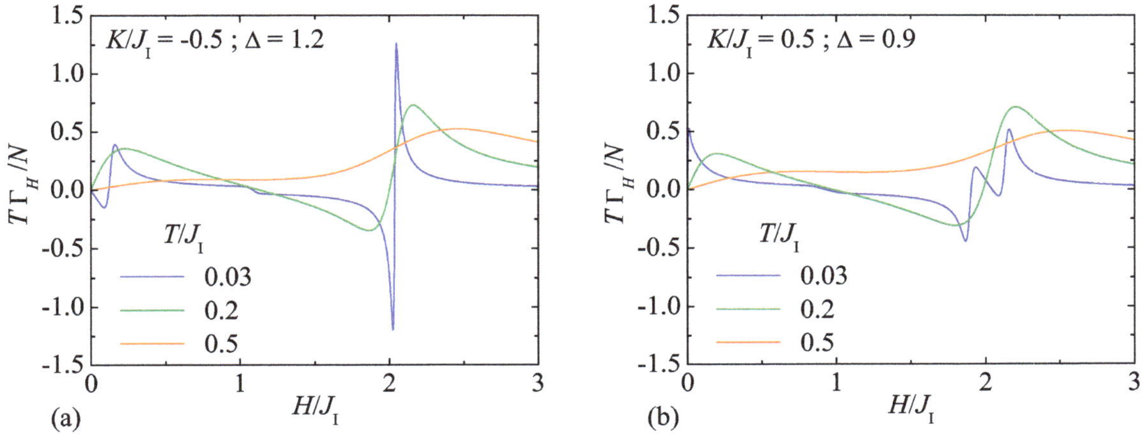


**Figure 7.** (Color online) The Grüneisen parameter multiplied by the temperature $T\Gamma_H$ versus the external magnetic field $H/J_{\mathrm{I}}$ for the set of parameters (a) $K/J_{\mathrm{I}} = -0.5$, $\Delta = 1.2$; (b) $K/J_{\mathrm{I}} = 0.5$, $\Delta = 0.9$, by assuming the temperatures $T/J_{\mathrm{I}} = 0.03, 0.2$ and $0.5$ in both cases.

of minima in low-entropy isentropes plotted in the $H-T$ plane. This leads to a pronounced cooling of the system during the adiabatic demagnetization in close vicinity of quantum phase transitions when low temperatures are reached. As a consequence, we have found large positive values of the adiabatic cooling rate (the Grüneisen parameter multiplied by the temperature) for magnetic fields slightly above critical points. In addition, we have concluded that the MCE observed just around field-induced ground-state phase transitions is extremely sensitive to the nature of the degeneracy of the model at these points. The most rapid cooling (approximately twice as fast as others) has been observed just around the field-induced ground-state phase transition QFI–SPP, where strong thermal excitations of the decorated Heisenberg spins are present at low temperatures due to breaking up the antisymmetric quantum superpositions of their up-down states at zero temperature, regardless of the nature of the Ising four-spin interaction. By contrast, the effect of Ising four-spin interaction on the adiabatic cooling rate of the system is the same in the vicinity of all field-induced phase transitions. Namely, the increasing Ising four-spin interaction (ferromagnetic as well as antiferromagnetic) accelerates the cooling of the system around phase boundaries during the adiabatic demagnetization.

The considered spin-1/2 Ising–Heisenberg diamond chain with the Ising four-spin interaction, thanks to their simplicity, has enabled the exact analysis of the MCE. Although to our knowledge there is no particular compound which can be described by the model investigated, our results might be useful in comparing the effects of ground-state phase transitions of different origin on the enhancement of the MCE. On the other hand, the comparison between theory and experiment may be resolved in future in connection with further progress in the synthesis of new magnetic chain compounds.

## References

1. Warburg E., Ann. Phys. Chem., 1881, **13**, 141.
2. Giauque W.F., MacDougall D.P., Phys. Rev., 1933, **43**, 768; doi:10.1103/PhysRev.43.768.
3. Strehlow P., Nuzha H., Bork E., J. Low Temp. Phys., 2007, **147**, 81; doi:10.1007/s10909-006-9300-y.
4. Zhitomirsky M.E., Honecker A., J. Stat. Mech., 2004, P07012; doi:10.1088/1742-5468/2004/07/P07012.
5. Zhitomirsky M.E., Tsunetsugu H., Phys. Rev. B, 2004, **70**, 100403; doi:10.1103/PhysRevB.70.100403.
6. Honecker A., Wessel S., Physica B, 2006, **378–380**, 1098; doi:10.1016/j.physb.2006.01.436.
7. Čanová L., Strečka J., Jaščur M., J. Phys.: Condens. Matter, 2006, **18**, 4967; doi:10.1088/0953-8984/18/20/020.
8. Derzhko O., Richter J., Eur. Phys. J. B, 2006, **52**, 23; doi:10.1140/epjb/e2006-00273-y.
9. Derzhko O., Richter J., Honecker A., Schmidt H.-J., Low Temp. Phys., 2007, **33**, 745; doi:10.1063/1.2780166.
10. Schnack J., Schmidt R., Richter J., Phys. Rev. B, 2007, **76**, 054413; doi:10.1103/PhysRevB.76.054413.
11. Schmidt B., Thalmeier P., Shannon N., Phys. Rev. B, 2007, **76**, 125113; doi:10.1103/PhysRevB.76.125113.
12. Pereira M.S.S., de Moura F.A.B.F., Lyra M.L., Phys. Rev. B, 2009, **79**, 054427; doi:10.1103/PhysRevB.79.054427.

13. Čanová L., Strečka J., Lučivjanský, Condens. Matter Phys., 2009, **12**, 353; doi:10.5488/CMP.12.3.353.
14. Honecker A., Wessel S., Condens. Matter Phys., 2009, **12**, 399; doi:10.5488/CMP.12.3.399.
15. Trippe C., Honecker A., Klümper A., Ohanyan V., Phys. Rev. B, 2010, **81**, 054402; doi:10.1103/PhysRevB.81.054402.
16. Lang M., Tsui Y., Wolf B., Jaiswal–Nagar D., Tutsch U., Honecker A., Removic–Langer K., Prokofiev A., Assmus W., Donath G., J. Low Temp. Phys., 2010, **159**, 88; doi: 10.1007/s10909-009-0092-8.
17. Ribeiro G.A.P., J. Stat. Mech., 2010, P12016; doi:0.1088/1742-5468/2010/12/P12016.
18. Jafari R., Eur. Phys. J. B, 2012, **85**, 167; doi:10.1140/epjb/e2012-20682-5.
19. Topilko M., Krokhmalskii T., Derzhko O., Ohanyan V., Eur. Phys. J. B, 2012, **85**, 278; doi:10.1140/epjb/e2012-30359-8.
20. Fisher M.E., Phys. Rev., 1959, **113**, 969; doi:10.1103/PhysRev.113.969.
21. Syoyi I., In: Phase Transition and Critical Phenomena. Vol. 1, Domb C., Green M.S. (Eds.), Academic Press, New York, 1972, 269–329.
22. Rojas O., Valverde J.S., de Sousa S.M., Physica A, 2009, **388**, 1419; doi:10.1016/j.physa.2008.12.063.
23. Kramers H.A., Wannier G.H., Phys. Rev., 1944, **60**, 252; doi:10.1103/PhysRev.60.252.
24. Baxter R.J., Exactly Solved Models in Statistical Mechanics, Academic Press, New York, 1982, 32–38.
25. Gálisová L., Phys. Status Solidi B, 2013, **250**, 187; doi:10.1002/pssb.201248260.
26. Zhu L., Garst M., Rosch A., Si Q., Phys. Rev. Lett., 2003, **91**, 066404; doi:10.1103/PhysRevLett.91.066404.
27. Garst M., Rosch A., Phys. Rev. B, 2005, **72**, 205129; doi:10.1103/PhysRevB.72.205129.

# 2

# Interaction of phonons at superfluid helium-solid interfaces

I.N. Adamenko, E.K. Nemchenko

V.N. Karazin Kharkiv National University, 4 Svobody Sqr., 61022 Kharkiv, Ukraine

A new method of obtaining the interaction Hamiltonian of phonons at superfluid helium-solid interface is proposed in the work. Equations of hydrodynamic variables are obtained in terms of second quantization if helium occupies a half-space. The contributions of all processes to the heat flux from solid to superfluid helium are calculated based on the obtained Hamiltonian. The angular distribution of phonons emitted by a solid is found in different processes. It is shown that all the exit angles of superfuild helium phonons are allowed. The obtained results are compared with experimental data and with previous theoretical works.

**Key words:** *phonon, angular distribution, heat flow, Kapitza gap, interface*

## 1. Introduction

Superfluid helium has a whole number of unique phenomena that take place at superfluid helium-solid interface. One of such phenomena is the thermal boundary resistance discovered by Kapitza P.L. [1]. It was discovered that there is a constant temperature difference between the contacting solid and superfluid helium when a solid emits heat. Since then, this phenomenon has been studied by different authors because so far there is no satisfactory agreement between experimental data and theoretical research.

The first theoretical explanation of Kapitza gap was given by Khalatnikov [2–4]. According to works [2–4], the heat flow occurs due to incident phonons in both superfluid helium and a solid. These phonons with difficulty pass through the interface due to acoustic mismatch of the media and due to the smallness of incident phonon angle in liquid helium above which total internal reflection occurs. Transition probability of phonon from one media to another which was obtained in [2–4] is proportional to interface impedance $\rho_{\mathrm{L}} c_{\mathrm{L}} / \rho_{\mathrm{S}} c_{\mathrm{S}}$. Critical angle is equal to $c_{\mathrm{L}} / c_{\mathrm{S}}$, where $c_{\mathrm{L}}$ and $c_{\mathrm{S}}$ are the velocities of sound of liquid and solid, respectively, $\rho_{\mathrm{L}}$ and $\rho_{\mathrm{S}}$ are densities of liquid and solid, respectively.

The results of many experiments obtained by various authors significantly differed from the calculated values of theories [2–4]. Particularly, experimental values of heat transfer rate of superfluid helium-solid interface are more often by two orders of magnitude larger than theoretical values in works [2–4].

Large experimental values of heat transfer rate mean that there are other mechanisms of heat transfer between superfluid helium and solid along with the so-called acoustic channel that was considered in [2–4]. To our best knowledge, all theoretical works dedicated to the the search for such mechanisms were based on the fact that the interface with superfluid helium surface of solid was not perfectly smooth and clean and contained roughness, various defects and monolayers.

The imperfection of an interface leads to the assumption that phonons could pass into a solid at any incident angles and not just in a narrow cone with a solid angle $(\ _{\mathrm{L}}/\ _{\mathrm{S}})^2$ which was formed by a critical angle following from Khalatnikov theory [2–4]. In this case, heat transfer rate may increase by $(c_{\mathrm{S}}/c_{\mathrm{L}})^2$ times. This value is of the order of $10^2$ for the superfluid helium-solid interface. This fact can reconcile the theory with experiments.

In this regard, numerous experiments were carried out by different authors in which the role of a solid surface in heat transfer between solid and superfluid helium was investigated. The results of such experiments are given in the review [5] and in later experimental works [6–9]. These experiments indicated that the condition of a solid surface does substantially alter the heat transfer coefficient so that it becomes larger than only an order of magnitude of the coefficient calculated in theory [2–4].

In order to understand the Kapitza gap problem, direct experiments [10–14] were performed in which energy and angular distribution of the emitted phonons by a solid in cold ($T < 100$ mK) superfluid helium were measured. In these experiments, phonon beams were emitted from a heated solid to superfluid helium that was almost at zero temperature (i.e., superfluid vacuum). As the heaters there were used both conductive metal films and cleaved surfaces of crystals that were almost perfect surfaces. It was shown in works [10–14] that even with almost perfect solid surface there are two channels of phonon transfer from solid to superfluid helium which are demonstrated in figure 1

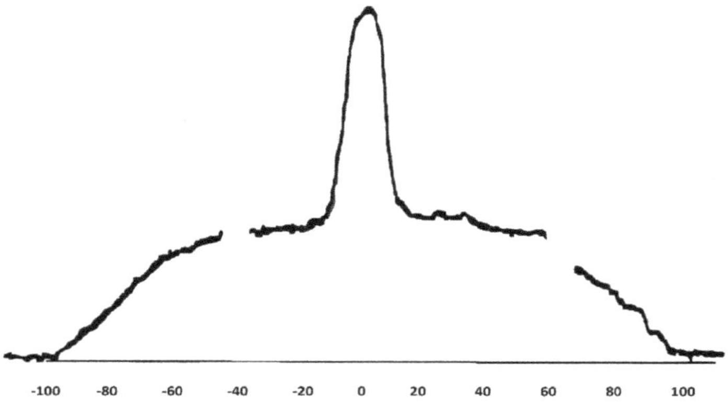


**Figure 1.** Angular distribution of heat flow from heated solid to superfluid that is observed in [10–14].

The first channel formed a sharp peak of phonons emitted in a narrow cone of angles whose axis was normal to the solid surface (see figure 1). The observed value in [11] of the angle of the cone coincides with the calculated values for different solids in the acoustic mismatch theory which was based on Khalatnikov theory [2–4]. This channel was called the acoustic channel.

The second channel, the so-called background channel, contained phonons emitted in all directions. Moreover, it was shown that the contribution of a background channel was an order of magnitude larger than the contribution of an acoustic channel during experimental data analysis.

In work [15] which was performed based on the results of experimental work [16] it was shown that the phonons emitted by a heated and rather rough gold surface in superfluid helium were also distributed through two channels observed in [10–14].

Accordingly, a question has arisen: what is the physical reason for the existence of such a large background channel at almost perfect surface solid? Great hope to explain the existence of the background channel and large observed values of the heat transfer coefficient in Kapitza gap experiments was entrusted to the processes in which there was a different number of phonons in the initial and the final states. These are the so-called inelastic interaction processes. A possible diagram of inelastic process can be found in the experimental work [12], where one phonon of a solid transforms into two phonons of a liquid that could pass at any angles to the interface. One of the possible inelastic processes that differs from the one illustrated in [12] was considered by Khalatnikov [4] who showed that the contribution of this process was relatively small. It is worth pointing out that the inelastic process considered in [4] does not contribute to the heat flux from solid to superfluid helium which is almost at zero temperature.

In this regard, consideration of all possible inelastic processes turns out to be relevant as well as the calculation of their contribution to the background channel. This is the focus of the present work.

The first attempts to solve the above mentioned problem was made in works [17, 18] in which it was suggested to create a microscopical theory of Kapitza gap at the He II-solid interface. However, it was a failure to create a self-consistent approach capable of yielding the results in accord with the acoustic theory [2–4] corresponding to elastic phonon processes. This is apparently connected with the calculations that were not brought to final analytical formulas and to specific numerical values in works [17] and [18].

The original results of constructing a unified self-consistent theory describing both elastic and inelastic processes at the superfluid helium-solid interface were presented at the QFS2012 conference[1] by the authors of this paper. These first results were published in the materials of the conference [19]. The contribution of inelastic processes to Kapitza gap was considered in the work [20].

The main goal of this paper is to investigate all possible inelastic processes that contribute to the heat flow from the solid to the superfluid helium and to consider the angular distribution of the emitted phonons in different processes.

## 2. Interaction Hamiltonian of helium phonons with an oscillating surface of a solid

For the interaction Hamiltonian of helium phonons with an oscillating surface of a solid, we calculate the density of the energy of superfluid helium in the presence of an oscillating interface. The obtained Hamiltonian will essentially differ from the Hamiltonians used in works [4, 17, 18] and will yield a correct result regarding the heat flow due to the elastic process that is equal to the result obtained in [3].

Oscillations of interface excite in helium oscillations of density $\rho_{\mathrm{i}}$ and velocity $\mathbf{v}_{\mathrm{i}}$ along with the intrinsic oscillations of $\rho$ and $\mathbf{v}$ in the liquid. In this case, the interaction energy is

$$E=\frac{1}{2}\left(\rho_{\mathrm{L}}+\rho+\rho_{\mathrm{i}}\right)\left(\mathbf{v}+\mathbf{v}_{\mathrm{i}}\right)^{2}+E_{\rho}\left(\rho_{\mathrm{L}}+\rho+\rho_{\mathrm{i}}\right), \tag{2.1}$$

where $E_\rho$ is the density functional.

To simplify the problem, we restrict ourselves to longitudinal phonons in the solid. The inclusion of transverse phonons does not cause fundamental difficulties, but all the calculations become more cumbersome and lead to the appearance of a factor $F$ in the final calculations that depends on the elastic constants of the solid. $F$ varies over small limits and remains of the order of 2 for different solids.

Now we reduce the equation (2.1) to the form of an expansion accurate to cubic terms in the small parameters $\rho_{\mathrm{i}}$, $\mathbf{v}_{\mathrm{i}}$, $\rho$ and $\mathbf{v}$:

$$E=E_{0,1}+\frac{1}{2}\rho_{\mathrm{L}}\left(\mathbf{v}+\mathbf{v}_{\mathrm{i}}\right)^{2}+\frac{c_{\mathrm{L}}^{2}}{2\rho_{\mathrm{L}}}\left(\rho+\rho_{\mathrm{i}}\right)^{2}+\frac{1}{2}\left(\rho+\rho_{\mathrm{i}}\right)\left(\mathbf{v}+\mathbf{v}_{\mathrm{i}}\right)^{2}+\frac{c_{\mathrm{L}}^{2}}{6\rho_{\mathrm{L}}^{2}}\left(2u-1\right)\left(\rho+\rho_{\mathrm{i}}\right)^{3}, \tag{2.2}$$

where $u=\frac{\rho_{\mathrm{L}}}{c_{\mathrm{L}}}\frac{\partial c_{\mathrm{L}}}{\partial \rho_{\mathrm{L}}}$ is the Gruneisen constant, which equals 2.84 for helium,

$$E_{0,1}=E_{\rho}\left(\rho_{\mathrm{L}}\right)+\left.\frac{\partial E_{\rho}\left(\rho_{\mathrm{t}}\right)}{\partial \rho_{\mathrm{t}}}\right|_{\rho_{\mathrm{t}}=\rho_{\mathrm{L}}}\left(\rho+\rho_{\mathrm{i}}\right) \tag{2.3}$$

is the sum of zero and the first terms of the expansion which does not contribute to the interaction of the liquid and the solid, $\rho_{\mathrm{t}}=\rho_{\mathrm{L}}+\rho+\rho_{\mathrm{i}}$.

Then, the contribution to the interaction of helium with a wall will yield a term that simultaneously contains parameters characterizing both the solid and the liquid. In this case, the interaction energy is as follows:

$$E_{\mathrm{int}}=\rho_{\mathrm{L}}\mathbf{v}\mathbf{v}_{\mathrm{i}}+\frac{c_{\mathrm{L}}^{2}}{\rho_{\mathrm{L}}}\rho\rho_{\mathrm{i}}+\frac{\rho}{2}\left(2\mathbf{v}\mathbf{v}_{\mathrm{i}}+\mathbf{v}_{\mathrm{i}}^{2}\right)+\frac{\rho_{\mathrm{i}}}{2}\left(2\mathbf{v}\mathbf{v}_{\mathrm{i}}+\mathbf{v}^{2}\right)+\frac{c_{\mathrm{L}}^{2}}{2\rho_{\mathrm{L}}^{2}}\left(2u-1\right)\rho\rho_{\mathrm{i}}\left(\rho+\rho_{\mathrm{i}}\right). \tag{2.4}$$

The first two terms in the equation (2.4) describe the two-phonon interactions and the remaining terms describe the three-phonon interactions (in terms of secondary quantization).

In this problem, the velocity and density of solid and liquid phonons are specified for a half space, whereas there are problems expanding them in Fourier series and with the subsequent use of the second quantization method. The following method for analytic continuation of the solutions is proposed to overcome these difficulties and make it possible to use the Fourier expansion and secondary quantization.

[1]QFS2012: International Conference on Quantum Fluids and Solids, 15–21 August 2012, Physics Department, Lancaster University, UK.

To this end, we carry out calculations on the entire axis $z$ that is perpendicular to the superfluid helium-solid interface. Moreover, due to boundary conditions at $z = 0$, $v_z$ is oddly extended to the entire space so that $v_z(z>0) = -v_z(z<0)$ and $v_x$, $v_y$, $\rho$ and $v_{iz}$ are evenly extended to the entire space.

However, we should note that the helium perturbations generated by oscillations of the interface, on the one hand, contribute to the energy of helium, and, on the other hand, are determined by the parameters which characterize the vibrations of a solid interface (amplitude and displacement velocity). The relationship between these parameters is given by standard boundary conditions for a normal component of the velocity at the solid-superfluid liquid interface, which is superfluid helium. Thus, parameters describing the vibrations of the interface after the second quantization and the change of helium energy caused by these vibrations will contain the creation and annihilation operators of solid phonons. In this respect, those perturbation operators of density and velocity of helium and velocity of interface vibrations are Hermitian after the second quantization. We get the final form of these operators:

$$\hat{\rho} = \rho_{\mathrm{L}} \sum_{k_z=0}^{+\infty} \sum_{\mathbf{k}_{||}} \frac{\mathrm{i}}{c_{\mathrm{L}}} \sqrt{\frac{\hbar\omega}{2\rho_{\mathrm{L}} V_{\mathrm{L}}}} \left(\hat{a}_{\mathbf{k}} - \hat{a}^{+}_{-\mathbf{k}} + \hat{a}_{-\mathbf{k}} - \hat{a}^{+}_{\mathbf{k}}\right) \left(\frac{\mathrm{e}^{\mathrm{i}k_z z} + \mathrm{e}^{-\mathrm{i}k_z z}}{\sqrt{2}}\right) \mathrm{e}^{\mathrm{i}\mathbf{k}_{||}\mathbf{r}_{||}},$$

$$\hat{v}_z = \sum_{k_z=0}^{+\infty} \sum_{\mathbf{k}_{||}} \sqrt{\frac{\hbar\omega}{2\rho_{\mathrm{L}} V_{\mathrm{L}}}} \mathrm{i}\frac{k_z}{k} \left(\hat{a}_{\mathbf{k}} + \hat{a}^{+}_{-\mathbf{k}} + \hat{a}_{-\mathbf{k}} + \hat{a}^{+}_{\mathbf{k}}\right) \left(\frac{\mathrm{e}^{\mathrm{i}k_z z} - \mathrm{e}^{-\mathrm{i}k_z z}}{\sqrt{2}}\right) \mathrm{e}^{\mathrm{i}\mathbf{k}_{||}\mathbf{r}_{||}},$$

$$\hat{v}_{iz} = \sum_{q_z=0}^{+\infty} \sum_{\mathbf{q}_{||}} \sqrt{\frac{\hbar\Omega}{2\rho_{\mathrm{S}} V_{\mathrm{S}}}} \mathrm{i}\frac{q_z}{q} \left(\hat{b}_{\mathbf{q}} - \hat{b}^{+}_{-\mathbf{q}} + \hat{b}_{-\mathbf{q}} - \hat{b}^{+}_{\mathbf{q}}\right) \left(\frac{\mathrm{e}^{\mathrm{i}b_z z} + \mathrm{e}^{-\mathrm{i}b_z z}}{\sqrt{2}}\right) \mathrm{e}^{\mathrm{i}\mathbf{q}_{||}\mathbf{r}_{||}}, \tag{2.5}$$

where $\mathbf{k}$ and $\mathbf{q}$ are wave vectors of helium and solid phonons, respectively, $\omega$ and $\Omega$ are frequencies of helium and solid phonons, respectively, $V_{\mathrm{L}}$ and $V_{\mathrm{S}}$ are volumes that liquid and solid occupies, $\hat{a}^{+}_{\mathbf{k}}\,(\hat{a}_{\mathbf{k}})$ and $\hat{b}^{+}_{\mathbf{q}}\,(\hat{b}_{\mathbf{q}})$ are operators of creation (annihilation) of helium and solid phonons, respectively; axis $z$ is directed perpendicular to the interface, and $\mathbf{k}_{||}$ and $\mathbf{q}_{||}$ tangential components of the wave vectors of helium and solid phonons, respectively. Equations (2.4) and (2.5) permit to submit Hamilton operator

$$\hat{H}_{\mathrm{int}} = \int_0^L \mathrm{d}z \int \mathrm{d}S E_{\mathrm{int}} \tag{2.6}$$

in terms of the second quantization. In equation (2.6), integration is over the volume of the liquid $V_{\mathrm{L}} = LS$, where $S$ is the area of the superfluid helium-solid interface. The Hamiltonian equation (2.6) describes the creation and annihilation of phonons at the He II-solid interface, which is caused by vibrations of the interface.

After these procedures, the Hamiltonian (2.6) will have the following form to within the cubic terms

$$\hat{H}_{\mathrm{int}} = \hat{H}^{(2)}_{\mathrm{int}} + \hat{H}^{(3)}_{\mathrm{int}}. \tag{2.7}$$

Here, the first term

$$\hat{H}^{(2)}_{\mathrm{int}} = \mathrm{i}c_{\mathrm{L}} \sqrt{\frac{\rho_{\mathrm{L}}}{\rho_{\mathrm{S}}}} \frac{\hbar S}{\sqrt{V_{\mathrm{L}} V_{\mathrm{S}}}} \sum_{\mathbf{k}} \sum_{\mathbf{q}} \frac{q_z}{q} \left(\hat{a}_{\mathbf{k}} + \hat{a}^{+}_{-\mathbf{k}} + \hat{a}_{-\mathbf{k}} + \hat{a}^{+}_{\mathbf{k}}\right)\left(\hat{b}_{\mathbf{q}} - \hat{b}^{+}_{-\mathbf{q}} + \hat{b}_{-\mathbf{q}} - \hat{b}^{+}_{\mathbf{q}}\right)\delta_{\mathbf{k}_{||},\mathbf{q}_{||}} \tag{2.8}$$

contains a single phonon annihilation (creation) operator and a single creation (annihilation) operator for the solid.

Thus, $\hat{H}^{(2)}_{\mathrm{int}}$ describes the conversion of a liquid (solid) phonon into a solid (liquid) phonon at the superfluid helium-solid interface. In this transition, the phonon retains its energy. We refer to this kind of a process as an elastic one. The second term in equation (2.7) has the form

$$\hat{H}^{(3)}_{\mathrm{int}} = \frac{\hbar^{3/2} S}{c_{\mathrm{L}} V_{\mathrm{L}} \sqrt{V_{\mathrm{S}} \rho_{\mathrm{S}}}} \sum_{\mathbf{k},\mathbf{q}} \sqrt{\omega_1 \omega_2 \Omega} \frac{k_{2z}}{k_2} \frac{q_z}{q} \frac{k_{2z}}{k_{2z}^2 - k_{1z}^2} \delta_{\mathbf{k}_{1||}+\mathbf{k}_{2||}+\mathbf{q}_{||},0} \left(\hat{a}_{\mathbf{k}_1} - \hat{a}^{+}_{-\mathbf{k}_1} + \hat{a}_{-\mathbf{k}_1} - \hat{a}^{+}_{\mathbf{k}_1}\right)$$
$$\times \left(\hat{a}_{\mathbf{k}_2} + \hat{a}^{+}_{-\mathbf{k}_2} + \hat{a}_{-\mathbf{k}_2} + \hat{a}^{+}_{\mathbf{k}_2}\right)\left(\hat{b}_{\mathbf{q}} - \hat{b}^{+}_{-\mathbf{q}} + \hat{b}_{-\mathbf{q}} - \hat{b}^{+}_{\mathbf{q}}\right) + \frac{\hbar^{3/2} \sqrt{\rho_{\mathrm{L}}} S}{c_{\mathrm{L}} V_{\mathrm{L}} \sqrt{V_{\mathrm{S}} \rho_{\mathrm{S}}}} \sum_{\mathbf{k},\mathbf{q}} \sqrt{\omega \Omega_1 \Omega_2} \frac{1}{k} \frac{q_{1z}}{q_1} \frac{q_{2z}}{q_2} \delta_{\mathbf{q}_{1||}+\mathbf{q}_{2||}+\mathbf{k}_{||},0}$$
$$\times \left(\hat{a}_{\mathbf{k}} + \hat{a}^{+}_{-\mathbf{k}} + \hat{a}_{-\mathbf{k}} + \hat{a}^{+}_{\mathbf{k}}\right)\left(\hat{b}_{\mathbf{q}_1} - \hat{b}^{+}_{-\mathbf{q}_1} + \hat{b}_{-\mathbf{q}_1} - \hat{b}^{+}_{\mathbf{q}_1}\right)\left(\hat{b}_{\mathbf{q}_2} - \hat{b}^{+}_{-\mathbf{q}_2} + \hat{b}_{-\mathbf{q}_2} - \hat{b}^{+}_{\mathbf{q}_2}\right), \tag{2.9}$$

where $\omega_{1,2}$ and $\mathbf{k}_{1,2}$ are frequencies and wave vectors of the superfluid helium phonons and $\Omega_{1,2}$ and $\mathbf{q}_{1,2}$ are frequencies and wave vectors of the solid. Equation (2.9) describes the processes in which there are different numbers of phonons in the initial and final states. We shall refer to these kinds of processes as inelastic phonon processes.

## 3. Heat flow through the superfluid helium-solid interface

In order to calculate the heat flow from a solid to a liquid, it is necessary to write down the probability of a phonon conversion process at the liquid helium-solid interface. From the Hamiltonian equations (2.7), (2.8) and (2.9), there are four possible three-phonon inelastic processes along with one elastic process. We enumerate these processes with a subscript $k$ equal to 0 for the elastic process and $1 \div 4$ for the four possible inelastic processes. Here are diagrams of all possible processes. The diagrams for reverse processes are obtained by reversing the directions of all the arrows in the diagram for a forward process.

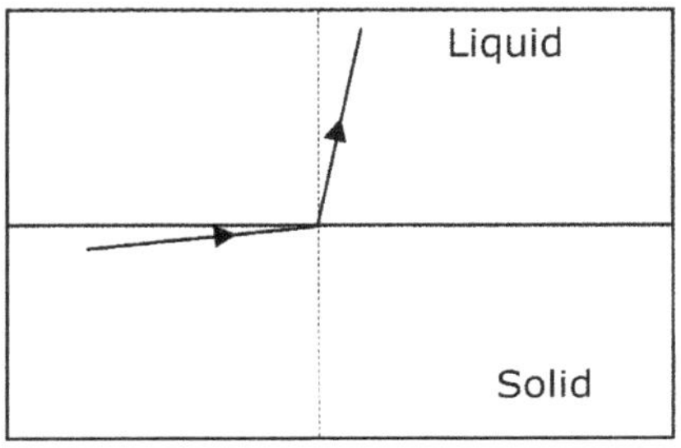


**Figure 2.** Diagram for a direct elastic phonon conversion process at a superfluid helium-solid interface $(k = 0)$.

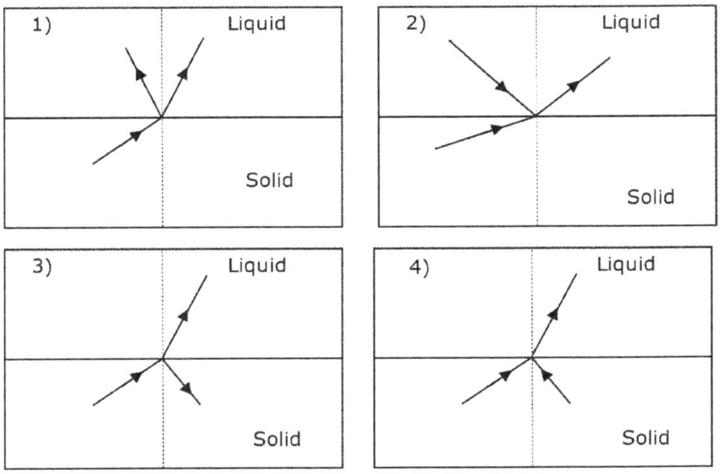


**Figure 3.** Diagrams for direct inelastic processes $(k = 1 \div 4)$.

The second inelastic process does not give a contribution to the heat flow from the heated solid to superfluid helium that is at zero temperature. Therefore, we will consider only the first, the third and the fourth inelastic processes along with the elastic process.

The probability $w_k$ of process $k$, which is determined by the matrix element $M_{fi}^{(k)} = \langle f | \hat{H}_{\text{int}} | i \rangle$ for a transition from the initial state $i$ to the final state $f$, if a particular process results from the Hamiltonian equation (2.7), is given by

$$w_k = \frac{2\pi}{\hbar S}\left|M_{fi}^{(k)}\right|^2 \delta\left(E_f - E_i\right). \tag{3.1}$$

Here, $E_f$ and $E_i$ are the total energy of the phonons in the final and initial states, respectively. The quantity (3.1) is the probability that phonons transfer from state $i$ into state $f$ per unit time through unit area of the interface surface.

The expression for a heat flux per unit time through unit area of the interface surface in the normalization of the operators to the energy of a single phonon for the $k$-th process that we have chosen is

$$W^{(k)} = \int w_k \sum_f \varepsilon_f \cos\theta_f \prod_f \left[1 + n\left(\varepsilon_f\right)\right] \mathrm{d}\Gamma_f \prod_i n\left(\varepsilon_i\right) \mathrm{d}\Gamma_i, \tag{3.2}$$

where the sum is taken over all the final phonons, while the products $\prod_f$ and $\prod_i$ are taken over all the final and initial phonons, respectively, $\varepsilon_f$ and $\varepsilon_i$ are the energies of the final and initial phonons, respectively, $n(\varepsilon)$ is the Bose distribution function, $\mathrm{d}\Gamma = \mathrm{d}^3 p \mathrm{d}^3 r / (2\pi\hbar)^3$ is the number of quantum states in an element of phase space, and $\theta_f$ is the exit angle for a final phonon with energy $\varepsilon_f$. Here, and in what follows, all the angles are reckoned from the normal to the superfluid helium-solid interface boundary.

We consider the heat flow due to an elastic process. The matrix element of this process is as follows:

$$M_{fi}^{(0)} = \frac{2\mathrm{i} c_{\mathrm{L}} \hbar S}{\sqrt{V_{\mathrm{L}} V_{\mathrm{S}}}} \sqrt{\frac{\rho_{\mathrm{L}}}{\rho_{\mathrm{S}}}} \frac{q_z}{q} \delta_{\mathbf{k}_{||}, \mathbf{q}_{||}}. \tag{3.3}$$

For the heat flow from a solid at temperature $T_{\mathrm{S}}$ into superfluid helium, which is at zero temperature, we begin with equations (3.1), (3.2), and (3.3) and obtain

$$W^{(0)} = \frac{4\pi^4}{15} \frac{\rho_{\mathrm{L}} c_{\mathrm{L}}}{\rho_{\mathrm{S}} c_{\mathrm{S}}^3} \frac{1}{3(2\pi)^2 \hbar^3} \left(k_{\mathrm{B}} T_{\mathrm{S}}\right)^4. \tag{3.4}$$

According to conservation of energy and conservation of tangential impulse component of phonon, it follows that in an elastic process, the heat flux (3.4) will fill a narrow cone of angles with solid angle $(\ _{\mathrm{L}}/c_{\mathrm{S}})^2$, whose axis is directed normal to the interface.

The first inelastic process, which corresponds to a transition from a state with one solid phonon to a state with two liquid phonons, is calculated in a standard way and is as follows:

$$M_{fi}^{(1)} = \frac{2\sqrt{2}\hbar^{\frac{3}{2}} S}{_{\mathrm{L}} V_{\mathrm{L}} \sqrt{V_{\mathrm{S}} \rho_{\mathrm{S}}}} \sqrt{\omega_1 \omega_2 \Omega} \left[\frac{k_{2z}^2}{k_2\left(k_{2z}^2 - k_{1z}^2\right)} - \frac{k_{1z}^2}{k_1\left(k_{1z}^2 - k_{2z}^2\right)}\right] \frac{q_z}{q} \delta_{\mathbf{k}_{1||} + \mathbf{k}_{2||} + \mathbf{q}_{||}, 0}. \tag{3.5}$$

On the assumption of (3.1), (3.2) and (3.5), the heat flow from a solid at temperature $T_{\mathrm{S}}$ into a liquid helium at zero temperature is as follows:

$$\begin{aligned} W^{(1)}(S \to L) \quad = \quad & \frac{8}{(2\pi)^4 \rho_s c_{\mathrm{L}}^4 c_{\mathrm{S}}^3 \hbar^6} \left(k_b T_{\mathrm{S}}\right)^8 \int \mathrm{d}x \mathrm{d}y \sin\theta \mathrm{d}\theta \sin\theta_1 \mathrm{d}\theta_1 \cos^2\theta \frac{1}{\mathrm{e}^x - 1}\left[y \frac{\cos\theta_1}{\cos\theta_2} + (x - y)\right] \\ & \times y^3 x^3 (x - y)\left[\frac{(x - y)\cos^2\theta_2 + y\cos^2\theta_1}{(x - y)^2\cos^2\theta_2 - y^2\cos^2\theta_1}\right]^2, \end{aligned} \tag{3.6}$$

where $x = \hbar\Omega / k_{\mathrm{B}} T_{\mathrm{S}}$, $y = \hbar\omega_1 / k_{\mathrm{B}} T_{\mathrm{S}}$ and $\hbar\omega_2 / k_{\mathrm{B}} T_{\mathrm{S}} = x - y$ are from the conservation of energy law, $\theta_{1,2}$ are the exit angles of liquid phonons, $\theta$ is the incident angle of a solid phonon. The numerical value of the dimensionless integral (3.6) is independent of temperature, but it does depend on the ratio of the speeds of sound of the solid and the liquid. For the value $c_{\mathrm{L}}/c_{\mathrm{S}} = 0.1$, that will be used further, dimensionless integral (3.6) is equal to $4.78 \cdot 10^2$. As will be shown below, this value weakly depends on the ratio $c_{\mathrm{L}}/c_{\mathrm{S}}$.

It should be noted that in integrals (3.6) and further, integration of the azimuthal angle is replaced by multiplication by $2\pi$ for simplicity. The limits of integration in these integrals and the function of the integration variables $\theta_2 = \theta_2(\theta, \theta_1, x, y)$ are determined by the conservation of energy and by tangential component of impulse laws. Equation for $\sin\theta_2$ is as follows:

$$\sin\theta_2 = \frac{1}{x - y}\left(y \sin\theta_1 + x \frac{c_{\mathrm{L}}}{_{\mathrm{S}}} \sin\theta\right). \tag{3.7}$$

The conditions of exit phonon angles with energies $\omega_1$ could be obtained from equation (3.7).

1. $0 \leqslant \theta_1 \leqslant \frac{\pi}{2}$, for $x \leqslant \frac{y}{2}\left(1 - \frac{c_L}{c_S}\sin\theta\right)$.
2. $0 \leqslant \theta_1 \leqslant \arcsin\left[\frac{y}{x}\left(1 - \frac{c_L}{c_S}\sin\theta\right) - 1\right]$, for $x \geqslant \frac{y}{2}\left(1 - \frac{L}{c_S}\sin\theta\right)$.

Consider the limiting cases of these conditions.

a) The elastic case: $\omega_1 = \Omega$, $\omega_2 = 0$. The condition on the exit angle is as follows:

$$0 \leqslant \theta_1 \leqslant \arcsin\left(\frac{c_L}{c_S}\sin\theta\right), \tag{3.8}$$

which coincides with the condition in the elastic process.

b) Weak-inelastic case: $\omega_1 = \Omega - \Delta$, $\omega_2 = \Delta$, considering that $\Delta \ll \Omega$, but $\Delta > \Omega\frac{c_L}{c_S}\sin\theta$. The condition takes in account the smallness of $c_L/c_S$

$$0 \leqslant \theta_1 \leqslant \frac{\Delta}{\Omega}. \tag{3.9}$$

c) Inelastic case: $\omega_1 = \omega_2 = \Omega$. The condition is as follows:

$$0 \leqslant \theta_1 \leqslant \frac{\pi}{2}. \tag{3.10}$$

This shows that if the energies of the created phonons are equal to each other, all the exit phonon angles are allowed in the first inelastic process. The ban on these angles is determined by the proximity of the liquid phonon energy to the energy of a solid phonon. According to the conditions on the incident phonon angles, the ratio of velocities $c_L/c_S$ gives a small contribution both to integration limits and to the value of integral (3.6).

Unlike the elastic process, the phonons which were born in this inelastic process will move in all directions relative to the normal to the interface. Then, the phonons that were emitted in all directions should be observed in the angular distribution of phonons emitted by a heated solid to superfluid helium along with a sharp acoustic peak (see figure 1). The rate of the heat flow due to the elastic (3.3) process and the first inelastic (3.6) process is as follows:

$$\frac{W^{(0)}}{W^{(1)}} = \frac{1}{1.08\cdot 10^4}\frac{\pi^6\rho_L c_L^5\hbar^3}{(k_B T_S)^4}. \tag{3.11}$$

Equation (3.11) shows that for $T_S = 5$ K, the heat flux through the superfluid helium-solid interface produced by the first inelastic process is 2.3 times greater than that produced by the elastic process. This value is by a factor of four smaller than the one observed experimentally [10–14]. For $T_S = 1$ K, the contribution of the heat flux from the first inelastic process is by a factor of 272 less than that from the elastic process. Thus, the first inelastic process cannot completely explain the relatively large experimentally observed [10–14] level of background emission.

Analogously, for the third process and the fourth process, the heat flow will be as follows:

$$W^{(3)} = \frac{32\rho_L}{(2\pi)^4\rho_S^2 c_S^4 c_L^3\hbar^6}(k_B T_S)^8\int \mathrm{d}x\mathrm{d}y\sin\theta\mathrm{d}\theta\sin\theta_1\mathrm{d}\theta_1\frac{1}{e^x-1}y^2x^3(x-1)\cos\theta\cos\theta_2\cos^2\theta_1, \tag{3.12}$$

$$W^{(4)} = \frac{32\rho_L}{(2\pi)^4\rho_S^2 c_S^4 c_L^3\hbar^6}(k_B T_S)^8\int \mathrm{d}x\mathrm{d}y\sin\theta\mathrm{d}\theta\sin\theta_1\mathrm{d}\theta_1\frac{1}{e^x-1}\frac{1}{e^{y-x}-1}y^2x^3(x-1)\cos\theta\cos\theta_2\cos^2\theta_1, \tag{3.13}$$

where $x = \hbar\Omega_1/k_B T_S$, $y = \hbar\omega/k_B T_S$ and $\hbar\Omega_2/k_B T_S = x - y$ from the conservation of energy law. Numerical values of integrals that are in (3.12) and (3.13) are $6.32\cdot 10^3$ and $8.53\cdot 10^3$, respectively. The ratio of contributions of the third and the second processes to the contribution of the first process due to (3.6), (3.12) and (3.13) are, respectively, as follows:

$$\frac{W^{(3)}}{W^{(1)}} = 5.21\frac{\rho_L c_L}{\rho_S c_S}, \qquad \frac{W^{(4)}}{W^{(1)}} = 7.07\frac{\rho_L c_L}{\rho_S c_S}. \tag{3.14}$$

The investigation of angular phonon distribution emitted by a heated solid in the third and the fourth processes similar to those that were presented for the first process shows that phonons are emitted in all directions to the superfluid helium. According to equation (3.14), the third and the fourth processes give contributes into the heat flow from solid to superfluid helium of the same order of magnitude. This contribution is by an order of magnitude less than contribution of the first process.

## 4. Conclusion

In this paper we have derived the interaction Hamiltonian (2.7)–(2.9) of phonons of superfluid helium with an oscillating solid interface. The phonon field has been quantized in the half space, which made it possible to write down this Hamiltonian in terms of annihilation and creation operators for phonons of the superfluid helium and of the solid.

The probabilities both of the elastic process and all of the inelastic processes have been calculated from the Hamiltonian. The derived equations allowed us to calculate the heat fluxes from the heated solid to the superfluid helium. The equation for the heat flow, owing to the elastic process, is the same as the result [3] obtained using the methods of classical acoustics.

It has been shown that all of the exit phonons angles in inelastic processes are allowed in a liquid helium, which was observed in experiments [10–14]. According to (3.14), the first inelastic process gives the main contribution to the background channel. The heat flow (3.6) from the solid heated to 5 K to the cold superfluid helium due to the first inelastic process is 2.3 times greater than the heat flow produced by the elastic process. This flow decreases as $T^4$ when the temperature is lowered. Namely, when the temperature of a solid increases, the contribution of the background radiation increases to the contribution of the elastic process, which corresponds to the behavior observed in experiments [10–14, 16]. The absolute values for the heat flux owing to the first inelastic process could only partially explain the big values of the background radiation, which was observed in experiments [10–14] (see figure 1). Calculated in [20] the contribution to the heat transfer coefficient owing to inelastic processes has also proved to be relatively small and could not explain the large values of the heat transfer coefficient, which was observed in the experiments on the Kapitza gap.

Thus, an inelastic process has only partially justified the expectations, and a new investigation of the heat transfer between solid and superfluid helium will be needed to reconcile the theory with the experiments.

## Acknowledgement

We thank A. F. G. Wyatt for useful discussions which led to the initiation of this work.

## References

1. Kapitza P.L., Zh. Eksp. Teor. Fiz., 1941, **11**, 1 (in Russian).
2. Khalatnikov I.M., Zh. Eksp. Teor. Fiz., 1952, **22**, 687 (in Russian).
3. Khalatnikov I.M., An Introduction to the Theory of Superfluidity, Addison-Wesley, Redwood, 1989.
4. Khalatnikov I.M., Theory of Superfluidity, Nauka, Moscow, 1971 (in Russian).
5. Swartz E.T., Pohl R.O., Rev. Mod. Phys., 1989, **61**, 605; doi:10.1103/RevModPhys.61.605.
6. Amrit J., Francois M.X., J. Low Temp. Phys., 2002, **128**, 113; doi:10.1023/A:1016341826786.
7. Amrit J., Thermeau J.P., J. Phys.: Conf. Ser., 2009, **150**, 032002; doi:10.1088/1742-6596/150/3/032002.
8. Amrit J., Phys. Rev. B, 2010, **81**, 054303; doi:10.1103/PhysRevB.81.054303.
9. Amrit J., Ramiere A., Low Temp. Phys., 2013, **39**, 752; doi:10.1063/1.4821076.
10. Wyatt A.F.G., Page G.J., Sherlock R.A., Phys. Rev. Lett., 1976, **36**, 1184; doi:10.1103/PhysRevLett.36.1184.
11. Page G.J., Sherlock R.A., Wyatt A.F.G., Ziebeck K.R.A., In: Phonon Scattering in Solids, L.J. Challis (Ed.), Plenum Press, New York, 1976.
12. Wyatt A.F.G., Crisp G.N., J. Phys. Colloques, 1978, **39**, C6-244; doi:10.1051/jphyscol:19786107.
13. Wyatt A.F.G., In: Nonequlubbrium Superconductivity, Phonons, and Kapitza Boundaries. NATO Advanced Study Institutes Series Vol. 65, Gray K. (Ed.), Plenum Press, New York, 1981, 31–72; doi:10.1007/978-1-4684-3935-9_2.

14. Wyatt A.F.G., Sherlock R.A., Allum D.R., J. Phys. C, 1982, **15**, 1897; doi:10.1088/0022-3719/15/9/012.
15. Adamenko I.N., Nemchenko K.E., Slipko V.A., Wyatt A.F.G., J. Low Temp. Phys., 2009, **157**, 509; doi:10.1007/s10909-009-0011-z;
16. Smith D.H.S, Wyatt A.F.G, Phys. Rev. B, 2007, **76**, 224519; doi:10.1103/PhysRevB.76.224519.
17. Sheard F.W., Bowley R.M., Tombs G.A., Phys. Rev. A, 1973, **8**, 3135; doi:10.1103/PhysRevA.8.3135.
18. Zhukov A., Phys. Scripta, 2004, **69**, 59; doi:10.1238/Physica.Regular.069a00059.
19. Adamenko I.N., Nemchenko E.K., J. Low Temp. Phys., 2013, **171**, 266; doi:10.1007/s10909-012-0754-9.
20. Adamenko I.N., Nemchenko E.K., Low Temp. Phys., 2013, **39**, 756; doi:10.1063/1.4821756.

3

# Effect of magnetic field on electron spectrum in spherical nano-structures

V. Holovatsky*, O. Voitsekhivska, I. Bernik
Chernivtsi National University, 2 Kotsiubynsky St., 58012 Chernivtsi, Ukraine

The effect of a magnetic field on the energy spectrum and on the wave functions of an electron in spherical nano-structures such as single quantum dot and spherical layer is investigated. It is shown that the magnetic field removes the spectrum degeneration with respect to the magnetic quantum number. An increasing magnetic field induction entails a monotonous character of electron energy for the states with $m \geqslant 0$ and a non-monotonous one for the states with $m < 0$. The electron wave functions of the ground state and several excited states are studied considering the effect of the magnetic field. It is shown that $1s$ and $1p$ states are degenerated in the spherical layer driven by a strong magnetic field. In the limit case, a series of size-quantized levels produce the Landau levels which are typical of bulk crystals.

**Key words:** *electron spectrum, quantum dot, spherical layer, magnetic field*

## 1. Introduction

The multilayered spherical nano-structures consisting of a core and a few spherical shells attract a particular attention of scientists. These structures are grown using the chemical colloidal method on the basis of CdS, CdSe, ZnS, ZnSe, HgS and other semiconductor materials [1–3]. Multilayered nano-structures have wide prospects for being utilized in medicine and electronics, for instance for the purpose of fabricating efficient biosensors and fluorescent labels [4, 5]. Multilayered spherical nano-structure containing two quantum wells formed by the core and spherical layer is called quantum-dot-quantum-well (QDQW). Such structures can be the basis for modern highly efficient white light sources [6]. In QDQW, a quasi-particle can be located in one of the quantum wells. The peculiarities of electron and hole location in multilayered spherical nano-structures have been theoretically studied in [7–10]. The impurities, external electrical and magnetic fields produce an effect on the location of quasi-particles in quantum wells [11–19] and consequently effect the optical properties of the structures. The on-center Coulomb impurity does not violate the spherical symmetry of the system and thus the Schrödinger equation for an electron or hole is solved exactly [11–13]. When the external fields are present, the spherical symmetry is violated and the calculation of an energy spectrum becomes more complicated due to the fitting conditions at the interfaces [14–16]. Therefore, in the majority of theoretical studies, the investigation of ground state energies of quasi-particles are performed within the variational method in the frames of the model of infinitely deep potential well [17, 18]. The effect of a magnetic field on the energies and on wave functions of the excited states of quasi-particles is still insufficiently investigated for a spherical quantum dot. The analogous problem for QDQW has not been studied at all.

In the case of high potential barrier of QDQW, the electron in low states does not penetrate through the interfaces of the system and quantum wells become decoupled [20]. The effect of a magnetic field on the states of electron located in the core or in the spherical layer of QDQW is different and can be investigated considering these two potential wells independently. Therefore, in this paper we study the

* E-mail: ktf@chnu.edu.ua

magnetic field effect on the energy spectrum and on the wave functions of an electron located in the quantum dot (QD) and in the spherical layer (SL), assuming that the potential barriers are infinite.

## 2. Schrödinger equation for the electron in spherical nano-structures driven by magnetic field

We consider the spherical QD with the radius $r_0$ and the spherical layer with the inner and outer radii $r_1$ and $r_2$, respectively, having impenetrable boundaries. The coordinate system is taken in such a way that its origin is in the center of the structure and Oz axis coincides with the direction of the magnetic field induction.

The potentials of size quantization for the electron are as follows:

$$U^{(0)}(r) = \begin{cases} 0, & r \leqslant r_0, \\ \infty, & r > r_0, \end{cases} \tag{1}$$

$$U^{(1)}(r) = \begin{cases} 0, & r_1 \leqslant r \leqslant r_2, \\ \infty, & r < r_1,\ r > r_2. \end{cases} \tag{2}$$

Here, $U^{(0)}$ is the potential energy of an electron in spherical QD and $U^{(1)}$ — in SL having impenetrable boundaries.

Schrödinger equations for the electron in these systems in a magnetic field are as follows:

$$\left[\frac{1}{2\mu}\left(\vec{p} - \frac{e}{c}\vec{A}\right)^2 + U^{(0,1)}(r)\right]\psi^{(0,1)}(\vec{r}) = E^{(0,1)}\psi^{(0,1)}(\vec{r}). \tag{3}$$

When $\vec{A} = [\vec{r} \times \vec{B}]/2$, the Hamiltonians become

$$H^{(0,1)} = -\frac{\hbar^2}{2\mu}\Delta + \frac{eB}{2c\mu}L_z + \frac{e^2B^2r^2\sin^2\theta}{8c^2\mu} + U^{(0,1)}(r), \tag{4}$$

where $L_z = -i\hbar\,\partial/\partial\varphi$.

Using the dimensionless magnitudes: $R^* = e^2/(2\varepsilon a^*)$ — effective Rydberg energy, $a^* = \hbar^2\varepsilon/(\mu e^2)$ — effective Bohr radius and parameter $\eta = \hbar\omega_c/(2R^*)$, where $\omega_c = Be/(\mu c)$ — cyclotron frequency, the Hamiltonian (4) is transformed into

$$H^{(0,1)} = -\Delta + \eta L_z + \frac{1}{4}(\eta r \sin\theta)^2 + U^{(0,1)}(r). \tag{5}$$

When $\eta = 0$, the Schrödinger equation with Hamiltonian (5) has the exact solutions

$$\Phi^{(0,1)}_{nlm}(r,\theta,\varphi) = R^{(0,1)}_{nl}(r)Y_{lm}(\theta,\varphi), \tag{6}$$

where $R^{(0)}_{nl}(r) = A^{(0)}_{nl}j_l(\chi_{nl}r/r_0)$, $R^{(1)}_{nl}(r) = A^{(1)}_{nl}j_l(k_{nl}r) + B^{(1)}_{nl}n_l(k_{nl}r)$, $A^{(0)}_{nl} = \sqrt{2}/[r_0^{3/2}j_{l+1}(\chi_{nl})]$, $B^{(1)}_{nl} = -A^{(1)}_{nl}j_l(k_{nl}r_1)/n_l(k_{nl}r_1)$, $j_l(z)$, $n_l(z)$ are Bessel spherical functions of the first and the second kind, respectively, $\chi_{nl}$ are the roots of Bessel spherical function $[j_l(\chi_{nl}) = 0]$, the values $k_{nl}$ are fixed by the condition $R^{(1)}_{nl}(r_2) = 0$ and the coefficients $A^{(1)}_{nl}$ are fixed by the normality condition.

The square term (with respect to the magnetic field) in the Hamiltonian (5) rapidly increases its contribution into the complete energy. Thus, it is impossible to use the perturbation method. In order to obtain the ground state energy using the variational method it is necessary to define the approximated wave function. In [21] it is written as follows:

$$\psi^{(0,1)}_{100}(\vec{r}) = e^{-g(\vec{r})}\Phi^{(0,1)}_{100}(\vec{r}), \tag{7}$$

where the function $g(\vec{r})$ is to ensure the compensation of a quadratic term after substitution of (7) into (5). The minimum condition for

$$F(\eta) = \frac{1}{4}(\eta r \sin\theta)^2 - \left|\vec{\nabla}g(\vec{r})\right|^2 + \Delta g(\vec{r}) \tag{8}$$

at $g(\vec{r})\big|_{\eta\to 0}=0$ fixes $g(\vec{r})=\eta r^2\sin^2\theta/4$. Herein, $F(\eta)$ linearly depends on $\eta$, thus, the variational function of the electron ground state can be written as follows:

$$\psi_{100}^{(0,1)}(\vec{r})=C\exp\left(-\eta r^2\sin^2\theta/4\right)\Phi_{100}^{(0,1)}(\vec{r})\exp(\lambda r), \tag{9}$$

where $C$ is the normality constant and $\lambda$ is the variational parameter. The form of the variational wave function is confirmed by physical considerations. The magnetic field directed along Oz axis deforms the wave function compressing it in perpendicular direction. This fact is represented by an angular dependence of the function (9). Variational parameter $\lambda$ and the energy of ground state are obtained from the minimum of the whole energy

$$E_{100}^{(0,1)}=\min_{\lambda}\left\langle\psi_{100}^{(0,1)}(\vec{r})\middle|H^{(0,1)}\middle|\psi_{100}^{(0,1)*}(\vec{r})\right\rangle. \tag{10}$$

In order to study the excited states, the orthonormality condition for the wave functions should be fulfilled, which makes the problem rather complicated. Therefore, we are going to use another method to solve it. We expand the wave function using a complete set of eigenfunctions of the electron in a spherical nano-structure without the magnetic field obtained as the exact solutions of Schrödinger equation [15]. When the magnetic field is applied, the spherical symmetry is violated and the orbital quantum number becomes inconvenient. The new states characterized by a magnetic quantum number $m$ are presented as a linear combination of the states $\Phi_{nlm}^{(0,1)}(\vec{r})$

$$\psi_{jm}^{(0,1)}(\vec{r})=\sum_{n}\sum_{l}c_{nl}\Phi_{nlm}^{(0,1)}(\vec{r}). \tag{11}$$

Substituting (11) into Schrödinger equation with Hamiltonian (5), we obtain a secular equation for the electron energy spectrum

$$\left|H_{nl,n'l'}^{(0,1)}-E_{jm}^{(0,1)}\delta_{n,n'}\delta_{l,l'}\right|=0, \tag{12}$$

where the matrix elements $H_{n'l',nl}^{(0,1)}$ have the form

$$H_{n'l',nl}^{(0,1)}=\left(E_{nl}^{0(0,1)}+m\eta\right)\delta_{n'n}\delta_{l'l}+\frac{\eta^2}{2}\left\{\alpha_{l,m}\delta_{l',l+2}+\beta_{l,m}\delta_{l',l}+\gamma_{l,m}\delta_{l',l-2}\right\}I_{n'l',nl}^{(0,1)}, \tag{13}$$

$$I_{n'l',nl}^{(0)}=A_{n'l'}^{(0)}A_{nl}^{(0)}\int_0^{r_0}r^4 j_{l'}(\chi_{n'l'}r/r_0)j_l(\chi_{nl}r/r_0)\mathrm{d}r,$$

$$I_{n'l',nl}^{(1)}=\int_{r_1}^{r_2}r^4\left[A_{n'l'}j_{l'}(k_{n'l'}r)+B_{n'l'}n_{l'}(k_{n'l'}r)\right]\left[A_{nl}j_l(k_{nl}r)+B_{nl}n_l(k_{nl}r)\right]\mathrm{d}r,$$

$$\alpha_{l,m}=-\sqrt{\frac{[(l+2)^2-m^2][(l+1)^2-m^2]}{(2l+5)(2l+3)^2(2l+1)}},$$

$$\beta_{l,m}=1-\frac{(l+1)^2-m^2}{(2l+1)(2l+3)}-\frac{l^2-m^2}{4l^2-1},\qquad \gamma_{l,m}=-\sqrt{\frac{(l^2-m^2)[(l-1)^2-m^2]}{(2l+1)(2l-1)^2(2l-3)}}.$$

Using the eigenvalues and eigenvectors of the matrix

$$F_{nl,n'l'}^{(0,1)}=H_{nl,n'l'}^{(0,1)}-E_{jm}^{(0,1)}\delta_{n,n'}\delta_{l,l'}, \tag{14}$$

we obtain the energy spectrum and wave functions of the electron in a spherical nano-structure driven by the magnetic field.

**Table 1.** Expansion coefficients of electron wave functions $\overset{(0)}{\psi}_{100}, \overset{(0)}{\psi}_{110}, \overset{(0)}{\psi}_{11-1}$ in QD and $\overset{(1)}{\psi}_{100}, \overset{(1)}{\psi}_{110}, \overset{(1)}{\psi}_{11-1}$ in SL at $B = 40$ T.

| | $\overset{(0)}{\psi}_{100}$ | $\overset{(1)}{\psi}_{100}$ | $\overset{(0)}{\psi}_{110}$ | $\overset{(1)}{\psi}_{110}$ | $\overset{(0)}{\psi}_{11-1}$ | $\overset{(1)}{\psi}_{11-1}$ |
|---|---|---|---|---|---|---|
| $c_{10}$ | 0.9986 | 0.7018 | – | – | – | – |
| $c_{11}$ | – | – | 0.9993 | 0.9151 | 0.9991 | 0.7949 |
| $c_{12}$ | 0.0378 | 0.6888 | – | – | – | – |
| $c_{13}$ | – | – | 0.0336 | 0.3969 | 0.0282 | 0.5923 |
| $c_{14}$ | 0.0007 | 0.1801 | – | – | – | – |
| $c_{15}$ | – | – | 0.0006 | 0.0705 | 0.0005 | 0.1306 |
| $c_{16}$ | 0 | 0.0232 | – | – | – | – |
| $c_{17}$ | – | – | 0 | 0,0069 | 0 | 0.0144 |
| $c_{20}$ | 0.0360 | 0.0044 | – | – | – | – |
| $c_{21}$ | – | – | 0.0161 | 0.0044 | 0.0331 | 0.0087 |
| $c_{22}$ | 0.0009 | 0.0014 | – | – | – | – |
| $c_{23}$ | – | – | 0.0010 | 0.0007 | 0.0007 | 0.0022 |
| $\sum_i c_i^2$ | 1.000 | 1.000 | 1.000 | 1.000 | 1.000 | 1.000 |

## 3. Analysis of the results

The computer calculations were performed using the physical parameters of CdSe semiconductor material: electron effective mass $\mu = 0.13\, m_e$ ($m_e$ — the mass of pure electron), dielectric constant $\varepsilon = 10.6$.

Expanding the wave functions (13) we took into account a sufficient number of terms, provided that the sum of squares of expansion coefficients was equal to a unit with the accuracy not less than 0.01%. In table 1 the expansion coefficients for the lowest states at $B = 40$ T are presented. Here one can see that the required accuracy is provided by 6 major terms. For convenience, we use the same quantum numbers characterizing the states of an electron in a nano-structure driven by a magnetic field as the ones without the field.

The dependencies of electron energy spectrum on the magnetic field induction in CdSe QD and SL are presented in figure 1.

From figure 1 one can see that the energy of $\psi_{100}$ state calculated by a variational method correlates well to the one obtained by the matrix method. Moreover, even in the case of a strong magnetic field, the

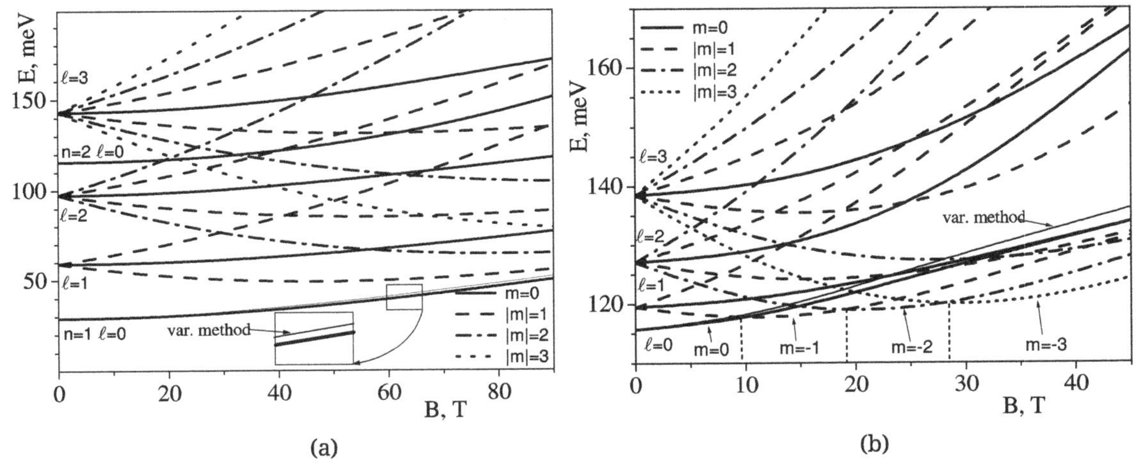


**Figure 1.** Electron energy spectrum as function of magnetic field induction in QD with $r_0 = 10$ nm (a) and SL with $r_1 = 10$ nm, $r_2 = 15$ nm (b).

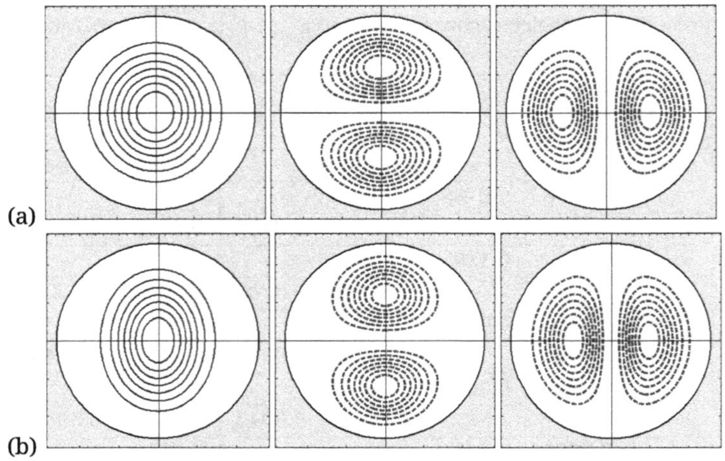

**Figure 2.** Distribution of probability density of electron location in QD with $r_0 = 10$ nm, at $B = 40$ T (a), $B = 80$ T (b) for the quantum states with $\psi^{(0)}_{100}$, $\psi^{(0)}_{110}$, $\psi^{(0)}_{11-1}$.

error for the electron energy in both nano-structures does not exceed 3%. Comparing the dependencies shown in figures 1 (a) and 1 (b), one can see that the magnetic field produces a greater effect on the energy states of an electron located in SL than on the energy states in QD. In both nano-structures, the degeneracy over the magnetic quantum number is removed. The energies of the states with $m \geqslant 0$ increase under the effect of the magnetic field. For the states with $m < 0$, the non-monotonous dependence of the energy on the magnetic field is caused by the linear and quadratic terms contributed by the magnetic field into the Hamiltonian (7).

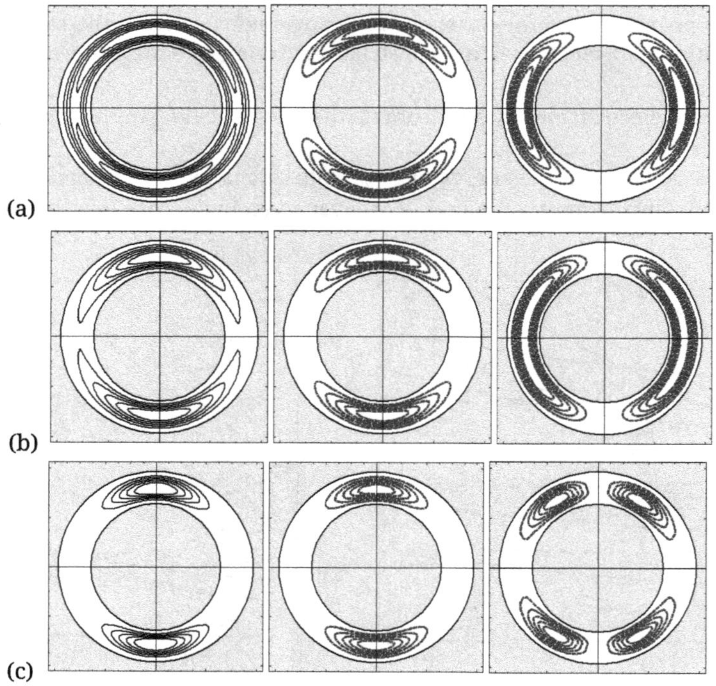

**Figure 3.** Distribution of probability density of electron location in SL with $r_1 = 10$ nm, $r_2 = 15$ nm at $B = 10$ T (a), $B = 20$ T (b) and $B = 40$ T (c) for the quantum states with $\psi^{(1)}_{100}$, $\psi^{(1)}_{110}$, $\psi^{(1)}_{11-1}$.

In the SL placed into a strong magnetic field, $1s$ and $1p$ states with $m = 0$ are degenerated, unlike in the QD. However, in zero-dimensional systems, the electron ground state is always non-degenerated. Consequently, when the magnetic field increases, the lowest states with $m = 0, -1, -2, \ldots$ successively play the role of the ground state. The ground state of an electron with a certain value of a magnetic quantum number transforms into the state with the other $m$ when the magnetic field intensity increases at an equal magnitude. The distance between the points of such a transition increases when SL radius becomes smaller. A similar behavior of the electron ground state energy was theoretically obtained and experimentally confirmed for the semiconductor quantum rings [22]. The oscillation of energies is known as Aharonov Bohm effect.

The distribution of probability density of electron location in QD and SL in different quantum states is presented in figures 2, 3, respectively.

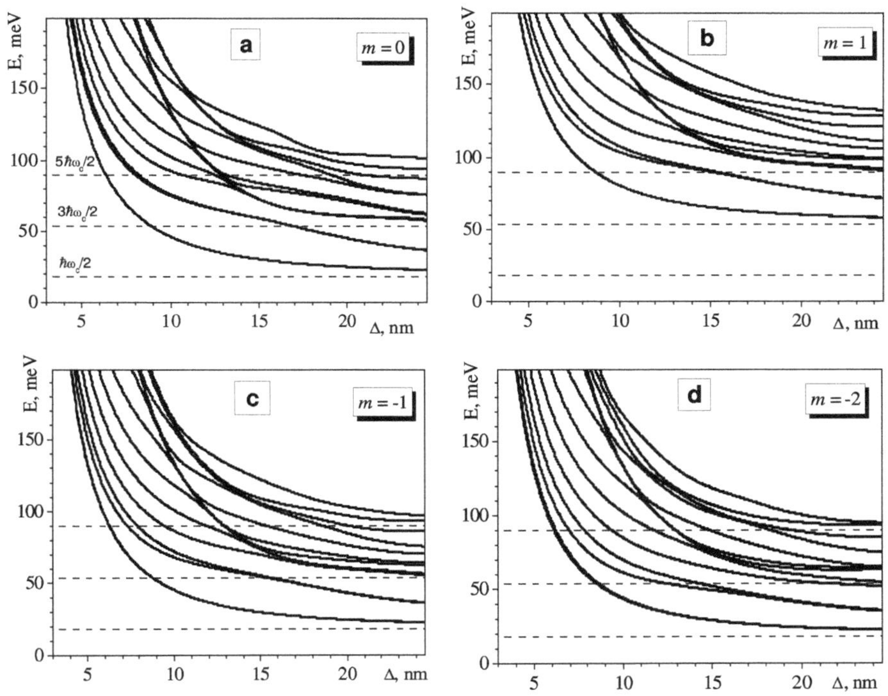


**Figure 4.** Electron energy spectrum as function of $\Delta = r_2 - r_1$ at $B = 40$ T for $m = 0$ (a), $m = 1$ (b), $m = -1$ (c) and $m = -2$ (d). Landau levels in bulk crystal are shown by dash lines.

Figures 2, 3, prove that the electron wave functions are deformed due to the effect of the strong magnetic field. When its induction increases, the angular probability increases near $\theta = 0, \pi$ and decreases near $\theta = \pi/2$. Herein, in the SL the wave function $\psi^{(1)}_{100}$ of $1s$ state, due to the deformation, becomes similar to the $\psi^{(1)}_{110}$ one of the excited $1p$ state. At $B = 40$ T, these states become indistinguishable both for the distribution of probability density [figure 3 (c)] and the energy [figure 1 (b)].

Computer calculations prove that in the limit case when the inner radius of SL $r_1$ diminishes at $r_2 =$ const and at the constant magnetic field induction, the electron energy spectrum coincides with the one for the QD with $r_0 = r_2$.

In limit cases, when the QD radius or SL thickness increases at $B =$ const due to a decrease of the quantum confining effect, the electron energy levels should coincide with Landau levels which are typical of a bulk crystal placed into the magnetic field. The process of the formation of Landau levels for the quantum states with $m = 0, 1, -1$ in the SL is presented in figure 4.

Figure 4 proves that the quantum confining effect diminishes when the sizes of the structure increase. This process is accompanied by a decrease of all energy levels and by the formation of Landau levels. For example, at $m = 0$, the lowest Landau level is formed by the set of levels with $n = 1$, the next one with

$n = 2$ and so on. A complicated dependence of an energy spectrum is observed during their formation due to the anti-crossing effect.

## 4. Summary

We studied the electron energy spectrum in QD and SL under the effect of a magnetic field. The problem is solved using the variational method and the method of electron wave function expansion over the set of eigenfunctions being the exact solutions of Schrödinger equation for the same structures without the magnetic field. The results obtained within the both methods are in good agreement. The variational method describes the lowest electron state with $m = 0$. However, the ground state in SL is formed by the states with $m = 0, -1, -2, \ldots$ consequently with the growth of the magnetic field induction. It is shown that the major contribution into the expansion of a wave function of an arbitrary electron state even in the strong magnetic field is performed by a few neighboring (over the energy) quantum states which are the exact solutions of Schrödinger equation for the electron when there is no magnetic field.

The wave functions of the electron in QD and SL are deformed under the effect of a magnetic field. The degeneration of an energy spectrum with respect to a magnetic quantum number is removed. The electron energies for the states with positive and negative values of a magnetic quantum number differently depend on the magnetic field induction: for the states with $m \geqslant 0$, the energy monotonously increases and for $m < 0$, the energy decreases at first and then, only when the magnetic field becomes strong enough, enhances.

It is proven that the effect of a magnetic field on the electron energy spectrum in SL is stronger than that in QD. Moreover, it is shown that in SL driven by the strong magnetic field, the neighboring states with the same $m$ become degenerated. For example, in the studied structure, $1s$ and $1p$ states are degenerated at $B > 30$ T. The degeneration of higher energy states takes place at a stronger magnetic field.

The validity of the obtained results is confirmed by the limit cases: when the quantum confining effect reduces at a constant magnetic field induction, the energy levels rebuild, saturate and form the respective Landau levels. When the inner radius of SL ($r_1$) decreases at $r_2 = \text{const}$, the electron energy spectrum becomes the same as in the QD having the radius $r_2$.

The results of the investigation make it possible to estimate the energies and the most probable place of electron location in QDQW. Different dependencies of electron energies in QD and SL on the magnetic field induction lead to their anticrossing in QDQW. This feature permits to change the quasiparticle location by a magnetic field for multilayered systems with penetrable interfaces.

## References

1. Eychmuller A., Mews A., Weller H., Chem. Phys. Lett., 1993, **208**, 59; doi:10.1016/0009-2614(93)80076-2.
2. Dorfs D., Eychmuller A., Z. Phys. Chem., 2006, **220**, 1539; doi:10.1524/zpch.2006.220.12.1539.
3. Little R., El-Sayed M., Bryant G., Burke S., Chem. Phys., 2001, **114**, 1813.
4. Frasco M.F., Chaniotakis N., Sensors, 2009, **9**, 7266; doi:10.3390/s90907266.
5. Liu Y.S., Sun Y., Vernier P.T., Liang C. H., Chong S.Y., Gundersen M.A., J. Phys. Chem. C. Nanomater Interfaces, 2007, **111**, 2872; doi:10.1021/jp0654718.
6. Demir H., Nizamoglu S., Mutlugun E., Ozel T., Sapra S., Gaponik N., Eychmuller A., Nanotechnology, 2008, **19**, 335203; doi:10.1063/1.2898892.
7. SalmanOgli A., Rostami A., J. Nanopart. Res., 2011, **13**, 1197; doi:10.1007/s11051-010-0112-2.
8. Tkach N., Voitsekhovska O., Holovatsky V., Mihalyova M., Izvestiya vuzov. Fizika, 1998, **12**, 58 (in Russian) [Russ. Phys. J., 1998, **41**, 1229; doi:10.1007/BF02514561].
9. Holovatsky V., J. Phys. Stud., 1998, **2**, 583 (in Ukrainian).
10. Tkach M., Holovatsky V., Voitsekhivska O., Mikhalyova M., Electrochemical Society Proceedings, 1998, **25**, 316.
11. Tkach M., Holovatsky V., Voitsekhivska O., Fiz. Teh. Pol., 2000, **34**, 602 (in Russian) [Semiconductors, 2000, **34**, 583; doi:10.1134/1.1188032].
12. Holovatsky V., Makhanets O., Voitsekhivska O., Physica E, 2009, **41**, 1522; doi:10.1016/j.physe.2009.04.027.
13. Boichuk V.I., Bilynskyi I.V., Leshko R.Ya., Voronyak L.Ya., Ukr. J. Phys., 2009, **54**, 1021.
14. Tas H., Sahin M., J. Appl. Phys., 2012, **111**, 083702; doi:10.1063/1.4751483.
15. Wu S., Wan L., J. Appl. Phys., 2012, **111**, 063711; doi:10.1063/1.3695454.

16. Rahmani K., Zorkani I., M. J. Condensed Matter., 2009, **11**, 35.
17. Chakraborty T., Apalkov V., Physica E, 2003, **16**, 253; doi:10.1016/S1386-9477(02)00674-4.
18. Xiao Z., J. Appl. Phys., 1999, **86**, 4509; doi:10.1063/1.371394.
19. Planelles J., Diaz J., Climente J., Jaskolski W., Phys. Rev. B, 2002, **65**, 245302; doi:10.1103/PhysRevB.65.245302.
20. Battaglia D., Blackman B., Peng X., J. Am. Chem. Soc., 2005, **127**, 10889-10897; doi:10.1021/ja0437297.
21. Jiang H., Phys. Rev. B., 1987, **35**, 9287; doi:10.1103/PhysRevB.35.9287.
22. Lorke A., Luyken R., Govorov A., Kotthaus J.J., Garcia M., Petroff P., Phys. Rev. B, 2000, **84**, 2223; doi:10.1103/PhysRevLett.84.2223.

# 4

# Complex conductivity in strongly fluctuating layered superconductors

B.D. Tinh, L.M. Thu, L.V. Hoa

Department of Physics, Hanoi National University of Education, 136 Xuanthuy, Caugiay, Hanoi, Vietnam

The time-dependent Ginzburg-Landau approach is used to calculate the complex fluctuation conductivity in layered type-II superconductor under magnetic field. Layered structure of the superconductor is accounted for by means of the Lawrence-Doniach model, while the nonlinear interaction term in dynamics is treated within self-consistent Gaussian approximation. In high-$T_c$ materials, large portion of the $H-T$ diagram belongs to vortex liquid phase. The expressions summing contributions of all the Landau levels are presented in explicit form which are applicable essentially to the whole phase and are compared to experimental data on high-$T_c$ superconductor $YBa_2Cu_3O_{7-\delta}$. Above the crossover to the "normal phase", our results agree with the previously obtained.

**Key words:** *time-dependent Ginzburg-Landau, complex conductivity, type-II superconductor*

## 1. Introduction

There has been a renewed interest in the effect of strong thermal fluctuations in layered high $T_c$ superconductors that exhibit Nernst effect [1] well above $T_c$. The experiments were interpreted microscopically as due to virtual (preformed) pairs [2], although can be described on the "mesoscopic" level using Ginzburg-Landau approach [3]. This indicates large renormalization of $T_c$ (of order 1) by thermal fluctuations in strongly layered materials like $Bi_2Sr_2CaCuO_{8+\delta}$ and $La_{2-x}Sr_xCuO_4$. Microscopic parameters of the material determine both the actual $T_c$ at which the Cooper pairs are coherent and the mean field critical temperature $T_c^{MF}$ until which incoherent Cooper pairs exist and effect the transport, magnetism and thermodynamics of the material. More recently strong diamagnetism was also observed [4], although it is still debated on experimental level. All these new results are concerned with DC transport.

The AC transport in strongly fluctuating type-II superconductors have been a subject for active research for many years, both theoretically and experimentally. While the seminal calculation of the enhancement of the DC conductivity in the normal phase due to virtual Cooper pair created by thermal fluctuations by Aslamazov and Larkin was done in the framework of the microscopic BCS theory [5], the approach becomes cumbersome in more complicated situations involving external magnetic field, layerred structure, etc. As usual in these circumstances (especially in the absense of a simple accepted microscopic model for high $T_c$ and other recently discovered "unconventional" strongly fluctuating superconductors), a more phenomenological Ginzburg-Landau approach adapted to incorporate thermal fluctuations turns out to be more effective [5, 6]. A general method to model the thermal fluctuations in dynamics is to add a random Langevin white noise to the time-dependent Ginzburg-Landau (TDGL) equations. The transport coefficients within this approach are obtained as a long time limit of a driven system. The model, therefore, becomes rather complicated and approximations should be made. In the gaussian fluctuations regime (in the normal phase not very close to criticality, where the quartic term in the GL free energy is dominant), the expressions for complex conductivity at zero magnetic field have been obtained very early on [7, 8]. This was expanded later by Dorsey and coworkers [9, 10] to include the critical fluctuations region by a variety of nonperturbative methods (Hartree approximation, large number of components $N$ limit, $\varepsilon$-expansion. The results were in line [9] with general physical scaling arguments by Fisher, Fisher, and Huse [11].

The complex conductivity in magnetic field in the normal phase was calculated using TDGL equation by Larkin and Varlamov [5]. The general expression valid for complex conductivity in layered superconductors under the assumption of gaussian fluctuations (neglecting the quartic in the order parameter term in the GL free energy) was also calculated and presented in [5] as a sum over all the Landau levels. On the opposite side of the phase diagram, namely in a strongly pinned case (vortex glass and Bragg glass), the same quantity was calculated using both macroscopic elastic theory [12] and the TGDL [13]. In yet another limit of the Abrikosov lattice phase of a clean superconductor, see [6], the complex conductivity was recently calculated [14] (within the lowest Landau level). The present work is complementary to all these in that we concentrate on the thermally depinned homogeneous phase marked as "vortex liquid" [6]. The complex conductivity in magnetic field in 2D and 3D was calculated using TDGL equation [15, 16]. In strongly layered high $T_c$ materials, this portion of the magnetic phase diagram is very large and consequently well studied experimentally [17–19]. In this region, thermal fluctuations are so strong that one cannot neglect the quartic in the order parameter term of the GL energy. This term, however, can be incorporated self-consistently into the framework of [5].

In this paper, the complex conductivity including all Landau levels is calculated in a layered superconductor under magnetic field in the vortex liquid phase by using TDGL approach with thermal fluctuations modelled by the Langevin white noise. We obtain an expression summing all Landau levels in an explicit form. The rest of the paper is organized as follows. In section 2, the Lawrence-Donich model in its time dependent form is briefly recalled and the main assumptions are specified. In section 3, the interaction term in dynamics is treated within self-consistent gaussian approximation sufficient for description of the vortex liquis. The complex conductivity calculation within the same approximation is the subject of section 4. The results are compared with experimental data on HTSC in section 5, while the work is summarized in section 6.

## 2. Thermal fluctuations in the time dependent GL Lawrence-Doniach model

Cooper pairing in layered superconductors can be described by the 2D distribution of the order parameter $\Psi_n(\mathbf{r})$ in each of the layers labeled by $n$. The Lawrence-Doniach version of the GL free energy includes the Josephson coupling between the layers [5]:

$$F_{\mathrm{GL}} = s' \sum_n \int \mathrm{d}^2 r \left( \frac{\hbar^2}{2m^*} |\mathbf{D}\Psi_n|^2 + \frac{\hbar^2}{2m_c d'^2} |\Psi_n - \Psi_{n+1}|^2 + a|\Psi_n|^2 + \frac{b'}{2}|\Psi_n|^4 \right). \tag{2.1}$$

Here, $s'$ is the order parameter effective "thickness" and $d' > s'$ is the distance between layers. The Lawrence-Doniach model approximates the paired electrons density of states by homogeneous infinitely thin planes separated by distance $d'$. For simplicity, we assume $a = \alpha T_c^{\mathrm{MF}}(t-1)$, $t^{\mathrm{MF}} \equiv T/T_c^{\mathrm{MF}}$, although this temperature dependence can be easily modified to better describe the experimental coherence length. The "mean field" critical temperature $T_c^{\mathrm{MF}}$ depends on UV cutoff and is often much larger than "renormalized" critical temperature $T_c$. This temperature is significantly higher than the measured critical temperature $T_c$ due to strong thermal fluctuations on the mesoscopic scale.

The covariant derivatives are defined by $\mathbf{D} \equiv \nabla + \mathrm{i}(2\pi/\Phi_0)\mathbf{A}$, where the vector potential describes a constant and homogeneous magnetic field $\mathbf{A} = (-By, 0)$ and $\Phi_0 = hc/e^*$ is the flux quantum with $e^* = 2|e|$. The two scales, the coherence length $\xi^2 = \hbar^2/(2m^*\alpha T_c)$, and the penetration depth, $\lambda^2 = c^2 m^* b'/(4\pi e^{*2}\alpha T_c)$ define the GL ratio $\kappa \equiv \lambda/\xi$, which is very large for HTSC. In the case of strongly type-II superconductors, the magnetization is by a factor $\kappa^2$ smaller than the external field for magnetic field larger than the first critical field $H_{c1}(T)$, so that we take $B \approx H$. The electric current, $\mathbf{J} = \mathbf{J}^{\mathrm{n}} + \mathbf{J}^{\mathrm{s}}$, includes both the Ohmic normal part

$$\mathbf{J}^{\mathrm{n}} = \sigma_n \mathbf{E}, \tag{2.2}$$

and the supercurrent

$$\mathbf{J}_n^{\mathrm{s}}(\mathbf{r}) = \frac{\mathrm{i}e^*\hbar}{2m^*} \left( \Psi_n^* \mathbf{D}\Psi_n - \Psi_n \mathbf{D}\Psi_n^* \right). \tag{2.3}$$

Since we are interested in a transport phenomenon, it is necessary to introduce a dynamics of the order parameter. The simplest one is a gauge-invariant version of the "type A" relaxational dynamics [20]. In the presence of thermal fluctuations, which on the mesoscopic scale are represented by a complex white noise, it reads:

$$\frac{\hbar^2\gamma'}{2m^*}D_\tau\Psi_n = -\frac{1}{s'}\frac{\delta F_{\mathrm{GL}}}{\delta\Psi_n^*}+\zeta_n, \tag{2.4}$$

with correlator

$$\left\langle \zeta_n(\mathbf{r},\tau),\zeta_m(\mathbf{r}',\tau')\right\rangle = \frac{\hbar^2\gamma' T}{m^* s'}\delta_{nm}\delta(\mathbf{r}-\mathbf{r}')\delta(\tau-\tau'). \tag{2.5}$$

Here, $D_\tau \equiv \partial/\partial\tau - \mathrm{i}(e^*/\hbar)\Phi$ is the covariant time derivative, with $\Phi = -E_\tau y$ being the scalar electric potential describing the driving force in a purely dissipative dynamics. The electric field is, to a good approximation in the vortex liquid phase coordinate, independent (at least for frequences below THz range, see argumentation in [21]), but is a monorchromatic periodic function of time

$$E_x = 0, \qquad E_y(\tau) = E\exp(-\mathrm{i}\omega\tau). \tag{2.6}$$

Throughout most of the paper we use the coherence length $\xi$ as a unit of length and $H_{c2} = \Phi_0/2\pi\xi^2$ as a unit of the magnetic field, with dimensionles field $b = B/H_{c2}$. In analogy to the coherence length, one defines a characteristic time scale: the GL "relaxation" time $\tau_{\mathrm{GL}} = \gamma'\xi^2/2$. Similarly, it is convenient to use the following unit of the electric field, $E_{\mathrm{GL}} = H_{c2}\xi/c\tau_{\mathrm{GL}}$, so that the dimensionless field is $\mathscr{E} = E/E_{\mathrm{GL}}$. The dynamical equation, equation (2.4), written in dimensionless units reads:

$$\left(D_\tau - \frac{1}{2}D^2\right)\psi_n + \frac{1}{2d^2}(2\psi_n - \psi_{n+1} - \psi_{n-1}) - \frac{1-t^{\mathrm{MF}}}{2}\psi_n + |\psi_n|^2\psi_n = \overline{\zeta}_n, \tag{2.7}$$

where $d = d'/\xi_z$ is dimensionless layer distance. The coherence length perpendicular to the layers is smaller compared to $\xi$ by the anisotropy parameter $\gamma$ [6].

The covariant time derivatives become $D_\tau = \partial/\partial\tau + \mathrm{i}\mathscr{E}(\tau)y$, the covariant derivatives are defined by $D_x = \partial/\partial x - \mathrm{i}by$, $D_y = \partial/\partial y$. The "mean field" critical temperature $T_c^{\mathrm{MF}}$ depends on the ultraviolet (UV) cutoff. This temperature is higher than the measured critical temperature $T_c$ due to strong thermal fluctuations on the mesoscopic scale, and it will be renormalized later. The dimensionless Langevin white-noise forces, $\overline{\zeta}_n = b'^{1/2}\left(2\alpha T_c^{\mathrm{MF}}\right)^{-3/2}\zeta_n$, are correlated through $\left\langle \overline{\zeta}_n^*(\mathbf{r},\tau)\overline{\zeta}_m(\mathbf{r}',\tau')\right\rangle = 2\eta t/s\delta_{nm}\delta(\mathbf{r}-\mathbf{r}')\delta(\tau-\tau')$, with a dimensionless fluctuation strength parameter related to the well known Ginzburg number [5, 6] by

$$\eta = \pi\sqrt{2\mathrm{Gi}}, \qquad \mathrm{Gi} = \frac{1}{2}\left(\frac{8e^2\kappa^2\xi T_c^{\mathrm{MF}}\gamma}{c^2\hbar^2}\right)^2. \tag{2.8}$$

The dimensionless current density is $\mathbf{J}^s = J_{\mathrm{GL}}\mathbf{j}^s$, where

$$\mathbf{j}_n^s = \frac{\mathrm{i}}{2}\left(\psi_n^*\mathbf{D}\psi_n - \psi_n\mathbf{D}\psi_n^*\right) \tag{2.9}$$

with $J_{\mathrm{GL}} = cH_{c2}/(2\pi\xi\kappa^2)$ being the unit of the current density. Consistently, the conductivity will be given in the units of

$$\sigma_{\mathrm{GL}} = \frac{J_{\mathrm{GL}}}{E_{\mathrm{GL}}} = \frac{c^2\gamma'}{4\pi\kappa^2}. \tag{2.10}$$

This unit is close to the normal state conductivity $\sigma_n$ in dirty limit superconductors [22]. In general, there is a factor $k$ of the order one relating the two: $\sigma_n = k\sigma_{\mathrm{GL}}$.

## 3. The Green's function of TDGL in Gaussian approximation

Let us first assume that the vortex liquid is not driven by the electric field. As mentioned above, the cubic term in the TDGL equation (2.7) can be treated in the self-consistent gaussian approximation (explained in detail in [23]) by replacing $|\psi_n|^2\psi_n$ with a linear one $2\left\langle|\psi_n|^2\right\rangle\psi_n$

$$\left(\frac{\partial}{\partial\tau} - \frac{1}{2}D^2 - \frac{b}{2}\right)\psi_n + \frac{1}{2d^2}(2\psi_n - \psi_{n+1} - \psi_{n-1}) + \varepsilon\psi_n = \overline{\zeta}_n. \tag{3.1}$$

Here, the value of the coefficient of the linear term,

$$\varepsilon = -\frac{1 - t^{\mathrm{MF}} - b}{2} + 2\left\langle |\psi_n|^2 \right\rangle, \tag{3.2}$$

is different from the noninteracting one.

The relaxational linearized TDGL equation with a Langevin noise, equation (3.1), is solved using the retarded (vanishing for $\tau < \tau'$) Green function (GF) $G^0_{k_z}(\mathbf{r},\tau;\mathbf{r}',\tau')$:

$$\psi_n(\mathbf{r},\tau) = \int_0^{2\pi/d} \frac{\mathrm{d}k_z}{2\pi} \mathrm{e}^{-ink_z d} \int \mathrm{d}\mathbf{r}' \int \mathrm{d}\tau' G^0_{k_z}(\mathbf{r},\tau;\mathbf{r}',\tau') \overline{\zeta}_{k_z}(\mathbf{r}',\tau'). \tag{3.3}$$

The GF satisfies

$$\left\{ \frac{\partial}{\partial \tau} - \frac{1}{2} D^2 - \frac{b}{2} + \frac{1}{d^2} [1 - \cos(k_z d)] + \varepsilon \right\} G^0_{k_z}(\mathbf{r},\mathbf{r}',\tau - \tau') = \delta(\mathbf{r} - \mathbf{r}')\delta(\tau - \tau'). \tag{3.4}$$

The GF is a Gaussian

$$G^0_{k_z}(\mathbf{r},\mathbf{r}',\overline{\tau}) = \theta(\overline{\tau}) C_{k_z} \exp\left[\frac{ib}{2} X (y + y')\right] \exp\left(-\frac{X^2 + Y^2}{2\beta}\right), \tag{3.5}$$

where $X = x - x'$, $Y = y - y'$, $\overline{\tau} = \tau - \tau'$. $\theta(\overline{\tau})$ is the Heaviside step function, $C$ and $\beta$ are coefficients.

Substituting equation (3.5) into equation (3.4), one obtains:

$$\beta = \frac{2}{b} \tanh(b\overline{\tau}/2), \tag{3.6}$$

$$C = \frac{b}{4\pi} \exp\left(-\left\{\varepsilon - \frac{b}{2} + \frac{1}{d^2}[1 - \cos(k_z d)]\right\} \overline{\tau}\right) \left[\sinh\left(\frac{b\overline{\tau}}{2}\right)\right]^{-1}. \tag{3.7}$$

The thermal average of the density of Cooper pairs can be expressed via the Green's functions:

$$\begin{aligned} \left\langle |\psi_n(\mathbf{r},\tau)|^2 \right\rangle &= 2\omega t \frac{d}{s} \int_0^{2\pi/d} \frac{\mathrm{d}k_z}{2\pi} \int \mathrm{d}\mathbf{r}' \int \mathrm{d}\tau' \left| G^0_{k_z}(\mathbf{r} - \mathbf{r}', \tau - \tau') \right|^2 \\ &= \frac{\omega t b}{2\pi s} \int_{\overline{\tau} = \tau_{\mathrm{c}}}^{\infty} \frac{f(\varepsilon, \overline{\tau})}{\sinh(b\overline{\tau})}, \end{aligned} \tag{3.8}$$

where

$$f(\varepsilon,\overline{\tau}) = \exp\left[\frac{2v^2}{b} \tanh\left(\frac{b\overline{\tau}}{2}\right)\right] \mathrm{e}^{-(2\varepsilon - b + v^2)\overline{\tau}} \mathrm{e}^{-2\overline{\tau}/d^2} I_0(2\overline{\tau}/d^2). \tag{3.9}$$

Substituting the density of Cooper pairs (3.8) into equation (3.2) and after renormalization [23], the latter takes the form

$$\varepsilon = -\frac{1 - b - t}{2} - \frac{\eta t}{\pi s} \int_0^{\infty} \mathrm{d}\overline{\tau} \ln[\sinh(b\overline{\tau})] \frac{\mathrm{d}}{\mathrm{d}\overline{\tau}} \left[\frac{g(\varepsilon,\overline{\tau})}{\cosh(b\overline{\tau})}\right] + \frac{\omega t}{\pi s} [\gamma_{\mathrm{E}} - \ln(bd^2)], \tag{3.10}$$

where $g(\varepsilon,\overline{\tau}) = \mathrm{e}^{-(2\varepsilon - b + 2/d^2)\overline{\tau}} I_0(2\overline{\tau} d^{-2})$ with $I_0(x) = (1/2\pi)\int_0^{2\pi} \mathrm{e}^{x\cos\theta} \mathrm{d}\theta$ being the modified Bessel function, $\overline{\tau} = \tau - \tau'$, $t = T/T_{\mathrm{c}}$, and $\gamma_{\mathrm{E}} = 0.577$ is Euler constant. As was discussed in [23] there are UV divergences in the intermediate steps, equation (3.2), that are regularized by a cutoff. There is a degree of arbitrariness in its choice and it was shown that in the TDGL theory, the cutoff having a dimension of time, $\tau_{\mathrm{c}}$, rather than energy, is most convenient. However, in the renormalized equation, equation (3.10), the cutoff does not appear: the mean field temperature $T_{\mathrm{c}}^{\mathrm{MF}}$ is now replaced by $T_{\mathrm{c}}$. Having determined the GF, one can use it to calculate the complex conductivity.

## 4. The complex conductivity of a layered superconductor

After the thermal averaging of the supercurrent density at time $\tau$, defined by equation (2.9), it can be expressed via the Green's functions as follows:

$$j_y^{\mathrm{s}}(\tau)=\mathrm{i}\eta t\frac{d}{s}\int_0^{2\pi/d}\frac{\mathrm{d}k_z}{2\pi}\int_{r',\tau'}G_{k_z}^{*}\left(\mathbf{r},\mathbf{r}',\tau-\tau'\right)\frac{\partial}{\partial y}G_{k_z}\left(\mathbf{r},\mathbf{r}',\tau-\tau'\right)+\text{c.c.},\tag{4.1}$$

where $G_{k_z}\left(\mathbf{r},\mathbf{r}',\tau-\tau'\right)$ is the Green's function of the linearized TDGL equation (2.7) in the presence of the scalar potential describing the electric field. The correction to the Green's function to linear order in the time dependent homogeneous electric field $\mathscr{E}(\tau)$ is as follows:

$$G_{k_z}(\mathbf{r},\mathbf{r}',\tau-\tau')=G_{k_z}^{0}(\mathbf{r},\mathbf{r}',\tau-\tau')-\mathrm{i}\int\mathrm{d}\mathbf{r}_1\int\mathrm{d}\tau_1 G_{k_z}^{0}(\mathbf{r},\mathbf{r}_1,\tau-\tau_1)\mathscr{E}(\tau_1)y_1G_{k_z}^{0}(\mathbf{r}_1,\mathbf{r}',\tau_1-\tau').\tag{4.2}$$

Substituting the full Green function (4.2) into expression (4.1), and performing the integrals, one obtains:

$$\begin{aligned}j_y^{\mathrm{s}}(\tau) &= \frac{b}{4\pi s}\frac{\eta t}{b^2+\omega^2}\mathscr{E}\int_0^{\infty}\mathrm{d}\overline{\tau}\exp\left[-\left(2\varepsilon-b+\frac{2}{d^2}\right)\overline{\tau}\right]I_0\left(\frac{2\overline{\tau}}{d^2}\right)\mathrm{csch}(b\overline{\tau})\\ &\quad\times\left\{b\cos(\tau\omega)\cos(b\overline{\tau})-b\cos[(\tau-\overline{\tau})\omega]\,\mathrm{csch}(b\overline{\tau})+\omega\sin(\tau\omega)\right\}.\end{aligned}\tag{4.3}$$

The real and imaginary parts of the complex conductivity after the Fourier transform

$$\sigma_{\mathrm{s}}(\omega)=\frac{j_{\mathrm{s}}(\omega)}{\mathscr{E}(\omega)}=\sigma_1(\omega)+\mathrm{i}\sigma_2(\omega),\tag{4.4}$$

therefore, are as follows:

$$\sigma_1(\omega)=\frac{b^2}{8\pi s}\frac{\eta t}{b^2+\omega^2}\int_0^{\infty}\mathrm{d}\overline{\tau}\exp\left[-\left(2\varepsilon-b+\frac{2}{d^2}\right)\overline{\tau}\right]I_0\left(\frac{2\overline{\tau}}{d^2}\right)\mathrm{csch}^2(b\overline{\tau})[\cosh(b\overline{\tau})-\cos(\omega\overline{\tau})],\tag{4.5}$$

$$\sigma_2(\omega)=\frac{b}{8\pi s}\frac{\eta t}{b^2+\omega^2}\int_0^{\infty}\mathrm{d}\overline{\tau}\exp\left[-\left(2\varepsilon-b+\frac{2}{d^2}\right)\overline{\tau}\right]I_0\left(\frac{2\overline{\tau}}{d^2}\right)\mathrm{csch}^2(b\overline{\tau})[\omega\sinh(b\overline{\tau})-b\sin(\omega\overline{\tau})].\tag{4.6}$$

This is the main result of the paper. The complex resistivity of the layered material is (after noting that superconducting layers constitute a fraction $s/d$ of the material)

$$\sigma(\omega)=\frac{s}{d}\sigma_{\mathrm{s}}(\omega)+\sigma_n(\omega).\tag{4.7}$$

Neglecting the normal part, one obtains resistivity:

$$\rho_{\mathrm{s}}(\omega)=\frac{d}{s\sigma_{\mathrm{s}}(\omega)}=\rho_1-\mathrm{i}\rho_2,\tag{4.8}$$

where

$$\rho_1=\frac{d}{s}\frac{\sigma_1}{\sigma_1^2+\sigma_2^2},\qquad \rho_2=\frac{d}{s}\frac{\sigma_2}{\sigma_1^2+\sigma_2^2}.\tag{4.9}$$

The difference between the conductivities in a zero field and in the field of $B$ is defined as follows:

$$\delta\sigma=\sigma(0)-\sigma(B).\tag{4.10}$$

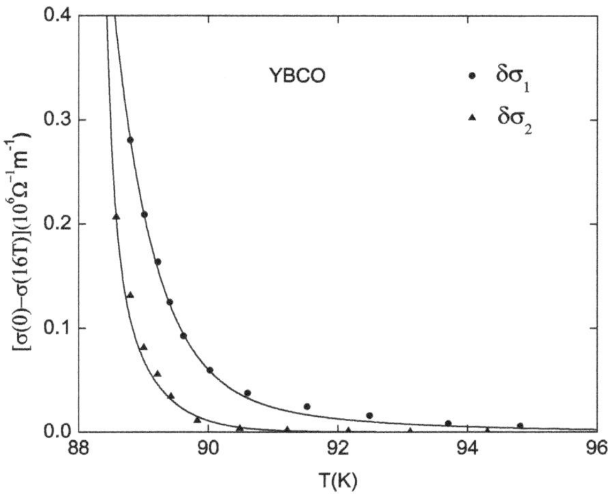


**Figure 1.** Points are the difference between the conductivities in zero field and in the field of 16 T ($\delta\sigma_1$ black circles, $\delta\sigma_2$ black triangles) of slightly underdoped YBCO. The solid lines are the theoretical values of resistivity for different temperatures at frequency $\omega/2\pi = 15.15$ GHz calculated from equation (4.10) with fitting parameters (see text).

# 5. Comparison with experiment

The experimental results by M.S. Grbić et al. [19], obtained from the the microwave absorption measurements at $\omega/2\pi = 15.15$ GHz on slightly underdoped $YBa_2Cu_3O_{7-\delta}$ (YBCO) with $T_c = 87$ K. The distance between the bilayers using the calculation is $d' = 11.68$ Å in [24]. In order to compare the fluctuation conductivity with experimental data in HTSC, one cannot use the expression of relaxation time $\gamma'$ in Bardeen-Cooper-Schrieffer theory which may be suitable for a low-$T_c$ superconductor. Instead of this, we use the factor $k$ as a fitting parameter. The comparison is presented in figure 1. The ac conductivity curves were fitted to equation (4.10) with the normal-state conductivity measured in [17] to be $\sigma_n = 3.3 \cdot 10^6$ $(\Omega\text{m})^{-1}$. The parameters we obtain from the fit are: $H_{c2}(0) = T_c dH_{c2}(T)/dT|_{T_c} = 178$ T (corresponding to $\xi = 13.6$ Å), the GL parameter $\kappa = 49.7$, the order parameter effective thickness $s' = 5.51$ Å, and the factor $k = \sigma_n/\sigma_{GL} = 0.86$, where we take $\gamma = 10$ for YBCO in [25]. Using those parameters, we obtain $\text{Gi} = 2.78 \cdot 10^{-3}$ (corresponding to $\eta = 0.176$). The order parameter effective thickness $s'$ can be taken to be equal to the layer distance (see in [26]) of the superconducting $CuO_2$ plane plus the coherence length $2\xi_c = 2\frac{\xi}{\gamma}$ due to the proximity effect: $3.18$ Å $+ 2\frac{13.6}{10}$ Å $= 4.54$ Å, roughly in agreement in magnitude with the fitting value of $s'$.

# 6. Discussion and conclusion

The complex conductivity was calculated in a layered type-II superconductor under magnetic field in the presence of strong thermal fluctuations on the mesoscopic scale in linear response. While in the normal state, the dissipation involves unpaired electrons, in the mixed phase it takes a form of the flux flow. Time dependent Ginzburg-Landau equations with thermal noise describing the thermal fluctuations are used to describe the vortex-liquid regime. The nonlinear term in dynamics is treated using the renormalized Gaussian approximation. Explicit expressions for the complex conductivity $\sigma_s$ and resistivity $\rho_s$ including all Landau levels were obtained, therefore the approach is valid for arbitrary values if the magnetic field is not too close to $H_{c1}(T)$.

The results were compared to the experimental data on HTSC. The results are in good qualitative and even quantitative agreement with experimental data on YBCO . The thermal fluctuation was included in the present approach, so the results should be applicable for above and below $T_c$.

## Acknowledgements

We are grateful to Baruch Rosenstein, Dingping Li for discussions. This work was supported by the National Foundation for Science and Technology Development (NAFOSTED) of Vietnam under Grant No. 103.02–2011.15.

# References

1. Xu Z.A., Ong N.P., Wang Y., Kakeshita T., Uschida S., Nature, 2000, **406**, 486; doi:10.1038/35020016.
2. Dorin V.V., Klemm R.A., Varlamov A.A., Buzdin A.I., Livanov D.V., Phys. Rev. B, 1993, **48**, 12951; doi:10.1103/PhysRevB.48.12951.
3. Ussishkin I., Sondhi S.L., Huse D.A., Phys. Rev. Lett., 2002, **89**, 287001; doi:10.1103/PhysRevLett.89.287001.
4. Li L., Wang Y., Komiya S., Ono S., Ando Y., Gu G.D., Ong N.P., Phys. Rev. B, 2010, **81**, 054510; doi:10.1103/PhysRevB.81.054510.
5. Larkin A., Varlamov A., Theory of Fluctuations in Superconductors, Clarendon Press, Oxford, 2005.
6. Rosenstein B., Li D., Rev. Mod. Phys., 2010, **82**, 109; doi:10.1103/RevModPhys.82.109.
7. Schmidt H., Z. Phys., 1968, **216**, 336; doi:10.1007/BF01391528.
8. Schmidt H., Z. Phys., 1970, **232**, 443; doi:10.1007/BF01395675.
9. Dorsey A.T., Phys. Rev. B, 1991, **43**, 7575; doi:10.1103/PhysRevB.43.7575.
10. Wickham R.A., Dorsey A.T., Phys. Rev. B, 2000, **61**, 6945; doi:10.1103/PhysRevB.61.6945.
11. Fisher D.S., Fisher M.P.A., Huse D.A., Phys. Rev. B, 1991, **43**, 130; doi:10.1103/PhysRevB.43.130.
12. Ong N.P., Wu H., Phys. Rev. B, 1997, **56**, 458; doi:10.1103/PhysRevB.56.458.
13. Maniv T., Rosenstein B., Shapiro I., Shapiro B.Ya., Phys. Rev. B, 2009, **80**, 134512; doi:10.1103/PhysRevB.80.134512.
14. Lin P.-J., Lipavsky P., Phys. Rev. B, 2009, **80**, 212506; doi:10.1103/PhysRevB.80.212506.
15. Tinh B.D., Thu L.M., Mod. Phys. Lett. B, 2012, **26**, 1250143; doi:10.1142/S0217984912501436.
16. Tinh B.D., Physica C, 2013, **485**, 10; doi:10.1016/j.physc.2012.10.005.
17. Tsuchiya Y., Iwaya K., Kinoshita K., Hanaguri T., Kitano H., Maeda A., Shibata K., Nishizaki T., Kobayashi N., Phys. Rev. B, 2001, **63**, 184517; doi:10.1103/PhysRevB.63.184517.
18. Hanaguri T., Tsuboi T., Tsuchiya Y., Sasaki K.I., Maeda A., Phys. Rev. Lett., 1999, **82**, 1273; doi:10.1103/PhysRevLett.82.1273.
19. Grbić M.S., Požek M., Paar D., Hinkov V., Raichle M., Haug D., Keimer B., Barišić N., Dulčić A., Phys. Rev. B, 2011, **83**, 144508; doi:10.1103/PhysRevB.83.144508.
20. Ketterson J.B., Song S.N., Superconductivity, Cambridge University Press, Cambridge, 1999.
21. Rosenstein B., Zhuravlev V., Phys. Rev. B, 2007, **76**, 014507; doi:10.1103/PhysRevB.76.014507.
22. Kopnin N., Vortices in Type-II Superconductors: Structure and Dynamics, Oxford University Press, Oxford, 2001.
23. Tinh B.D., Li D., Rosenstein B., Phys. Rev. B, 2010, **81**, 224521; doi:10.1103/PhysRevB.81.224521.
24. Yan Y., Blanchin M.G., Phys. Rev. B, 1991, **43**, 13717; doi:10.1103/PhysRevB.43.13717.
25. Li D., Rosenstein B., Phys. Rev. B, 2002, **65**, 220504(R); doi:10.1103/PhysRevB.65.220504.
26. Poole C.P.(Jr.), Farach H.A., Creswick R.J., Prozorov R., Superconductivity, Academic Press, Amsterdam, 2007.

# 5

# A semiflexible polymer chain under geometrical restrictions: Only bulk behaviour and no surface adsorption

P.K. Mishra*

Department of Physics, DSB Campus, Kumaun University, Nainital-263 002 (Uttarakhand), India

We analyse the conformational behaviour of a linear semiflexible homo-polymer chain confined by two geometrical constraints under a good solvent condition in two dimensions. The constraints are stair shaped impenetrable surfaces. The impenetrable surfaces are lines in a two dimensional space. The infinitely long polymer chain is confined in between such two ($A$ and $B$) surfaces. A lattice model of a fully directed self-avoiding walk is used to calculate the exact expression of the partition function, when the chain has attractive interaction with one or both the constraints. It has been found that under the proposed model, the chain shows only a bulk behaviour. In other words, there is no possibility of adsorption of the chain due to restrictions imposed on the walks of the chain.

**Key words:** *polymer adsorption, bulk behaviour, geometrical constraints, exact results*

## 1. Introduction

Biopolymers (*DNA* and *proteins*) are soft objects and, therefore, such molecules can be easily squeezed into the spaces that are much smaller than the natural size of the molecules. For example, *actin* filaments in *eukaryotic* cell or *protein* encapsulated in *Ecoli* [1–3] are the examples of confined molecules that may serve as the basis for understanding molecular processes occurring in the living cells. The conformational properties of single bio-polymers have attracted considerable attention in recent years due to the development of single molecule based experiments [4–9]. The entropy of a molecule having an excluded volume interaction gets modified due to the presence of geometrical restrictions. Therefore, geometrical constraints can modify conformational properties and the adsorption desorption transition behaviour of the confined polymer molecules.

The behaviour of a linear and flexible polymer molecule under good solvent condition, confined to different geometries, has been studied for the past few years [10–17]. Theoretical investigations of a semiflexible polymer chain under confined geometry also find considerable attention in recent years, see [18–25] and references quoted therein. For example, Whittington and his coworkers [10–17] used directed self-avoiding walk models to study the behaviour of a flexible polymer chain confined between two parallel walls on a square lattice and calculated the force diagram for a surface interacting polymer chain. Rensburg et al. [17] performed numerical studies using an isotropic self-avoiding walk model and showed that the force diagram obtained for surface interacting polymer chains confined in between two parallel plates have a qualitatively similar phase diagram obtained by Brak et al. [10–13] for a directed self-avoiding walk model of the problem.

However, in the present investigation, we consider an infinitely long-linear semiflexible polymer chain confined in between one dimensional two stair shaped impenetrable surfaces (geometrical constraints) under good solvent conditions and we discuss the conformational behaviour of the chain. Such

* pkmishrabhu@gmail.com

an investigation may be useful to understand the behaviour of a macromolecule near a membrane as well as the behaviour of *DNA* in micro-arrays and electrophoresis.

To analyze the conformational behaviour of such semiflexible chains we have chosen a fully directed self-avoiding walk model introduced by Privmann and coworkers [26, 27] and have used a generating function technique to solve the model analytically for different values of the spacing between the constraints. The result so obtained is used to discuss the possibility of an adsorption phase transition behaviour of the polymer chain on the stair shaped geometrical constraints. Since the constraint is an attractive surface, it contributes an energy $\epsilon_s$ ($< 0$) for each step of the fully directed self-avoiding walk touching the constraint. This leads to an increased probability defined by a Boltzmann weight $\omega = \exp(-\epsilon_s / k_B T)$ of stepping on the constraint ($\epsilon_s < 0$ or $\omega > 1$, $T$ is temperature and $k_B$ is the Boltzmann constant). The polymer chain gets adsorbed on the constraint at an appropriate value of $\omega$ or $\epsilon_s$. Therefore, the transition between an adsorbed to a desorbed phase is marked by a critical value of adsorption energy $\epsilon_s$ or $\omega_c$.

In this paper, we analytically solve the fully directed self-avoiding walk model to calculate the exact expression of the partition function of the chain when the chain has an attractive interaction either with one or both of the geometrical constraints. The results so obtained are compared with the case when the adsorption of a semiflexible polymer chain occurs on a flat surface [28–34].

The paper is organized as follows: In section 2, a square-lattice model of fully directed self-avoiding walk is described for an infinitely long and linear semiflexible homo-polymer chain confined in between the constraints for a given value of spacing between the constraints. In subsection 2.1, we discuss the possibility of an adsorption transition of the polymer chain when constraint $A$ has an attractive interaction with the semiflexible polymer chain. Subsection 2.2 is devoted to a discussion of the adsorption of a semiflexible polymer chain on the constraint $B$. While in subsection 2.3 the expression of the partition function of the polymer chain is obtained for the case when the chain has an attractive interaction with both the constraints. Finally, in section 3 we summarize and discuss the results obtained.

## 2. Model and method

A model of fully directed self-avoiding walks [26, 27] on a square lattice is used to investigate the possibility of an adsorption transition of an infinitely long linear semiflexible homopolymer chain on geometrical constraints, when the chain is confined in between two impenetrable stair shaped surfaces under a good solvent condition (as shown schematically in figure 1). The directed walk model is restrictive in the sense that the angle of bending has a unique value, that is 90° for a square lattice and the directivity of the walk amounts to a certain degree of stiffness in the walks of the chain because different directions of the space are not treated equally. Since the directed self-avoiding walk model can be solved analytically, it gives exact values of the partition function of the polymer chain. We consider a fully directed self-avoiding walk (FDSAW) model. Therefore, the walker is allowed to take steps along $+x$, and $+y$ directions on a square lattice in between the constraints.

The walks of the chain start from a point $O$ located on the impenetrable surface-$A$, and the walker moves in the space in between the two surfaces [as we have shown schematically in figure 1, a walk of the polymer chains confined in between two surfaces for a value of separation $n(=3)$ between them].

The stiffness of the chain is accounted for by associating a Boltzmann weight with the bending energy for each turn in the walk of the polymer chain. The stiffness weight is $k = \exp(-\beta\epsilon_b)$; where $\beta = 1/k_B T$ is the inverse of the temperature, $\epsilon_b (> 0)$ is the energy associated with each bend in the walk of the chain, $k_B$ is the Boltzmann constant and $T$ is temperature. For $k = 1$ or $\epsilon_b = 0$, the chain is said to be flexible and for $0 < k < 1$ or $0 < \epsilon_b < \infty$ the polymer chain is said to be semiflexible. However, when $\epsilon_b \to \infty$ or $k \to 0$, the chain has the shape of a rigid rod.

The partition function of a surface interacting semiflexible polymer chain can be written as follows:

$$Z(\omega, k) = \sum_{N=0}^{N=\infty} \sum_{\text{all walks of } N \text{ steps}} g^N \omega^{N_s} k^{N_b}, \tag{2.1}$$

where $N_b$ is the total number of bends in a walk of $N$ steps (monomers), $N_s$ is number of monomers in a $N$ step walk ($N_b \leqslant N - 1$ and $N_s \leqslant N$), lying on the surface, $g$ is the step fugacity of each monomer of the chain, and $\omega$ is the Boltzmann weight of the monomer-surface attraction energy.

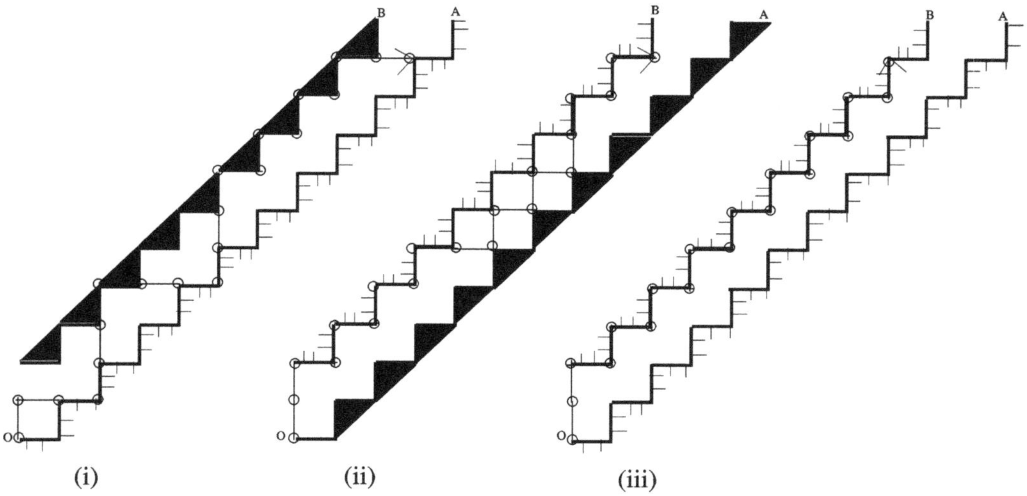


**Figure 1.** This figure shows a walk of an infinitely long linear semiflexible polymer chain confined in between two constraints (impenetrable stair-shaped surface). All walks of the chain start from a point $O$ on the constraint. We show three different cases viz. (i), (ii) and (iii) having separation $(n)$ between the constraints along the axis three monomers (steps). The separation between the constraints are defined on the basis of how many steps a walker can successively move at maximum along any of the $+x$ or $+y$ directions. In the case 1 (i), the constraint $A$ has an attractive interaction with the monomers of the chain, in 1 (ii) only constraint $B$ has an attractive interaction with the monomers of the chain while in 1 (iii) both constraints are shown to have an attractive interaction with the monomers of the polymer chain.

## 2.1. A semiflexible polymer chain interacting with constraint $A$

The partition function of an infinitely long linear semiflexible polymer chain confined in between the constraints (as shown schematically in figure 1 (i) and having an attractive interaction with the constraint $A$ can be calculated using the generating function technique. The components (as shown in figure 2) of the partition function $Z_3^A(k,\omega_1)$ (here we have used the suffix three because in figure 1 (i) case, the maximum step that a walker can move successively in one particular direction is three and $\omega_1$ is the Boltzmann weight of the attraction energy between monomers, and thus the constraint $A$) of the chain can be written as follows:

$$X_1^A = s_1 + ks_1 Y_3^A, \tag{2.2}$$

where $s_1 = \omega_1 g$.

$$X_2^A = g + g\left(X_1^A + kY_2^A\right), \tag{2.3}$$

$$X_3^A = g + g\left(X_2^A + kY_1^A\right), \tag{2.4}$$

$$Y_1^A = g + kgX_3^A, \tag{2.5}$$

$$Y_2^A = g + g\left(kX_2^A + Y_1^A\right), \tag{2.6}$$

and

$$Y_3^A = s_1 + s_1\left(kX_1^A + Y_2^A\right). \tag{2.7}$$

On solving equations (2.2)–(2.7), we find the expression for $X_1^A(k,\omega_1)$ and $Y_2^A(k,\omega_1)$. In obtaining the expression for $X_1^A(k,\omega_1)$ and $Y_2^A(k,\omega_1)$, we have solved a matrix of $2n \times 2n$ $(n = 3)$, for the present case i.e., figure 1 (i). Thus, we have an exact expression of the partition function for an infinitely long linear semiflexible polymer chain confined between the constraints and having an attractive interaction with the constraint $A$ [as shown in 1 (i)]. This is written as follows:

$$Z_3^A(k,\omega_1) = X_1^A(k,\omega_1) + Y_2^A(k,\omega_1) = -\frac{u_1 + u_2 + u_3 + 2k^4 s_1^2 g^4 - 2k^5 s_1^2 g^4}{-1 + k^6 s_1^2 g^4 + u_4}, \tag{2.8}$$

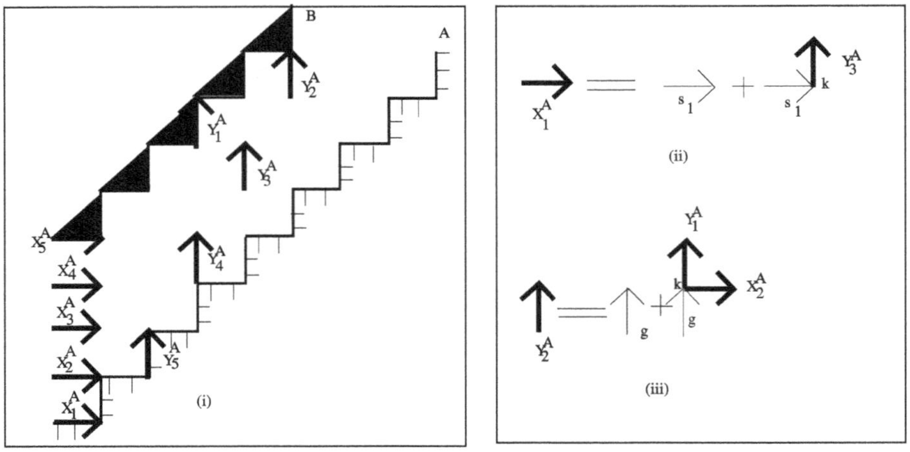


**Figure 2.** The components of the partition function is shown graphically in this figure. Term $X_m^A$ ($3 \leqslant m \leqslant n$) indicates the sum of Boltzmann weight of all the walks having the first step along $+x$ direction and suffix $n$ indicates maximum number of steps that a walker can successively take along $+x$ direction. Similarly, we have defined $Y_m^A$, where the first step of the walker is along $+y$ direction. In this figure, (ii) and (iii) graphically represents the recursion relation for equations (2.2) and (2.6), respectively.

where

$$u_1 = s_1 - ks_1^2 - g - ks_1^2 g + k^2 s_1^2 g - g^2 - kg^2 - ks_1 g^2 + 2k^2 s_1 g^2 - ks_1^2 g^2,$$
$$u_2 = -k^2 s_1^2 g^2 + 3k^3 s_1^2 g^2 - kg^3 + k^2 g^3 - k^2 s_1^2 g^3 + 2k^3 s_1^2 g^3 - k^4 s^2 g^3,$$
$$u_3 = -kg^4 + k^3 g^4 - ks_1 g^4 + k^2 s_1 g^4 + k^3 s_1 g^4 - k^4 s_1 g^4 - 2k^2 s_1^2 g^4 + 2k^3 s_1^2 g^4,$$

and

$$u_4 = -k^4 \left[g^4 + 2s_1^2 \left(g^2 + g^4\right)\right] + k^2 \left[g^2 \left(2 + g^2\right) + s_1^2 \left(1 + g^2 + g^4\right)\right].$$

From the singularity of the partition function, $Z_3^A(k, \omega_1)$, we obtain the critical value of the Boltzmann's weight for the monomer-constraint $A$ attraction energy,

$$\omega_{c1} = \frac{\sqrt{1 - 2k^2 g^2 - k^2 g^4 + k^4 g^4}}{\sqrt{k^2 g^2 + k^2 g^4 - 2k^4 g^4 + k^2 g^6 - 2k^4 g^6 + k^6 g^6}}.$$

This is required for the adsorption of an infinitely long linear semiflexible polymer chain on the constraint $A$. We obtain the value of $\omega_{c1} = 1$, when we substitute the value of $g_c$ in the expression of $\omega_{c1}$ corresponding to all possible values of $k[= \exp(-\beta\epsilon_b)]$ or the bending energy $\epsilon_b$ for which an infinitely long linear semiflexible polymer chain can be polymerized in between the constraints. It shows the existence of only one singularity $g_c$ of the partition function equation (2.8) and it corresponds to the bulk behaviour of the chain. There is no possibility of an adsorption transition of the chain on constraint $A$.

## 2.2. A semiflexible polymer chain interacting with constraint *B*

The partition function of an infinitely long linear semiflexible polymer chain confined in between the constraints [as shown schematically in figure 1 (ii)] and having an attractive interaction with the constraint $B$ is calculated following the method discussed in the above subsection. The components of the partition function $Z_3^B(k, \omega_2)$ (where $\omega_2$ is Boltzmann weight of attraction energy between the monomers of the chain and the constraint $B$) of the chain can be written as follows:

$$X_1^B = g + kgY_3^B, \tag{2.9}$$
$$X_2^B = g + g\left(X_1^B + kY_2^B\right), \tag{2.10}$$
$$X_3^B = s_2 + s_2\left(X_2^B + kY_1^B\right), \tag{2.11}$$

where $s_2 = \omega_2 g$.

$$Y_1^B = s_2 + ks_2 X_3^B, \tag{2.12}$$

$$Y_2^B = g + g\left(kX_2^B + Y_1^B\right), \tag{2.13}$$

and

$$Y_3^B = g + g\left(kX_1^B + Y_2^B\right). \tag{2.14}$$

On solving equations (2.9)–2.14), we find an expression for $X_1^B(k,\omega_2)$ and $Y_2^B(k,\omega_2)$. In obtaining the expression for $X_1^B(k,\omega_2)$ and $Y_2^B(k,\omega_2)$, we have to solve a matrix of $2n \times 2n$ [$n = 3$, for figure 1 (ii) case]. Thus, we obtain an exact expression of the partition function for an infinitely long linear semiflexible polymer chain confined between the constraints and having an attractive interaction with the constraint $B$ [as shown in figure 1 (ii)] which is as follows:

$$Z_3^B(k,\omega_2) = X_1^B(k,\omega_2) + Y_2^B(k,\omega_2) = -\frac{-g(s_2(1+kg^2-k^2g^2)+2u_5+ks_2^2u_6)}{-1+k^6s_2^2g^4+u_7}, \tag{2.15}$$

where

$$u_5 = 1 + k^2(-1+g)g^2 - k^3g^3 + kg(1+g),$$
$$u_6 = 1 + g + g^2 - 2k^3(-1+g)g^2 + 2k^4g^3 - k^2g\left(2+3g+3g^2\right) + 2k\left(-1+g^3\right),$$
$$u_7 = -k^4\left[g^4 + 2s_2^2\left(g^2+g^4\right)\right] + k^2\left[g^2\left(2+g^2\right) + s_2^2\left(1+g^2+g^4\right)\right].$$

From the singularity of the partition function, $Z_3^B(k,\omega_2)$, we obtain a critical value for the monomer-constraint $B$ attraction energy,

$$\omega_{c2} = \frac{\sqrt{1-2k^2g^2-k^2g^4+k^4g^4}}{\sqrt{k^2g^2+k^2g^4-2k^4g^4+k^2g^6-2k^4g^6+k^6g^6}} = \omega_{c1},$$

required for adsorption of an infinitely long linear semiflexible polymer chain on the constraint $B$. In this case too, we find $\omega_{c2} = 1$, for all possible values of the bending energy or stiffness of the semiflexible polymer chain and further there is no possibility for the existence of a new singularity of the partition function i.e. equation (2.15). Therefore, the adsorption of the chain on constraint $B$ is impossible.

## 2.3. A semiflexible polymer chain interacting with both the constraints $A$ and $B$

The partition function of an infinitely long linear semiflexible polymer chain confined in between the constraints [as shown schematically in figure 1 (iii)] and having an attractive interaction with both the constraints ($A$ and $B$) is calculated following the method discussed in the above subsections. The components of the partition function $Z_3^C(k,\omega_3,\omega_4)$ of the chain can be written as follows:

$$X_1^C = s_3 + ks_3 Y_3^C, \tag{2.16}$$

where $s_3 = \omega_3 g$.

$$X_2^C = g + g\left(X_1^C + kY_2^C\right), \tag{2.17}$$

$$X_3^C = s_4 + s_4\left(X_2^C + kY_1^C\right), \tag{2.18}$$

here, $s_4 = \omega_4 g$.

$$Y_1^C = s_4 + ks_4 X_3^C, \tag{2.19}$$

$$Y_2^C = g + g\left(kX_2^C + Y_1^C\right), \tag{2.20}$$

and

$$Y_3^C = s_3 + s_3(kX_1^C + Y_2^C). \tag{2.21}$$

On solving equations (2.16)–(2.21), we get the expression for $X_1^C(k,\omega_3,\omega_4)$ and $Y_2^C(k,\omega_3,\omega_4)$. In obtaining the expression for $X_1^C(k,\omega_3,\omega_4)$ and $Y_2^C(k,\omega_3,\omega_4)$, we have solved a matrix of $2n\times 2n$ [$n=3$, for figure 1 (iii) case]. Thus, we have an exact expression for the partition function of an infinitely long linear semiflexible polymer chain confined between the constraints and having an attractive interaction with the constraints [as shown in figure 1 (iii)]. This is written as follows:

$$Z_3^C(k,\omega_3,\omega_4) = X_1^C(k,\omega_3,\omega_4) + Y_2^C(k,\omega_3,\omega_4) = -\frac{-(gu_8+s_3u_9)+u_{10}+u_{11}}{-1+k^6s_3^2s_4^2g^2+u_{12}+u_{13}}, \tag{2.22}$$

where

$$u_8 = 1+s_4+kg-(-1+k)ks_4^2(1+g+kg),$$
$$u_9 = 1-k^2s_4^2g^2+k^4s_4^2g^2+k\left(1+s_4^2\right)g^2-k^2\left[g^2s_4^2\left(1+g^2\right)\right],$$
$$u_{10} = ks_3^2\left[1+g+s_4gk^2s_4^2(1-2g)g+2k^4s_4^2g^2+kg(-1-s_4+2g)\right],$$
$$u_{11} = ks_3^2\left\{kgs_4^2(1+2g)-k^2\left[2g^2+s_4^2\left(1+2g+2g^2\right)\right]\right\},$$
$$u_{12} = k^2\left\{g^2+s_4^2\left(1+g^2\right)+s_3^2\left[1+\left(1+s_4^2\right)g^2\right]\right\},$$

and

$$u_{13} = -k^4\left\{s_4^2g^2+s_3^2\left[g^2+s_4^2\left(1+2g^2\right)\right]\right\}.$$

From the singularity of the partition function, $Z_3^C(k,\omega_3,\omega_4)$, we obtain a critical value of the monomer-constraint attraction energy,

$$\omega_{c3} = \frac{\sqrt{1-k^2g^2-k^2g^2\omega_4^2-k^2g^4\omega_4^2+k^4g^4\omega_4^2}}{\sqrt{k^2g^2+k^2g^4-k^4g^4-k^4g^6\omega_4^2-2k^4g^6\omega_4^2+k^2g^6\omega_4^2+k^6g^6\omega_4^2}}, \tag{2.23}$$

required for the adsorption of an infinitely long linear semiflexible polymer chain on the constraints $A$, when both the constraints have an attractive interaction with the chain.

On substitution of the value of $\omega_4$ in equation (2.23) to get the value of $\omega_{c3}=1$,

$$\omega_{c4} = \frac{\sqrt{1-2k^2g^2-k^2g^4+k^4g^4}}{\sqrt{k^2g^2+k^2g^4-2k^4g^4+k^2g^6-2k^4g^6+k^6g^6}} = \omega_{c2}.$$

The method discussed above can be used for different values of $n$. The size of the matrix needed to solve for the partition function of the chain confined in between the constraints is $2n\times 2n$. We have calculated the exact expressions of the partition function for $n$ ($3\leqslant n\leqslant 19$).

We have found that the adsorption transition point of an infinitely long linear semiflexible polymer chain on the constraint $A$, $B$ and simultaneously on both the constraints $A$ and $B$ has the value of unity. The equation (2.23) has only a singularity that corresponds to the polymerization of an infinitely long linear homopolymer chain in between the constraints. Therefore, there is no possibility of the adsorption-desorption phase transition in the proposed model. This fact is true for the chosen values of $k$ or the bending energy (as checked for $3\leqslant n\leqslant 19$) for which an infinitely long polymer chain can be polymerized in between the constraints.

## 2.4. General expressions of the recursion relations

In this subsection, we should like to express the recursion relations with the least possible number of equations. This method is useful in solving a matrix of $n\times n$ rather than $2n\times 2n$ as discussed in the subsections 2.1–2.3. For instance, equations (2.16)–(2.21) can be written as follows:

$$\begin{aligned} W_1^n &= s_3+ks_3^2+kgs_3^3+\cdots+kg^{n-2}s_3^2s_4+k^2s_3^2W_1^n \\ &\quad +k^2gs_3^2W_2^n+k^2g^2s_3^2W_3^n+\cdots+k^2g^{n-2}s_3^2s_4W_n^n, \end{aligned} \tag{2.24}$$

$$\begin{aligned} W_m^n &= g+kg^2+\cdots+kg^{n+1-m}s_4+gW_{m-1}^n+k^2g^2W_m^n \\ &\quad +k^2g^3W_{m+1}^n+\cdots+k^2g^{n+1-m}W_{n-1}^n+k^2g^{n+1-m}s_4W_n^n, \end{aligned} \tag{2.25}$$

where $(1 < m < n)$ and $W_m^n = 0$, when $m < 1$.

$$W_n^n = s_4 + ks_4^2 + s_4 W_{n-1}^n + k^2 s_4 W_n^n. \tag{2.26}$$

The equations (2.24)–(2.26) can be used to express recursion relations, $X_n^{\mathrm{C}}$ (for all values of the chosen $n$), and mutual exchange of $s_3$ with $s_4$ will result in the recursion relations $Y_n^{\mathrm{C}}$ for the chosen values of $n$. The partition function of the chain can now be written as follows:

$$Z_n^{\mathrm{C}}(k, \omega_3, \omega_4) = W_1^n + W_{n-1}^n, \tag{2.27}$$

where $W_1^n$ is the sum of the Boltzmann weights of all walks starting from a point $O$ lying on the constrant $A$ and having the first step along $+x$ direction, while $W_{n-1}^n$ is the sum of the Boltzmann weights of all the walks starting from point $O$ and with the first step along $+y$ direction.

However, substituting $s_4 = g$ and $s_3 = s_1$, we have recursion relations and a partition function for the case 1 (i), as shown in figure 1 and when we substitute $s_3 = g$ and $s_4 = s_2$, recursion relations and partition function for the case 1 (ii) of figure 1 were found by us.

If the constraints are assumed to be neutral, the recursion relations can be written for any given value of $n$ as follows:

$$\begin{aligned} W_m^n &= g + kg^2 + kg^3 + \dots + kg^{n+2-m} + gW_{m-1} \\ &\quad + k^2 g^2 W_m + k^2 g^3 W_{m+1} + \dots + k^2 g^{n+2-m} W_n^n, \end{aligned} \tag{2.28}$$

where $1 \leqslant m \leqslant n$ and $W_0^n = 0$.

## 3. Summary and conclusions

We have considered an infinitely long linear semiflexible homopolymer chain confined in between two impenetrable stair shaped surfaces (constraint) in two dimensions under good solvent condition. We have used a fully directed self-avoiding walk model to study the adsorption phase transition behaviour of the polymer chain on any of the two constraints ($A$ and $B$) and simultaneous adsorption of the polymer chain on both the constraints ($A$ and $B$). The generating function technique is used to solve the model analytically and an exact expression of the partition function of the surface interacting semiflexible polymer chain is obtained for different values of spacing $(3 \leqslant n \leqslant 19)$ between the constraints.

We find in the case 1 (i), 1 (ii) and 1 (iii) that the bulk behaviour of the polymer chain occurs on the constraints for the values $\omega_{c1} = \omega_{c2} = \omega_{c3} = 1$ for all possible values of $k$ or the bending energy of the chain for which an infinitely long linear semiflexible polymer chain can be polymerized in between the constraints. The critical value of $\omega$ is unity for all cases considered and for different values of spacing between the constraints $(3 \leqslant n \leqslant 19)$. This result is obvious because the walks of the chain are directed along the constraint(s), therefore, the partition function of the chain is dominated by the walks lying on the constraints, and the bulk behaviour is observed on the constraints. We have shown the results for a few values of $n = 3, 7, 11, 16$ in the table 1 for the case 1 (i), when the chain interacts with the constraint $A$. The chain is grafted to the constraint $A$ for the case 1 (i), 1 (ii) and 1 (iii), as shown in figure 1. An infinitely long linear chain is polymerized in between the two constraints $A$ and $B$, when $g = g_{\mathrm{c}}$.

However, in the case of adsorption of an infinitely long linear semiflexible polymer chain on a flat surface, the adsorption transition point is found to depend on the bending energy or stiffness of the chain. In this case, the partition function of the surface interacting chain has two singularities. One singularity corresponds to the bulk behaviour i.e., polymerization of an infinitely long linear chain and the other singularity corresponds to adsorption transition of the chain on the surface [28–34].

We have also expressed general expressions of the recursion relations, when the chain has an attractive interaction with any or both the constraints and when the constraints are assumed to be neutral. It has been found that polymerization of an infinitely long flexible polymer chain is not possible for separations ($n$) between constraints 3, 6 and 8. In the case of $n = 3$, the imaginary part of the critical value of step fugacity is negligible. However, for other values of separation between the constraints, i.e., $n = 6$ and 8 the imaginary part in the critical value of step fugacity is reasonable and cannot be ignored. We plan to discuss these issues in a another paper to be submitted elsewhere in due time.

**Table 1.** This table shows the values of $g_c$ and $s_c(=\omega_{c1}g_c)$ for different values of separation $(n)$ between the constrains, for the case 1 (i), as shown in figure 1. The value of $s_c = g_c$ indicates that $\omega_{c1} = 1$.

| | $n=3$ | | $n=7$ | | $n=11$ | | $n=16$ | |
|---|---|---|---|---|---|---|---|---|
| $k$ | $g_c$ | $s_c$ | $g_c$ | $s_c$ | $g_c$ | $s_c$ | $g_c$ | $s_c$ |
| 0.1 | – | – | – | – | – | – | 0.99303 | 0.99303 |
| 0.2 | – | – | – | – | 0.92215 | 0.92215 | 0.88185 | 0.88185 |
| 0.3 | – | – | 0.89545 | 0.89545 | 0.83077 | 0.83077 | 0.80227 | 0.80227 |
| 0.4 | – | – | 0.80962 | 0.80962 | 0.76023 | 0.76023 | 0.73867 | 0.73867 |
| 0.5 | 0.93879 | 0.93879 | 0.74186 | 0.74186 | 0.70259 | 0.70259 | 0.68558 | 0.68558 |
| 0.6 | 0.84709 | 0.84709 | 0.68613 | 0.68613 | 0.65400 | 0.65400 | 0.64017 | 0.64017 |
| 0.7 | 0.77358 | 0.77358 | 0.63907 | 0.63907 | 0.61222 | 0.61222 | 0.60072 | 0.60072 |
| 0.8 | 0.71293 | 0.71293 | 0.59857 | 0.59857 | 0.57576 | 0.57576 | 0.56603 | 0.56603 |
| 0.9 | 0.66182 | 0.66182 | 0.56325 | 0.56325 | 0.54359 | 0.54359 | 0.53524 | 0.53524 |
| 1.0 | 0.61803 | 0.61803 | 0.53208 | 0.53208 | 0.51496 | 0.51496 | 0.50771 | 0.50771 |

## Acknowledgements

The financial support received from Department of Science and Technology, New Delhi (SR/FTP/PS-122/2010) thankfully acknowledged. The author also would like to thank Professor D. Dhar, TIFR, Mumbai (India) and the anonymous referee for useful corrections in the earlier version of the manuscript.

## References

1. Vacha M., Habuchi S., NPG Asia Materials, 2010, **2**, 134; doi:10.1038/asiamat.2010.135.
2. Köster S., Steinhauser D., Pfohl T., J. Phys.: Condens. Matter, 2005, **17**, S4091; doi:10.1088/0953-8984/17/49/006.
3. Morrison G., Thirumalai D., J. Chem. Phys., 2005, **122**, 194907; doi:10.1063/1.1903923.
4. Henrichson S.E., Misakian M., Bobertson B., Kasianowicz J.J., Phys. Rev. Lett., 2000, **85**, 3057; doi:10.1103/PhysRevLett.85.3057.
5. Structure and Dynamics of Confined Polymers, Kasianowicz J.J., Kellermayer M., Deamer D. (Eds.), Kluwer Academic Publishers, Dordrecht, 2002.
6. Reccius C.H., Mannion J.T., Cross J.D., Craighead H.G., Phys. Rev. Lett., 2000, **95**, 268101; doi:10.1103/PhysRevLett.95.268101.
7. Reisner W., Morton K.J., Riehn R., Wang Y.M., Yu Z., Rosen M., Sturm J.C., Chou S.Y., Frey E., Austin R.H., Phys. Rev. Lett., 2005, **94**, 196101; doi:10.1103/PhysRevLett.94.196101.
8. Nykypanchuk D., Strey H.H., Hoagland D.A., Macromolecules, 2005, **38**, 145; doi:10.1021/ma048062n.
9. Ichikawa M., Matsuzawa Y., Yoshikawa K., J. Phys. Soc. Jpn., 2005, **74**, 1958; doi:10.1143/JPSJ.74.1958.
10. DiMarzio E.A., Rubin R.J., J. Chem. Phys., 1971, **55**, 4318; doi:10.1063/1.1676755.
11. De Gennes P.G., Scaling Concepts in Polymer Physics, Cornell University Press, Ithaka, 1979.
12. Eisenriegler E., Polymer Near Surfaces, World Scientific, Singapore, 1993.
13. Brak R., Owczarek A.L., Rechnitzer A., Whittington S.G., J. Phys. A: Math. Gen., 2005, **38**, 4309, doi:10.1088/0305-4470/38/20/001.
14. Klushin L.I., Polotsky A.A., Hsu H.P., Makelov D.A., Binder K., Skvortsov A.M., Phys. Rev. E, 2013, **87**, 022604; doi:10.1103/PhysRevE.87.022604.
15. Janse van Rensburg E.J., Orlandini E., Whittington S.G., J. Phys. A: Math. Gen., 2006, **39**, 13869; doi:10.1088/0305-4470/39/45/003.
16. Brak R., Iliev G.K., Rechnitzer A., Whittington S.G., J. Phys. A: Math. Theor., 2007, **40**, 4415; doi:10.1088/1751-8113/40/17/001.
17. Janse van Rensburg E.J., Orlandini E., Owczarek A.L., Rechnitzer A., Whittington S.G., J. Phys. A: Math. Gen., 2005, **38**, L823; doi:10.1088/0305-4470/34/29/301.
18. Harnau L., Reineker P., Phys. Rev. E, 1999, **60**, 4671; doi:10.1103/PhysRevE.60.4671.
19. Bicout D.J., Burkhardt T.W., J. Phys. A: Math. Gen., 2001, **34**, 5745; doi:10.1088/0305-4470/34/29/301.
20. Yang Y., Burkhardt T.W., Gompper G., Phys. Rev. E, 2007, **76**, 011804; doi:10.1103/PhysRevE.76.011804.

21. Levi P., Mecke K., Europhys. Lett., 2007, **78**, 38001; doi:10.1209/0295-5075/78/38001.
22. Wagner F., Lattanzi G., Frey E., Phys. Rev. E, 2007, **75**, 050902; doi:10.1103/PhysRevE.75.050902.
23. Cifra P., Benkova Z., Bleha T., J. Phys. Chem. B, 2009, **113**, 1843; doi:10.1021/jp806126r.
24. Cifra P., J. Chem. Phys., 2009, **131**, 224903; doi:10.1063/1.3271830.
25. Cifra P., Bleha T., Eur. Phys. J. E, 2010, **32**, 273; doi:10.1140/epje/i2010-10626-y.
26. Privman V., Frisch H.L., J. Chem. Phys., 1988, **88**, 469; doi:10.1063/1.454626.
27. Privman V., Svrakic N.M., Directed Models of Polymers, Interfaces, and Clusters: Scaling and Finite-Size Properties, Springer, Berlin, 1989.
28. Mishra P.K., Kumar S., Singh Y., Physica A, 2003, **323**, 453; doi:10.1016/S0378-4371(02)01993-3.
29. Mishra P.K., J. Phys.: Condens. Matter, 2010, **22**, 155103; doi:10.1088/0953-8984/22/15/155103.
30. Mishra P.K., Acad. Arena, 2009, **1**, No. 6, 1.
31. Mishra P.K., New York Sci. J., 2010, **3**, No. 1, 32.
32. Mishra P.K., Phase Transitions, 2010, **83**, 47; doi:10.1080/01411590903537588.
33. Mishra P.K., Phase Transitions, 2011, **84**, 291; doi:10.1080/01411594.2010.534657.
34. Sintex T., Sumithra K., Straube E., Macromolecules, 2001, **34**, 1352; doi:10.1021/ma000493s.

# 6

# Conformational properties of polymers in anisotropic environments

K. Haydukivska, V. Blavatska

Institute for Condensed Matter Physics of the National Academy of Sciences of Ukraine,
1 Svientsitski St., 79011 Lviv, Ukraine

We analyze the conformational properties of polymer macromolecules in solutions in presence of extended structural obstacles of (fractal) dimension $\varepsilon_d$ causing the anisotropy of environment. Applying the pruned-enriched Rosenbluth method (PERM), we obtain numerical estimates for scaling exponents and universal shape parameters of polymers in such environments for a wide range $0 < \varepsilon_d < 2$ in space dimension $d = 3$. An analytical description of the model is developed within the des Cloizeaux direct polymer renormalization scheme. Both numerical and analytical studies qualitatively confirm the existence of two characteristic length scales of polymer chain in directions parallel and perpendicular to the extended defects.

**Key words:** *polymers, scaling, disorder, renormalization group, computer simulations*

## 1. Introduction

Many physical objects are characterized by anisotropy of structure: real magnetic crystals often contain extended defects in the form of linear dislocations, disclinations or planar regions of different phases [1–5]; in polymer systems, the understanding of the behavior of macromolecules in solutions having spatial anisotropy caused by the presence of fibrous obstacles is of great importance, e.g., in gels [6], intra- and extracellular environment [7, 8], or in the vicinity of planes (membranes) [9].

The analytical description of crystalline materials with extended defects attracts a lot of interest [1–5, 10–14]. In particular, Dorogovtsev [1] proposed the model of a $d$-dimensional spin system with quenched nonmagnetic defects in the form of $\varepsilon_d$-dimensional objects, which are randomly distributed over the remaining $d - \varepsilon_d$ dimensions. The anisotropy of the system brings about two different characteristic length scales (correlation lengths $\xi_{||}$ and $\xi_{\perp}$), reflecting the macroscopic properties of the system along the directions "parallel" to the $\varepsilon_d$-dimensional defect and along the "perpendicular" directions:

$$\xi_{||} \sim |t|^{-\nu_{||}}, \qquad \xi_{\perp} \sim |t|^{-\nu_{\perp}}, \tag{1.1}$$

where $t$ is the reduced distance to the critical temperature $t = (T - T_c)/T_c$, $\nu_{\perp}$ and $\nu_{||}$ are universal critical exponents.

A number of conformational properties of long flexible polymer chains in solutions can also be described within the critical exponent formalism: for example, the averaged mean-squared distance between two ends of a chain obeys the scaling law:

$$\langle R^2 \rangle \sim N^{2\nu}, \tag{1.2}$$

where $N$ is the number of monomers in the chain and $\nu$ is an universal quantity that does not depend on chemical properties of the macromolecule, but only on space dimension $d$ (e.g., the phenomenological Flory theory [15] gives $\nu(d) = 3/(d+2)$). Thus, for polymers in $d = 1$-dimensional space, this exponent takes on the maximal value of 1 (completely stretched chain), in $d = 2$, one restores the exact value

3/4 [16] and in the 3-dimensional case, the Flory theory nicely agrees, e.g., with the analytical result $\nu = 0.5882 \pm 0.0011$ [17]. For space dimension $d \geqslant 4$, the polymer behaves as an idealized Gaussian chain with $\nu = 1/2$. The relation of the polymer size exponent (1.2) to the correlation length critical index of the $m$-component spin vector model in the formal limit $m \to 0$ was provided by P.-G. de Gennes (the well-known de Gennes limit [15]).

Note that the Flory theory is applicable only in the case of polymers in a pure environment, but in reality most of the solutions contain impurities (obstacles), that interact with the macromolecules. These obstacles can be very small or penetrate through the whole space; randomly distributed or obeying some correlations on a mesoscopic scale [18, 19]. It was shown [20–22] that in the environment having a weak concentration of quenched point-like obstacles, the macromolecules behave like in pure solutions, namely the value of the critical exponent $\nu$ in (1.2) is the same as in the idealized case of a pure solvent. Only when the concentration of defects is close to the percolation threshold where an incipient percolation cluster of fractal structure emerges in the system, the scaling properties of polymers are modified in a non-trivial way [20–40]. It is appropriate to mention that according to a simple generalization of the Flory formula $\nu(d_f) = 3/(d_f + 2)$ [21], there should be a different behavior for polymers on spaces with fractal dimension $d_f < d$, e.g., on percolation clusters.

The conformational properties are also influenced in a non-trivial way when the position of one obstacle particle affects the other, i.e., there are correlations in the spatial distribution of impurities. In particular, these correlations often express themselves by a power law behavior $\sim |r|^{-a}$ [41] with $a < d$, where $r$ is the distance between two obstacles. This type of disorder has a direct interpretation for integer values of $a$, namely, the cases $a = d-1$ ($a = d-2$) describe straight lines (planes) of impurities of random orientation, whereas non-integer values of $a$ are interpreted in terms of impurities organized in fractal structures. This type of disorder leads to a new universality class of polymers [42].

Herein above we were speaking about polymers in disordered but isotropic environment. It means that all the observable statistical properties of the molecules are the same when explored in different spatial directions. An interesting question concerns a quantitative change of these properties when there are some selected directions in the system: the case of spatial anisotropy. In this concern, a widely discussed model of directed self-avoiding walks (DSAW) [43] may describe the properties of macromolecules in an applied external field. This causes an elongation of the molecule along the field direction [44, 45] and leads to the existence of two characteristic length scales in parallel and perpendicular directions. The anisotropic properties of the environment can also be caused by the presence of extended obstacles correlated in $\varepsilon_d$-dimensions, similar to those discussed previously for the spin systems. These may be ordered colloid particles, gel fibers or biological species in a cell environment. In these systems we should also expect a different behavior along the chosen direction and perpendicular to it. Baumgartner and Muthukumar [46] discussed a model of polymer chains in an environment having impurities in the form of absorbing parallel cylinders. They predicted that the polymer chain is elongated in the direction parallel to the cylinders and, correspondingly, the parallel component of the end-to-end distance is governed by an exponent $\nu_{||} = 1$, whereas in normal direction there will be no dependence of the end-to-end distance component on the number of monomers at all (exponent $\nu_{\perp} = 0$). This gives a reason to expect an anisotropic behavior for polymers in solution having impurities correlated in $\varepsilon_d$ dimensions [47].

In the present paper, we study the conformational properties of polymers in the environment where an anisotropy is caused by the presence of $\varepsilon_d$-dimensional defects of parallel orientation, which are randomly distributed in the remaining $d - \varepsilon_d$ dimensions, both using numerical simulations based on a discrete lattice model (Section II) and an analytical description within des Cloizeaux direct polymer renormalization scheme (Section III). We end up by giving conclusions and an outlook.

## 2. Numerical studies

### 2.1. The model

Our goal is to investigate the conformational properties of polymers in anisotropic environments in the presence of obstacles that are ordered in some subspace. For this purpose, we choose a lattice model of a long flexible polymer chain — the model of self-avoiding random walks (SAW), which is established

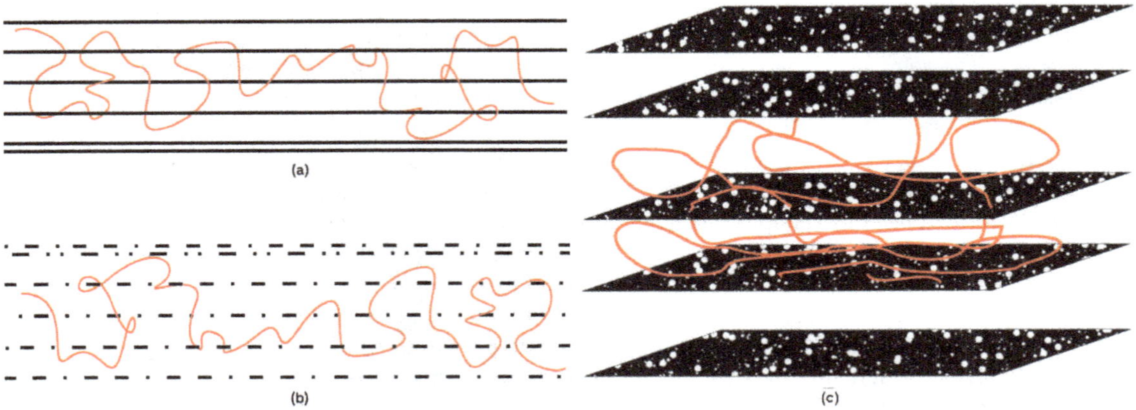


**Figure 1.** (Color online) Schematic representation of polymer chain in the environment having structural obstacles in the form of lines (a), partially penetrable lines (b) and partially penetrable planes (c).

to perfectly capture the universal properties of polymers in good solvent with the excluded volume effect. We deal with the cubic lattice since it is known that universal properties do not depend on the lattice type [51] and the cubic one is the most simple and easy to realize.

The simplest types of extended obstacles that can be chosen for our purposes are spacial objects in the form of lines ($\varepsilon_d = 1$) of parallel orientation (see figure 1 (a)), spreading throughout the lattice in some chosen direction, since they should obviously lead to a different behavior in directions parallel and perpendicular to them. The case of homogeneous planes is not of interest since they divide the space into small restricted regions, and thus the problem is reduced to that of polymers in confined geometries.

We start with an $xy$-plane of a lattice with concentration $c$ of randomly chosen sites containing point-like defects and then we build lines perpendicular to this plane in $z$-direction through chosen sites. In figure 2 one can see a projection of the SAW trajectory on this plane for different concentrations of lines. As one can see at high concentrations $c$ of such obstacles, the SAW trajectory appears to collapse in small restricted regions. In this case, a long polymer chain will behave like a one-dimensional rod, extended

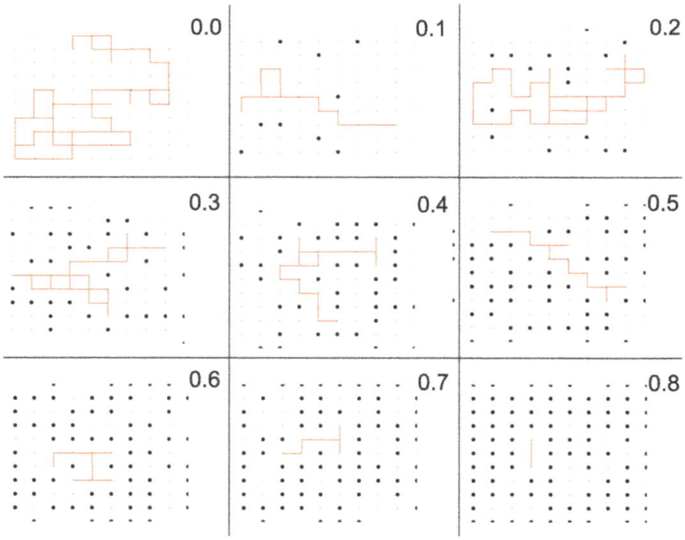


**Figure 2.** (Color online) Projections of SAW trajectories on the $xy$-plane of a lattice at different concentrations $c$ of impurity lines extended throughout the system in $z$ direction.

in the direction parallel to the lines of defects. Thus, we expect a crossover to an extended regime for an increasing defect concentration.

We also consider more interesting situations, namely a set of partially penetrable lines (figure 1 (b)) and planes (figure 1 (c)). To this end, with some fixed probability $p$, we randomly choose the sites on the constructed lines (planes) of defects and treat them as "open" (allowed for SAW trajectory). Note that one can, roughly speaking, treat these objects as fractals. For example, the fractal (cluster) dimension of such a line with concentration $p$ of the lacking sites can be estimated from the well-known relation between the linear size of an object and "the number of particles": $\varepsilon_d \cong \ln[(1-p)\cdot L]/\ln(L)$ (here, $L$ is the length of a line). Indeed, at $p=0$, one has a Euclidian line (like in figure 1 (a)) and simply restores $\varepsilon_d = 1$, whereas with increasing $p$, the line can be treated as a set of disconnected sites (points) and the fractal dimension of this object gradually tends to 0.

## 2.2. The method

For our purposes we use the Pruned-Enriched Rosenbluth Method (PERM) [52]. It is based on the original Rosenbluth-Rosenbluth algorithm of growing chains with population control parameters [53]. On each step $n$, the chain has a weight $W_n$ given by:

$$W_n = \prod_{i=1}^{n} m_i, \tag{2.1}$$

where $m_i$ is the number of possibilities to perform the next step, which varies from 0 to $2d-1$ due to the fact that the chain may not cross itself. This value is also reduced by the presence of impurities.

When the chain of total length $N$ is constructed, a new one is started from the same starting point, until the desired number of chain configurations are obtained. In this manner, all observables should be averaged over an ensemble of different chain configurations $M$:

$$\langle(\ldots)\rangle = \frac{1}{Z_N}\sum_{k=1}^{M} W_N^k(\ldots), \qquad Z_N = \sum_{k=1}^{M} W_N^k, \tag{2.2}$$

here, $W_N^k$ is the weight of the $k$-th configuration of the $N$-step trajectory.

It is also necessary to average over different configurations of disorder:

$$\overline{(\ldots)} = \frac{1}{p}\sum_{k=1}^{p}(\ldots), \tag{2.3}$$

where $p$ is the number of replicas (the number of different realizations of the disorder). For our purposes, we consider about $10^5$ chains for each of the 400 replicas.

Weight fluctuations are reduced by using population control (pruning and enrichment). It means that with probability of 1/2 we reject the chains having low weight and enrich the statistics by replication increasing the number of high weighted configurations. To do this, we use lower and upper bound weights, that are updated at each step according to [54–56]: $W_n^> = C(Z_n/Z_1)(c_n/c_1)^2$ and $W_n^< = 0.2W_n^>$, where $c_n$ is the number of chains of length $n$ created, and the parameter $C$ controls the pruning-enrichment statistics; it is chosen in such a way to allow one to receive in average 10 chains of total length $N$ in each tour.

## 2.3. Results

We concentrate on the conformational properties of long flexible polymer chains, in particular, in the critical exponents governing the behavior of the effective linear size of the macromolecule with respect to the number of monomers (1.2). Let us note that in the anisotropic case we expect to have two different exponents rather than one:

$$\begin{aligned} \overline{\langle R_{||}^2\rangle} &= (z_N - z_0)^2 \sim N^{2\nu_{||}}, \\ \overline{\langle R_{\perp}^2\rangle} &= (x_N - x_0)^2 + (y_N - y_0)^2 \sim N^{2\nu_{\perp}}, \end{aligned} \tag{2.4}$$

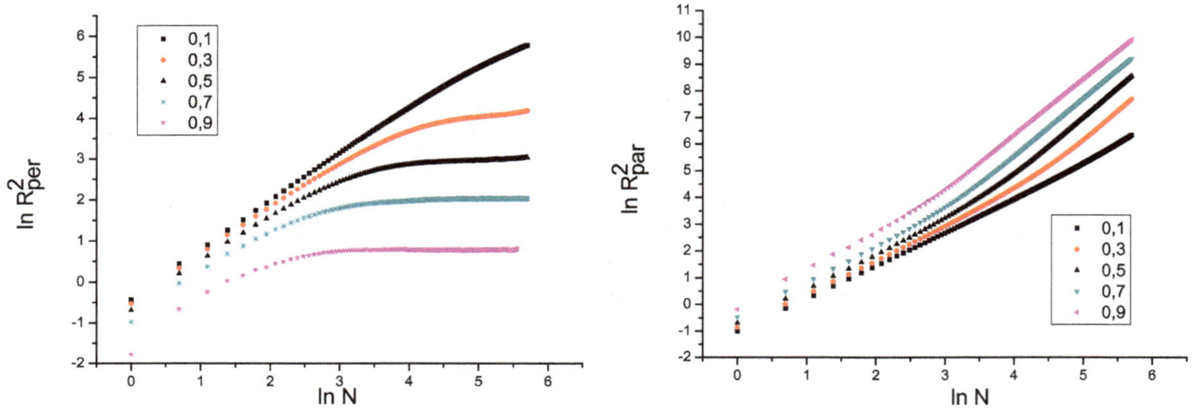


**Figure 3.** (Color online) Parallel (a) and perpendicular (b) components of the end-to-end distance of SAW as a function of chain length in a double logarithmic scale at various concentrations of defects in the form of parallel lines.

so that $R^2 = R_{\parallel}^2 + R_{\perp}^2$. The parallel and perpendicular components are expected to behave in a different way [46].

We start by considering the case of the anisotropy caused by the presence of obstacles in the form of impenetrable parallel lines [47]. In this case, we took chains up to $N = 300$ monomers and find the components of the end-to-end distance vector, performing double averaging according to (2.3). The results are presented in figure 3. It is clearly observed that the behavior of the two components differ from each other: the parallel component grows with an increase of concentration of obstacles, whereas the perpendicular component collapses indicating the stretching of the polymer chain in longitudinal direction. As one can see, there is a crossover between the two types of behavior: one for short chains (less than 40 monomers) and then the other one for long polymers. We expect this crossover to take place when the averaged end-to-end distance $R^2$ of the SAW trajectory is comparable to the distance between the lines of impurities. This is similar to the polymer behavior in a restricted space, for example a cylinder, when short chains (with end-to-end distance smaller than the radius of cylinder) behave like 3-dimensional, and long chains as 1-dimensional [15]. In our case, we observe a crossover to such a behavior when the concentration of lines that penetrate the system is close to a percolative concentration of point-like defects on a simple square lattice. At smaller concentrations, applying the least-square fitting of the data observed, we obtain the estimates for the two critical exponents $\nu_{\parallel}$ and $\nu_{\perp}$ (see figure 4), which coincide

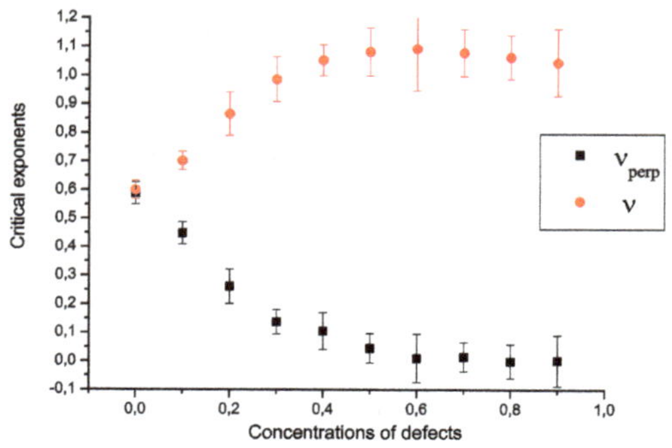


**Figure 4.** (Color online) Critical exponents $\nu_{\parallel}$ and $\nu_{\perp}$ governing the components of the end-to-end distance of SAWs, parallel and perpendicular to the defects in the form of parallel lines, as a function of the concentration of defects.

at $c = 0$, where we restore the corresponding value of the SAW exponent on a pure lattice. As one can see, the exponent governing the scaling of a parallel component of the end-to-end distance is larger than the pure one and gradually reaches the maximal value of 1 with an increasing concentration of disorder, while $\nu_\perp$ is smaller and gradually tends to zero. It gives us the right to say that polymers in anisotropic space are more elongated than polymers in isotropic environments.

Let us check the fact of possible elongation by analyzing the shape properties. All information concerning the shape measure of a chain is given by the gyration tensor with its components defined by:

$$Q_{\alpha\beta} = \frac{1}{N^2} \sum_{i=1}^{N} \sum_{i=1}^{N} (x_i^\alpha - x_j^\alpha)(x_i^\beta - x_j^\beta), \qquad \alpha, \beta = 1, \ldots, 3, \tag{2.5}$$

where $x^\alpha, x^\beta$ are the components of the position vector $\vec{r}_i$ of the $i$-th monomer. Eigenvalues of this tensor provide a full information on the shape of the polymer. To receive these in simulations we need to solve a cubic equation for each of the $10^5$ chains. Thus, we are interested in calculating the rotationally invariant shape characteristics, such as asphericity and prolateness, defined in $d = 3$ as combinations of gyration tensor components according to [48–50]:

$$A_3 = \frac{3}{2} \frac{\mathrm{Tr}\,\hat{Q}^2}{(\mathrm{Tr}\,Q)^2}, \qquad S_3 = 27 \frac{\det \hat{Q}^2}{(\mathrm{Tr}Q)^2}, \tag{2.6}$$

with $\hat{Q} = Q - \hat{I}\mathrm{Tr}Q/3$ ($\hat{I}$ being a unity matrix). Asphericity is normalized in such a way that it changes the value from 0 for spherical configuration to 1 for a completely stretched rod-like structure. Prolateness obeys the inequality: $-0.25 \leqslant S \leqslant 2$, it is negative for oblate configurations and positive for prolate ones; a value 2 corresponds to a rod-like (completely prolate) state.

In figures 5 and 6 we present our data for $\overline{\langle A_3 \rangle}$ and $\overline{\langle S_3 \rangle}$, averaged over realizations of disorder at various fixed concentrations of defects. At $c = 0$, in both cases we restore the corresponding values on a pure lattice. One can easily see in figures 5 (a) and 6 (a) that both quantities are growing with an increasing defect concentration and gradually reach the corresponding values of rod-like structures.

Next, we consider an interesting situation, where point-like defects in the lattice are aligned in some particular direction (say, $z$) forming partially penetrable lines (see figure 1 (b)). To investigate the influence of such a type of space anisotropy on the conformational properties of flexible polymer chains, we randomly choose some concentration $p$ of sites on lines of defects (constructed as described previously), and treat them as "open" (allowed to SAW trajectory). As a result, we obtain aligned fractal-like objects with dimension $0 < \varepsilon_d < 1$. In such problems, we have two parameters: the concentration $c$ of obstacles in the form of parallel lines, and the probability $p$ of a SAW trajectory to penetrate through this line.

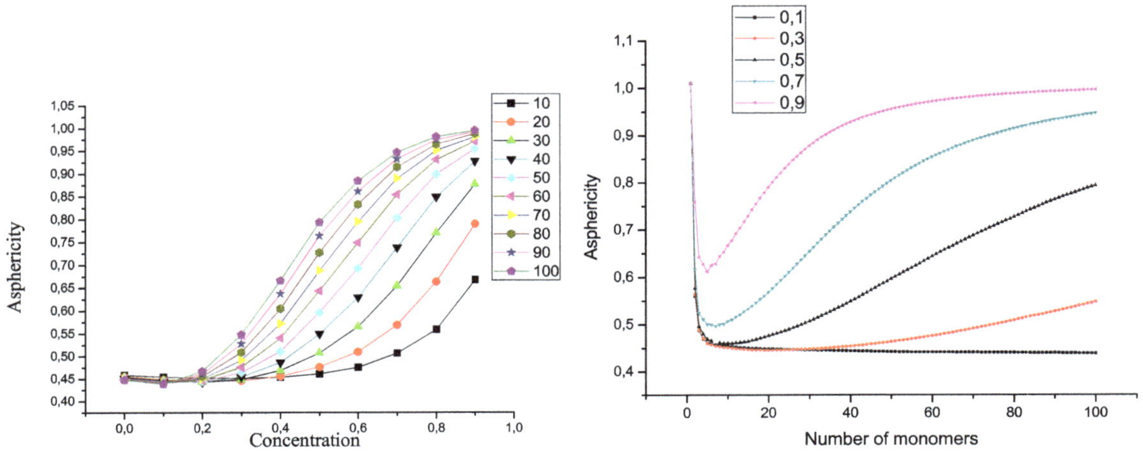


**Figure 5.** (Color online) Asphericity of SAW trajectories as a function of the concentration of defects at various numbers of monomers (a). Asphericity of SAW trajectories as a function of the number of monomers at various fixed concentrations of defects (b).

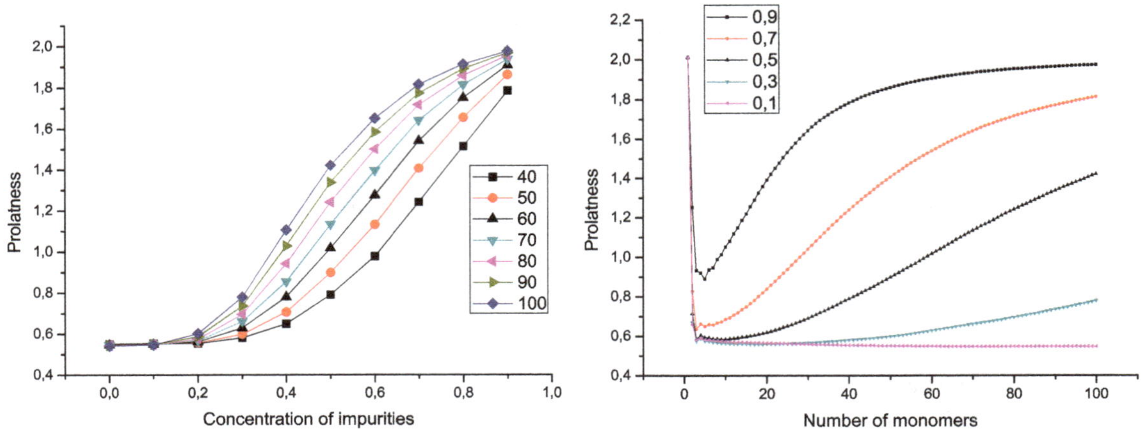


**Figure 6.** (Color online) Prolateness of SAW trajectories as a function of the concentration of defects at various numbers of monomers (a). Prolateness of SAW trajectories as a function of the number of monomers at various fixed concentrations of defects (b).

Performing simulations for chains up to $N = 100$ steps and applying the least-square fits for the data obtained for parallel and perpendicular components of the end-to-end distances of polymer chains, we received the estimates for critical exponents $\nu_{\parallel}$ and $\nu_{\perp}$ (see figure 7 (a)) as functions of these two parameters. Again, the exponent governing the scaling of a parallel component of the end-to-end distance is larger than the pure one and gradually reaches the maximal value of 1, while the other one is lower and gradually tends to zero. This tendency is kept even at a rather high probability of a growing trajectory to penetrate the line (up to the concentration $p$ of "open" sites close to a critical percolation concentration). When $p$ is close to 1, the spatial anisotropy disappears and both exponents gradually reach the corresponding value of the pure lattice case.

Finally, we consider another possible model of anisotropic environment caused by the presence of structural defects in the form of partially penetrable planes of parallel orientation (see figure 1 (c)). We start with homogeneous planes of concentration $c$, randomly distributed in $z$-direction, and again randomly choose some concentration $p$ of sites on these planes and treat them as "open"(allowed to SAW trajectory). As a result, we obtain fractal-like objects having dimension $1 < \varepsilon_d < 2$. Performing simula-

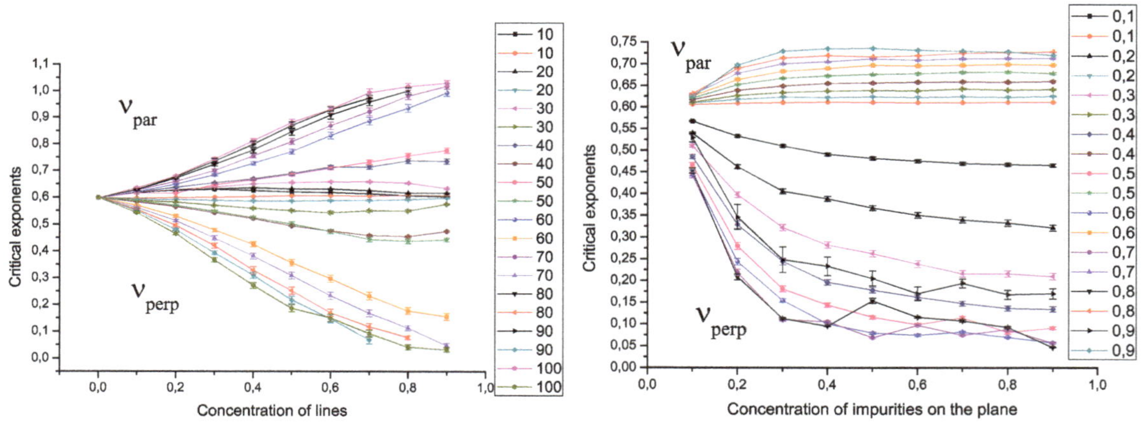


**Figure 7.** (Color online) Critical exponents $\nu_{\parallel}$ and $\nu_{\perp}$ of SAW on a lattice with partially penetrable lines of defects as a function of the concentration of lines at various probabilities to penetrate through these lines (a). Critical exponents $\nu_{\parallel}$ and $\nu_{\perp}$ of SAW on a lattice with partially penetrable planes of defects as a function of probabilities to penetrate these lines at various concentrations of planes (b).

tions for chains up to $N = 100$ steps and applying the least-square fits for data obtained for parallel and perpendicular components of the end-to-end distances of polymer chains, we receive estimates for the critical exponents $\nu_\parallel$ and $\nu_\perp$ (see figure 7 (b)) as functions of these two parameters. The exponent governing the scaling of the parallel component of the end-to-end distance gradually changes from the value on three-dimensional pure lattice (at $c = 0$) to that in two dimensions with a growing concentration of impurity planes. This can be treated as a crossover to a restricted geometry regime of polymers confined between two planes. The exponent $\nu_\perp$ gradually tends to zero.

## 3. Analytical approach

### 3.1. The model

We deal with flexible polymers in an environment with extended impurities correlated in $\varepsilon_d$-dimensions and randomly distributed in the remaining space. Let us start with a continuous x'model, where a polymer chain is presented as a path of length (or surface) $S$ parameterized by $\vec{r}(s)$, with $s = 0\ldots S$ [57]. An effective Hamiltonian of the system is given by:

$$H = \frac{1}{2}\int_0^S \left(\frac{\mathrm{d}\vec{r}(s)}{\mathrm{d}s}\right)^2 \mathrm{d}s + u\int_0^S \mathrm{d}s' \int_0^{s'} \mathrm{d}s'' \delta\left(\vec{r}(s') - \vec{r}(s'')\right) \mathrm{d}s + \int_0^S V(\vec{r}(s))\,\mathrm{d}s. \tag{3.1}$$

Here, the first term describes the chain connectivity, the second term reflects the short-range repulsion between monomers due to the excluded volume effect with coupling constant $u$, and the last term arises due to the interaction between the monomers of the polymer chain and the structural defects in the environment given by potential $V$. We work in the formalism of partition functions:

$$Z = \int \mathrm{D}\vec{r}\mathrm{e}^{-H},$$

where $\int \mathrm{D}\vec{r}$ denotes integration over different paths.

Dealing with systems that display randomness of structure, one usually encounters two types of ensemble averaging, treated as annealed and quenched disorder [58, 59]. In general, the critical behavior of disordered systems with annealed and quenched averaging is quite different. However, for the polymer systems it has been shown [28, 60–63], that the distinction between quenched and annealed averages for an infinitely long single polymer chain is negligible, and in performing analytical calculations for quenched polymer systems one may restrict the problem to the simpler case of annealed averaging. In this paper, we deal with annealed averaging over disorder because it is technically simpler. After averaging the partition sum over realizations of disorder and including only up to the second moment of cumulant expansion, we receive:

$$\begin{aligned}\overline{\exp\left\{\int_0^S V(\vec{r}(s'))\mathrm{d}s'\right\}} &= \exp\left\{\int_0^S \overline{V(\vec{r}(s'))}\mathrm{d}s'\right\} \\ &\times \exp\left\{\frac{1}{2}\int_0^S\int_0^{s'} \overline{V(\vec{r}(s'))V(\vec{r}(s''))} - \overline{V(\vec{r}(s'))}^2 \mathrm{d}s'\mathrm{d}s''\right\},\end{aligned}$$

where $\overline{V(\vec{r}(s'))}$ gives the average concentration of impurities $\rho$, and for the second moment we assume [1, 46]:

$$\overline{V(\vec{r}(s'))V(\vec{r}(s''))} = v\delta^{d-\varepsilon_d}\left(r_{d-\varepsilon_d}(s') - r_{d-\varepsilon_d}(s'')\right), \tag{3.2}$$

which reflects the fact that the impurities are correlated in $\varepsilon_d$ dimensions and uncorrelated in the remaining space. Omitting the terms $\rho S + \frac{1}{2}\rho^2 S^2$, which give a trivial constant shift, we obtain an averaged partition function $\overline{Z} = \int d\vec{r} e^{-H^{\text{eff}}}$ with an effective Hamiltonian:

$$H^{\text{eff}} = \frac{1}{2}\int_0^S \left(\frac{d\vec{r}(S)}{ds}\right)^2 ds + u\int_0^S ds' \int_0^{s'} ds'' \delta\left(\vec{r}(s') - \vec{r}(s'')\right) ds$$
$$-v\int_0^S ds' \int_0^{s'} ds'' \delta^{d-\varepsilon_d}\left(r_{d-\varepsilon_d}(s') - r_{d-\varepsilon_d}(s'')\right). \quad (3.3)$$

Note that the last term in (3.3) describes an effective attractive interaction between monomers in the direction perpendicular to the extended obstacles governed by a coupling constant $v$.

## 3.2. The method

Within the model, all the parameters depend on the polymer area $S$ and on dimensionless coupling constants $\{z_a\} = a(2\pi)^{-(d_a)/2} S^{4-(d_a)/2}$ (here $d_a$ is dimension of coupling constant $a$) in a way that when $\{z_a\} = 0$, one restores the case of idealized Gaussian chain without any interactions between monomers. To calculate the universal properties of the model we need to find such values of parameters that lead to physical values of the universal characteristics. For that reason, we use the direct renormalization technique proposed by des Cloiseaux [51].

Within this scheme, renormalized coupling constants are defined by:

$$g_a(\{z_a\}) = -[\chi_1(\{z_a\})]^{-4} Z_{z_a}(S,S)[2\pi\chi_0\{z_a\}S]^{-(2-\varepsilon_a/2)}, \quad (3.4)$$

where $\varepsilon_a$ is the deviation from the upper space dimension for the coupling constant $z_a$. $\chi_0 = R^2/Sd$ is the swelling factor that governs the behavior of the end-to-end distance of the polymer in solution. It can be presented as a perturbation theory series over the coupling constants:

$$\chi_0(\{z_a\}) = 1 + \sum_a z_a \cdot f_a(d_a), \quad (3.5)$$

where the first term (1) corresponds to an ideal Gaussian chain and the others give corrections caused by interactions in the system. $f_a(d_a)$ is the factor that depends only on the dimension of the corresponding coupling constant. This factor allows us to estimate the critical exponent $\nu$ using the relation:

$$2\nu - 1 = S\frac{\partial}{\partial S}\chi_0(\{z_a\}).$$

The factor $\chi_1$ is connected with the partition function $Z(S)/Z^0(S) = [\chi_1(\{z_a\})]^2$. Here, $Z^0(S)$ is the partition function of an idealized Gaussian chain, $Z_{z_a}(S,S)$ is the partition function of two interacting polymer chains.

Renormalized coupling constants given by the equation (3.4) tend to constant values or the so-called fixed points as the polymer area tends to infinity. The fixed points are defined as common zeros of the flow equations:

$$W_a = 2S\frac{\partial}{\partial S} z_a^*(\{z_a\}). \quad (3.6)$$

To find the fixed points of the model, one need to express $z_a$ in terms of $g_a$ and find the common zeros of (3.6) and then choose those that are stable and physical in the region of interest for the parameters $\varepsilon_a$.

## 3.3. Results

We start with the restricted partition function

$$\widetilde{Z}(\vec{k},S) = \overline{\left\langle e^{i\vec{k}(\vec{r}(S)-\vec{r}(0))}\right\rangle}, \quad (3.7)$$

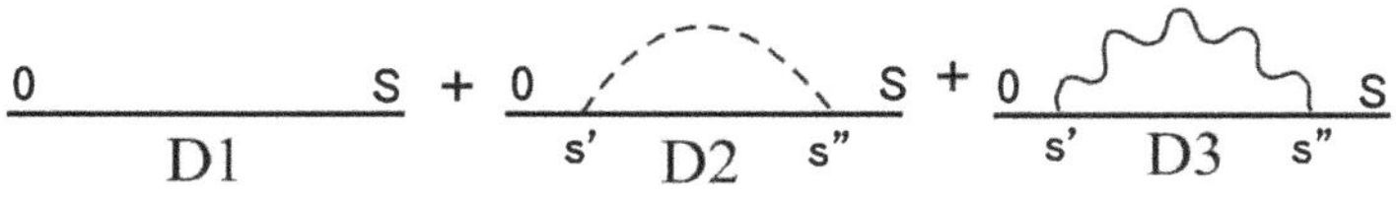


**Figure 8.** Diagrammatic presentation of the contributions to the restricted partition function (3.7) up to the first order of perturbation theory expansion in the coupling constants.

where $\overline{\langle(\ldots)\rangle}$ means averaging with the hamiltonian (3.3). We consider the evaluation of the expression (3.7) by performing the perturbation theory expansion in coupling constants $u$, $v$. The terms in this expansion can be presented diagrammatically as shown in figure 8. The first diagram describes the zeroth order approach corresponding to the idealized Gaussian chain without any interaction between the monomers. Solid line on the diagrams presents the polymer chain, the dashed line describes the excluded volume interaction between monomers governed by the coupling $u$, and the wavy line presents the attractive interaction caused by the presence of the impurities governed by the coupling $v$. Let us consider the expressions corresponding to the second and third diagram:

$$D2 = -u \int \mathrm{d}^d \vec{q} \int_0^S \mathrm{d}s' \int_0^{s'} \mathrm{d}s'' \mathrm{e}^{-\frac{q^2}{2}(s''-s')} \mathrm{e}^{-\frac{k^2}{2}(S-s''+s')},$$

$$D3 = v \int \mathrm{d}^{d-\varepsilon_d} \vec{q} \int_0^S \mathrm{d}s' \int_0^{s'} \mathrm{d}s'' \mathrm{e}^{-\frac{q^2}{2}(s''-s')} \mathrm{e}^{-\frac{k^2}{2}(S-s''+s')}. \qquad (3.8)$$

It is necessary to point out that in the expression for $D3$, the integration is performed only in subspace $d-\varepsilon_d$ due to the fact that the interaction $v$ acts only in this subspace. Using the Poisson formula to integrate over the wave vector $\vec{q}$ we receive:

$$D2 = -u \frac{1}{(2\pi)^{d/2}} \int_0^S \mathrm{d}s' \int_0^{s'} \mathrm{d}s'' (s''-s')^{\frac{d}{2}} \mathrm{e}^{-\frac{k^2}{2}(S-s''+s')},$$

$$D3 = v \frac{1}{(2\pi)^{(d-\varepsilon_d)/2}} \int_0^S \mathrm{d}s' \int_0^{s'} \mathrm{d}s'' (s''-s')^{\frac{d-\varepsilon_d}{2}} \mathrm{e}^{-\frac{k^2}{2}(S-s''+s')}. \qquad (3.9)$$

Expanding the exponents over $\vec{k}$ and then integrating over areas we finally receive:

$$D2 = -z_u \frac{1}{\left(1-\frac{d}{2}\right)\left(2-\frac{d}{2}\right)} + z_u \frac{k^2 S}{2} \frac{2}{\left(1-\frac{d}{2}\right)\left(2-\frac{d}{2}\right)\left(3-\frac{d}{2}\right)},$$

$$D3 = z_v \frac{1}{\left(1-\frac{d-\varepsilon_d}{2}\right)\left(2-\frac{d-\varepsilon_d}{2}\right)} - z_v \frac{k^2_{d-\varepsilon_d} S}{2} \frac{2}{\left(1-\frac{d-\varepsilon_d}{2}\right)\left(2-\frac{d-\varepsilon_d}{2}\right)\left(3-\frac{d-\varepsilon_d}{2}\right)}, \qquad (3.10)$$

where $z_u = u(2\pi)^{-d/2} S^{2-d/2}$ and $z_v = v(2\pi)^{-(d-\varepsilon_d)/2} S^{2-(d-\varepsilon_d)/2}$ are dimensionless coupling constants.

Collecting all contributions from the considered diagrams one receives an expression for the partition function of the model by keeping terms that do not depend on $\vec{k}$:

$$\overline{Z(S)} = 1 - \frac{z_u}{\left(1-\frac{d}{2}\right)\left(2-\frac{d}{2}\right)} - \frac{z_v}{\left(1-\frac{d-\varepsilon_d}{2}\right)\left(2-\frac{d-\varepsilon_d}{2}\right)}.$$

The expressions for the components of the end-to-end distance of the polymer chain can be estimated using the identities:

$$\overline{\langle R^2_{d-\varepsilon_d}\rangle} = -2\frac{1}{Z(S)}\left[\frac{\mathrm{d}}{\mathrm{d}\vec{k}_{d-\varepsilon_d}} \tilde{Z}(\vec{k},S)\right]_{\vec{k}=0}, \qquad \overline{\langle R^2_{\varepsilon_d}\rangle} = -2\frac{1}{Z(S)}\left[\frac{\mathrm{d}}{\mathrm{d}\vec{k}_{\varepsilon_d}} \tilde{Z}(\vec{k},S)\right]_{\vec{k}=0}.$$

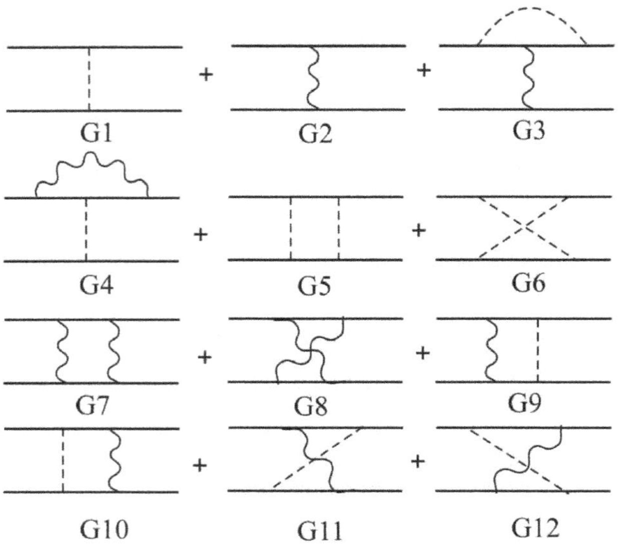


**Figure 9.** Diagramatic presentation of the contributions into the partition function $Z(S,S)$ of two interacting polymer chains up to the second order of expansion in the coupling constants.

We distinguish between the components in subspaces $\varepsilon_d$ and $d-\varepsilon_d$, corresponding to components of the end-to-end distance in directions parallel and perpendicular to extended defects:

$$\overline{\langle R^2_{d-\varepsilon_d}\rangle} = S(d-\varepsilon_d)\left[1+\frac{z_u}{\left(2-\frac{d}{2}\right)\left(3-\frac{d}{2}\right)}-\frac{z_v}{\left(2-\frac{d-\varepsilon_d}{2}\right)\left(3-\frac{d-\varepsilon_d}{2}\right)}\right], \tag{3.11}$$

$$\overline{\langle R^2_{\varepsilon_d}\rangle} = S\varepsilon_d\left[1+\frac{z_u}{\left(2-\frac{d}{2}\right)\left(3-\frac{d}{2}\right)}\right]. \tag{3.12}$$

References (3.11) and (3.12) confirm the existence of two characteristic lengths for polymers in anisotropic environments. The presence of extended defects makes the polymer radius shrink in transverse direction due to the attractive interactions between monomers governed by the coupling $v$, whereas in parallel direction, the increase of the effect of repulsive interactions (as consequence of the increase of monomer density) is responsible for the elongation of the polymer chains.

Calculating contributions to the partition function of two interacting polymer chains one may use a diagrammatic representation (see figure 9). Note that only those diagrams are taken into account which contain at least one interaction line. The first few diagrams, those with one interaction acting between two polymers ($G1-G4$), can be gathered and presented as $-uS^2Z(S)^2 - vS^2Z(S)^2$. Performing the dimensional analysis of the contributions, produced by different diagrams, we find two distinct classes of graphs. The first class of graphs produces terms which behave like $[S]^{\frac{4-d}{2}}$, the sum of all such terms gives contributions into the function denoted by $Z_u(S,S)$. The diagrams of the second class behave like $[S]^{\frac{4-d+\varepsilon_d}{2}}$ and thus give contributions into the function $Z_v(S,S)$. As a result, the "two polymer function" can be presented in the form: $Z(S,S) = Z_u(S,S) + Z_v(S,S)$, where $Z_u(S,S)$ and $Z_v(S,S)$ are given by the expressions:

$$\begin{aligned} Z_u(S,S) \quad = \quad & -uS^2\left[1+2\frac{z_u}{\left(1-\frac{d}{2}\right)\left(2-\frac{d}{2}\right)}-2\frac{z_v}{\left(1-\frac{d-\varepsilon_d}{2}\right)\left(2-\frac{d-\varepsilon_d}{2}\right)}\right. \\ & \left.+2z_u\frac{2^{4-d/2}-10+d}{\left(1-\frac{d}{2}\right)\left(2-\frac{d}{2}\right)\left(3-\frac{d}{2}\right)\left(4-\frac{d}{2}\right)}-4z_v\frac{2^{4-(d-\varepsilon_d)/2}-10+d}{\left(1-\frac{d-\varepsilon_d}{2}\right)\left(2-\frac{d-\varepsilon_d}{2}\right)\left(3-\frac{d-\varepsilon_d}{2}\right)\left(4-\frac{d-\varepsilon_d}{2}\right)}\right], \end{aligned}$$

$$Z_v(S,S) = vS^2\Bigg[1+2\frac{z_u}{\left(1-\frac{d}{2}\right)\left(2-\frac{d}{2}\right)}-2\frac{z_v}{\left(1-\frac{d-\varepsilon_d}{2}\right)\left(2-\frac{d-\varepsilon_d}{2}\right)} -2z_v\frac{2^{4-(d-\varepsilon_d)/2}-10+d}{\left(1-\frac{d-\varepsilon_d}{2}\right)\left(2-\frac{d-\varepsilon_d}{2}\right)\left(3-\frac{d-\varepsilon_d}{2}\right)\left(4-\frac{d-\varepsilon_d}{2}\right)}\Bigg]. \quad (3.13)$$

The swelling factor in our model reads:

$$\chi^0 = \frac{R^2}{Sd} = \frac{\varepsilon_d}{d}R^2_{\varepsilon_d} + \frac{d-\varepsilon_d}{d}R^2_{d-\varepsilon_d} = \left[1+\frac{z_u}{\left(2-\frac{d}{2}\right)\left(3-\frac{d}{2}\right)}+\frac{d-\varepsilon_d}{d}\frac{z_v}{\left(2-\frac{d-\varepsilon_d}{2}\right)\left(3-\frac{d-\varepsilon_d}{2}\right)}\right]. \quad (3.14)$$

The renormalized coupling constants can be presented in the form:

$$g_u = \chi_1^{-4}\chi_0^{-2+\varepsilon/2}Z_u(S,S),$$
$$g_v = \chi_1^{-4}\chi_0^{-2+\delta/2}Z_v(S,S).$$

The corresponding flow equations read:

$$W[g_u] = \varepsilon g_u - 8g_u^2 + 12g_ug_v,$$
$$W[g_v] = -\delta g_v - 8g_v^2 + 4g_ug_v,$$

here, $\varepsilon = 4-d$, $\delta = \varepsilon+\varepsilon_d$. The coordinates of fixed points can be found as common zeros of functions $W[g_u]$,$W[g_v]$:

$$g_u^* = 0, \qquad g_v^* = 0, \quad (3.15)$$
$$g_u^* = \varepsilon/8, \qquad g_v^* = 0, \quad (3.16)$$
$$g_u^* = 0, \qquad g_v^* = -\delta/8, \quad (3.17)$$
$$g_u^* = \varepsilon/2-(3/4)\delta, \qquad g_v^* = \varepsilon/4-\delta/2. \quad (3.18)$$

The first fixed point describes the case of an idealized Gaussian chain without any interactions between monomers. Expression (3.16) corresponds to the case of a polymer chain with short-range excluded volume interactions in a pure solvent. The fixed points (3.17) and (3.18) describe, correspondingly, the Gaussian chain and the chain with excluded volume effect in the anisotropic environment. However, since both of them are associated with attractive interactions between monomers due to the presence of defects, these fixed points appear to be unstable in the physical region of the parameters ($\varepsilon > 0$ and $\varepsilon_d > 0$) and thus cannot provide estimates of scaling exponents. Note that a similar problem of the absence of stable and physically accessible fixed points also exists in the case of uncorrelated point-like impurities [24]. The latter was solved by absorbing the interaction with disorder into the excluded volume interaction due to a special symmetry [64]. However, this does not work in the present case of extended defects.

## 4. Conclusions

We analyzed the influence of anisotropy of the environment caused by the presence of impurities correlated in $\varepsilon_d$ dimensions, on conformational size and shape characteristics of long flexible polymer chains. The integer values of $\varepsilon_d$ have direct physical interpretation and describe extended defects, e.g., in the form of lines or planes of parallel orientation ($\varepsilon_d = 1$ or 2, correspondingly). In this case, it is obvious that one should distinguish between two characteristic length scales, in directions parallel and perpendicular to such extended defects. Non-integer values of $\varepsilon_d$ may correspond to complex defects of fractal nature.

Applying the numerical simulations based on the model of self-avoiding random walks on a regular cubic lattice, we considered three cases: impurities in the form of parallel lines ($\varepsilon_d = 1$), fractal-like structures with $0 < \varepsilon_d < 1$ (which can be treated as partially penetrable lines) and fractal-like structures with $1 < \varepsilon_d < 2$ (partially penetrable planes). In the first case, we found that components of the effective linear size of polymer chain, that are either parallel or perpendicular to the lines of impurities, behave differently and their scaling is governed by two distinct scaling exponents $\nu_\parallel$ and $\nu_\perp$ (see equation (2.4)). The exponent governing the scaling of a parallel component of the end-to-end distance gradually reaches the maximal value of 1 with increasing of concentration of defects, while $\nu_\perp$ gradually tends to zero. Analyzing the influence of disorder in the form of partially penetrable lines on scaling properties of polymers, we again found the existence of two distinct exponents $\nu_\parallel$ and $\nu_\perp$ (see figure 7 (a)). This tendency (and thus the anisotropy) surprisingly persists even at high probability of the polymer chain to penetrate through such "line" (which corresponds to $\varepsilon_d$ close to 0). Considering structural defects in the form of partially penetrable planes of parallel orientation (see figure 1 (c)), we found that the exponent $\nu_\parallel$ gradually changes from the value found earlier for the three-dimensional pure lattice to that in two dimensions with growing concentration of impurity planes. This can be treated as a crossover to a restricted geometry regime of the polymer confined between two homogeneous planes. The exponent $\nu_\perp$ gradually tends to zero.

Our analytical studies were performed within the frames of the direct polymer renormalization approach using the double $\varepsilon$, $\varepsilon + \varepsilon_d$ expansion. In particular, we found expressions for the components of the end-to-end distance of polymer chain (3.11), (3.12). The presence of extended defects makes the polymer radius shrink in transverse direction due to attractive interactions between monomers governed by the coupling $v$, whereas in parallel direction the increase of the effect of repulsive interactions (as a consequence of the increase of monomer density) is responsible for the elongation of the polymer chain. We conclude that the presence of extended defects correlated in $\varepsilon_d$ dimensions makes the polymer chain elongated in the direction parallel to these extended impurities, which confirms the existence of two characteristic lengths for polymers in anisotropic environments.

## Acknowledgements

This work was supported in part by the FP7 EU IRSES projects N269139 "Dynamics and Cooperative Phenomena in Complex Physical and Biological Media" and N295302 "Statistical Physics in Diverse Realizations".

## References

1. Dorogovtsev S.M., Fiz. Tverd. Tela (Leningrad), 1980, **22**, 321.
2. Yamazaki Y., Holz A., Ochiai M., Fukuda Y., Physica A, 1988, **150**, 576; doi:10.1016/0378-4371(88)90257-9.
3. Yamazaki Y., Holz A., Ochiai M., Fukuda Y., Phys. Rev. B, 1986, **33**, 3460; doi:10.1103/PhysRevB.33.3460.
4. Yamazaki Y., Holz A., Ochiai M., Fukuda Y., Physica A, 1986, **136**, 303; doi:10.1016/0378-4371(86)90255-4.
5. Yamazaki Y., Ochiai M., Holz A., Fukuda Y., Phys. Rev. B, 1986, **33**, 3474; doi:10.1103/PhysRevB.33.3474.
6. Stylianopoulos T., Diop-Frimpong B., Munn L.L., Jain R.K., Biophys. J., 2010, **99**, 3119; doi:10.1016/j.bpj.2010.08.065.
7. Xiao F., Nicholson C., Hrabe J., Hrabětova S., Biophys. J., 2008, **95**, 1382; doi:10.1529/biophysj.107.124743.
8. Verkman A.S., Phys. Biol., 2013, **10**, 045003; doi:10.1088/1478-3975/10/4/045003.
9. Cannell D.S. Rondelez F., Macromolecules, 1980, **13**, 1599; doi:10.1021/ma60078a046.
10. Boyanovsky D., Cardy J.L., Phys. Rev. B, 1982, **33**, 154; doi:10.1103/PhysRevB.26.154.
11. Lawrie I.D., Prudnikov V.V., J. Phys. C: Solid State Phys., 1984, **17**, 1655; doi:10.1088/0022-3719/17/10/007.
12. Lee J.C., Gibbs R.L., Phys. Rev. B, 1992, **45**, 2217; doi:10.1103/PhysRevB.45.2217.
13. De Cesare L., Phys. Rev. B, 1994, **49**, 11742; doi:10.1103/PhysRevB.49.11742.
14. Korzhenevskii A.L., Herrmanns K., Schirmacher W., Phys. Rev. B, 1996, **53**, 14834; doi:10.1103/PhysRevB.53.14834.
15. De Gennes P.G., Scaling Concepts in Polymer Physics, Cornell University Press, Ithaca, 1979.
16. Nienhuis B., Phys. Rev. Lett., 1982, **49**, 1062; doi:10.1103/PhysRevLett.49.1062.
17. Guida R., Zinn-Justin J., J. Phys. A: Math. Gen., 1998, **31**, 8104; doi:10.1088/0305-4470/31/40/006.

18. Sahimi M., Flow and Transport in Porous Media and Fractured Rock, VCH, Weinheim, 1995.
19. Dullen A.L., Porous Media: Fluid Transport and Pore Structure, Academic, New York, 1979.
20. Kim Y., J. Phys. A: Math. Gen., 1987, **20**, 1293; doi:10.1088/0305-4470/20/5/039.
21. Kremer K., Z. Phys. B: Condens. Matter, 1981, **49**, 149; doi:10.1007/BF01293328.
22. Lee S.B., Nakanishi H., Phys. Rev. Lett., 1988, **61**, 2022; doi:10.1103/PhysRevLett.61.2022.
23. Lee S.B., Nakanishi H., Kim Y., Phys. Rev. B, 1989, **33**, 9561; doi:10.1103/PhysRevB.39.9561.
24. Chakrabarti B.K., Kertesz K., Z. Phys. B: Condens. Matter, 1981, **44**, 221; doi:10.1007/BF01297178.
25. Sahimi M., J. Phys. A: Math. Gen., **17**, 1984, L379; doi:10.1088/0305-4470/17/7/002.
26. Lam P.M., Zhang Z.Q., Z. Phys. B: Condens. Matter, 1984, **56**, 155; doi:10.1007/BF01469696.
27. Lyklema J.W., Kremer K., Z. Phys. B: Condens. Matter, 1984, **55**, 41; doi:10.1007/BF01307499.
28. Cherayil B.J., J. Chem. Phys., 1990, **92**, 6246; doi:10.1063/1.458349.
29. Roy A.K., Blumen A., J. Stat. Phys., 1990, **59**, 1581; doi:10.1007/BF01334765.
30. Lam P.M., J. Phys. A: Math. Gen., 1990, **23**, L831; doi:10.1088/0305-4470/23/16/010.
31. Kim Y., Phys. Rev. A, 1992, **45**, 6103; doi:10.1103/PhysRevA.45.6103.
32. Nakanishi H., Lee S.B., J. Phys. A: Math. Gen., 1991, **24**, 1355; doi:10.1088/0305-4470/24/6/026.
33. Vanderzande C., Komoda A., Phys. Rev. A, 1992, **45**, R5335; doi:10.1103/PhysRevA.45.R5335.
34. Ordemann A., Porto M., Roman H.E., Havlin S., Bunde A., Phys. Rev. E, 2000, **61**, 6858; doi:10.1103/PhysRevE.61.6858.
35. Janssen H.-K., Stenull O., Phys. Rev. E, 2007, **75**, 020801R; doi:10.1103/PhysRevE.75.020801.
36. Grassberger P., J. Phys. A: Math. Gen., 1993, **26**, 1023; doi:10.1088/0305-4470/26/5/022.
37. Lee S.B., J. Korean Phys. Soc., 1996, **29**, 1.
38. Blavatska V., Janke W., Europhys. Lett., 2008, **82**, 66006; doi:10.1209/0295-5075/82/66006.
39. Blavatska V., Janke W., Phys. Rev. Lett., 2008, **101**, 125701; doi:10.1103/PhysRevLett.101.125701.
40. Blavatska V., Janke W., J. Phys. A: Math. Theor., 2009, **42**, 015001; doi:10.1088/1751-8113/42/1/015001.
41. Weinrib A., Halperin B.I., Phys. Rev. B, 1983, **27**, 413; doi:10.1103/PhysRevB.27.413.
42. Blavats'ka V., von Ferber C., Holovatch Yu., Phys. Rev. E, 2001, **64**, 041102; doi:10.1103/PhysRevE.64.041102.
43. Vanderzande C., Lattice Models of Polymers, Cambridge University Press, Cambridge, 1998.
44. Bhattacharjee S.M., Physica A, 1992, **186**, 183; doi:10.1016/0378-4371(92)90374-Y.
45. Baram A., Stern P.S., J. Phys. A: Math. Gen., 1985, **17**, 1835; doi:10.1088/0305-4470/18/10/036.
46. Baumgartner A., Muthukumar M., In: Advances in Chemical Physics. Polymeric Systems, Vol. XCIV, Prigogine I., Rice S.A. (Eds.), John Wiley & Sons, New York, 1996, 625–709.
47. Blavatska V., Haydukivska K., Eur. Phys. J. – Spec. Top., 2013, **216**, 191; doi:10.1140/epjst/e2013-01742-2.
48. Šolc K., Stockmayer W.H., J. Chem. Phys., 1971, **54**, 2756; doi:10.1063/1.1675241.
49. Šolc K., J. Chem. Phys., 1971, **55**, 335; doi:10.1063/1.1675527.
50. Aronovitz J.A., Nelson D.R., J. Phys., 1986, **47**, 1445; doi:10.1051/jphys:019860047090144500.
51. Des Cloizeaux J., Jannink G., Polymers in Solutions: Their Modelling and Structure, Clarendon Press, Oxford, 1990.
52. Grassberger P., Phys. Rev. E, 1997, **56**, 3682; doi:10.1103/PhysRevE.56.3682.
53. Rosenbluth M.N., Rosenbluth A.W., J. Chem. Phys., 1955, **23**, 356; doi:10.1063/1.1741967.
54. Hsu H.P., Mehra V., Nadler W., Grassberger P., J. Chem. Phys., 2007, **118**, 444; doi:10.1063/1.1522710.
55. Bachmann M., Janke W., Phys. Rev. Lett., 2003, **91**, 208105; doi:10.1103/PhysRevLett.91.208105.
56. Bachmann M., Janke W., J. Chem. Phys., 2004, **120**, 6779; doi:10.1063/1.1651055.
57. Edwards S.F., Proc. Phys. Soc. London, 1965, **85**, 613; doi:10.1088/0370-1328/85/4/301.
58. Brout R., Phys. Rev., 1959, **115**, 824; doi:10.1103/PhysRev.115.824.
59. Folk R., Holovatch Yu., Yavors'kii T., Phys.-Usp., 2003, **46**, 169; doi:10.1070/PU2003v046n02ABEH001077.
60. Wu D., Hui K., Chandler D., J. Chem. Phys., 1991, **96**, 835; doi:10.1063/1.462469.
61. Ippolito I., Bideau D., Hansen A., Phys. Rev. E, 1998, **57**, 3656; doi:10.1103/PhysRevE.57.3656.
62. Patel D.M., Fredrickson G.H., Phys. Rev. E, 2003, **68**, 051802; doi:10.1103/PhysRevE.68.051802.
63. Blavatska V., J. Phys.: Condens. Matter, 2013, **25**, 505101; doi:10.1088/0953-8984/25/50/505101.
64. Kim Y., J. Phys. C: Solid State Phys., 1983, **16**, 1345; doi:10.1088/0022-3719/16/8/005.

# 7

# Scaling functions and amplitude ratios for the Potts model on an uncorrelated scale-free network

M. Krasnytska[1,2]

[1] Institute for Condensed Matter Physics of the National Academy of Sciences of Ukraine, 1 Svientsitskii St., 79011 Lviv, Ukraine

[2] Institut Jean Lamour, CNRS/UMR 7198, Groupe de Physique Statistique, Universite de Lorraine, BP 70239, F-54506 Vandœuvre-lés-Nancy Cedex, France

We study the critical behaviour of the $q$-state Potts model on an uncorrelated scale-free network having a power-law node degree distribution with a decay exponent $\lambda$. Previous data show that the phase diagram of the model in the $q, \lambda$ plane in the second order phase transition regime contains three regions, each being characterized by a different set of critical exponents. In this paper we complete these results by finding analytic expressions for the scaling functions and critical amplitude ratios in the above mentioned regions. Similar to the previously found critical exponents, the scaling functions and amplitude ratios appear to be $\lambda$-dependent. In this way, we give a comprehensive description of the critical behaviour in a new universality class.

**Key words:** *Potts model, complex networks, scaling, universality*

## 1. Introduction

The concept of universality plays a fundamental role in the theory of critical phenomena [1–3]. A lot of systems manifest similar behaviour near the critical point. The universality class does not depend on the local parameters but on the global ones, i.e., dimensionality, symmetry, nature of interaction, etc. If several systems are in the same universality class, they share, besides the values of the critical exponents, identical critical amplitude ratios and scaling functions [4].

The goal of our study is to analyze an universal content of the critical behaviour of the Potts model on a scale-free network near the critical point, in particular, to quantify it in terms of scaling functions and universal amplitude ratios. A lot of studies were devoted to the analysis of the critical behaviour of spin models on complex networks [5]. In this case, the disorder of an underlying structure is modelled in terms of a random graph. In the present work we will consider the $q$-state Potts model on uncorrelated scale-free network having a power-law node degree distribution with exponent $\lambda$. Similar to the lattice systems, for the Potts model on the uncorrelated scale-free networks one may observe either the 1st or the 2nd order phase transition. However, now the order of the phase transition depends, besides the $q$ value, on the node degree distribution decay exponent $\lambda$ [6–8]. The second order phase transition regime is characterized by power law dependencies of thermodynamic functions as functions of temperature and magnetic field in the vicinity of critical point. Critical exponents governing this transition depend on $\lambda$, which plays the role of a global variable for models on a network, like the dimension $d$ in the case of a lattice.

Depending on the particular value of $q$, the Potts model has been suited to describe various real and model systems. Besides the Ising model at $q = 2$, it also describes percolation at $q \to 1$ [9, 10]. The spanning treelike percolation with a geometric phase transition is described by a zero-state $q = 0$ Potts model [11]. Also in $q = 0$ limit, the Potts model can be used for a description of the Kirchhoff's rules via the resistor network models [12]. Subsequently, there has been shown the equivalence between the zero-state Potts model and Abelian sandpile models in case of arbitrary finite graphs [13]. Sandpile models describe processes in neural networks, fracture, hydrogen bonding in liquid water. Another particular case of Potts model at $q = 1/2$ is a spin glass model [14, 15]. The case of $0 \leqslant q < 1$ is used to describe gelation and vulcanization processes in branched polymers [16]. Other examples concern the application of the Potts model for larger values of $q$. Three-component $q = 3$ Potts model is used to describe a cubic ferromagnet with three axes in a diagonal magnetic field [17], an adsorption of 4He atoms on graphite in two dimensions [18], transition of helium films on graphite substrate [19], etc. The 4-state Potts model also describes the effect of absorbtion on surfaces [20]. The Potts model at large $q$ is used to simulate the processes of intercellular adhesion and cancer invasion [21], see also [22].

In this paper we will complete the analysis of the Potts model on an uncorrelated scale-free network [6] by calculating scaling functions and universal amplitude ratios. Recently, [23] the scaling functions and universal amplitude ratios were obtained for the Ising model on a scale free network. Here we will generalize these expressions for the Potts model case.

The structure of this paper is as follows. In the next section we write down the main relations of the scaling theory, expressions for thermodynamic functions in a scaling form and universal amplitude ratios. Section 3 is a short overview of our previous work, where the critical behavior of the Potts model on uncorrelated scale-free network was considered. In particular, it was shown that the phase diagram in the second order phase transition regime contains three regions, each being characterized by a different set of critical exponents. In section 4, we complete these results by finding analytic expressions for the scaling functions and critical amplitude ratios in the above mentioned regions. Similar to the previously found critical exponents, the scaling functions and amplitude ratios appear to be $\lambda$-dependent. In this way, we give a comprehensive description of the critical behaviour in a new universality class. Analytic expressions are summarized in table 2. In the last section we summarize the obtained results.

## 2. Main relations of scaling theory

In this paper we will be interested in the universal features of a system that are manifested in the vicinity of the critical (i.e., second order phase transition) point. Critical exponents, that govern the power-law behaviour of different observables near the critical point belong to such characteristics. Temperature driven phase transition into magnetically ordered state being taken for definiteness, one observes the power-law asymptotics near the critical point $T = T_c$, $h = 0$ [4, 24]. In particular, at $h = 0$ the (dimensionless) order parameter $m$, isothermal susceptibility $\chi_T$, specific heat $c_h$ and magnetocaloric coefficient $m_T$ are[1] power law functions of $\tau = |T - T_c|/T_c$:

$$m = B_- \tau^{\beta}, \qquad \chi_T = \Gamma_{\pm} \tau^{-\gamma}, \qquad c_h = \frac{A_{\pm}}{\alpha} \tau^{-\alpha}, \qquad m_T = B_T^{\pm} \tau^{-\omega} \quad \text{at} \quad h = 0. \tag{2.1}$$

Here, indices $\pm$ refer to the way the critical temperature is approached, $T - T_c \to 0^{\pm}$. In turn, directly at $T = T_c$ (i.e., $\tau = 0$) the following power law field dependencies hold:

$$m = D_c^{-1/\delta} h^{1/\delta}, \qquad \chi = \Gamma_c h^{-\gamma_c}, \qquad c_h = \frac{A_c}{\alpha_c} h^{-\alpha_c}, \qquad m_T = B_T^c h^{-\omega_c} \quad \text{at} \quad \tau = 0. \tag{2.2}$$

The above formulas (2.1), (2.2) introduce critical exponents and critical amplitudes that we are interested in in this study. Unlike the critical exponents, the critical amplitudes are non-universal, being dependent on the microscopic features of the system. However, their certain combinations appear to be universal as well [4]. In particular, in this study we will be interested in the following universal critical amplitude ratios:

$$R_{\chi}^{\pm} = \Gamma_{\pm} D_c B_-^{\delta-1}, \quad R_c^{\pm} = \frac{A_{\pm}\Gamma_{\pm}}{\alpha B_-^2}, \quad R_A = \frac{A_c}{\alpha_c} D_c^{-(1+\alpha_c)} B_-^{-2/\beta}, \quad A_+/A_-, \quad \Gamma_+/\Gamma_-. \tag{2.3}$$

[1]The magnetocaloric coefficient is defined by the mixed derivative of the free energy over magnetic field and temperature, $m_T = -T(\partial m/\partial T)_h$. It measures the heat released by the system upon an isothermal increase of the magnetic field due to the magnetocaloric effect (see, e.g., [23] and references therein).

The above quoted power law scaling in the behaviour of various thermodynamic functions, universality and scaling relations between critical exponents and amplitude ratios are the manifestations of special properties of the thermodynamic potential in the vicinity of critical point. In particular, the scaling hypothesis for the Helmholtz free energy $F(\tau, m)$ states that this thermodynamic potential is a generalized homogeneous function [25] and can be written as follows:

$$F(\tau, m) \approx \tau^{2-\alpha} f_{\pm}(x), \tag{2.4}$$

with the scaling variable $x = m/\tau^{\beta}$ and scaling function $f_{\pm}(x)$, signs + and − correspond to $T > T_c$ and $T < T_c$, respectively. The principal content of equation (2.4) is that $F(\tau, m)$ as a function of two variables can be mapped onto a single variable scaling function $f_{\pm}(x)$. It may be shown that all thermodynamic potentials are generalized homogeneous functions, provided one of them possesses such a property [25].

Based on the expression for the free energy one can also represent the thermodynamic functions in terms of appropriate scaling functions. In particular, magnetic and entropic equations of state read:

$$h(m, \tau) = \tau^{\beta\delta} H_{\pm}(x), \tag{2.5}$$

$$S(m, \tau) = \tau^{1-\alpha} \mathscr{S}(x), \tag{2.6}$$

with the scaling functions $H_{\pm}(x)$ and $\mathscr{S}(x)$. In turn, the scaling functions for the heat capacity, isothermal susceptibility, and magnetocaloric coefficient are defined via (see e.g., [23]):

$$c_h(m, \tau) = (1 \pm \tau)\tau^{-\alpha} \mathscr{C}_{\pm}(x), \tag{2.7}$$

$$\chi_T(m, \tau) = \tau^{-\gamma} \chi_{\pm}(x), \tag{2.8}$$

$$m_T(m, \tau) = (1 \pm \tau)\tau^{\beta-\gamma} \mathscr{M}_{\pm}(x). \tag{2.9}$$

Scaling functions are reachable in experiments and MC simulations. Together with critical exponents and critical amplitude ratios they constitute quantitative characteristics of a given universality class. In the rest of this paper we will complete the previous description of the critical behaviour of the Potts model on an uncorrelated scale-free network by calculating its amplitude ratios and scaling functions in the vicinity of the second order phase transition.

## 3. Potts model on an uncorrelated scale-free network

The $q$-state Potts model can be considered as one of the possible generalizations of the Ising model, where the spin variable can have $q$ possible states [26]. The Potts model Hamiltonian reads:

$$-H = \frac{1}{2}\sum_{i,j} J_{ij}\delta_{n_i,n_j} + h\sum_i \delta_{n_i,0}, \qquad (n_i = 0, 1, \ldots, q-1), \tag{3.1}$$

here $q$ is the number of Potts states, $h$ is a local external magnetic field directed along the 0-th component of the Potts variable $n_i$ (the Potts state on the node $i$). We consider the case where all spins are located on the nodes of a random graph (complex network) and are connected with each other in an appropriate way. The latter is determined by the adjacency matrix $J_{ij}$ with the elements $J_{ij} = 1$ if there exists a link between the nodes $i$ and $j$ and $J_{ij} = 0$ otherwise. One of the important characteristics of a network is its node degree distribution $P(k)$: a probability that the randomly chosen node has a degree (number of links) $k$. We will consider the case of Potts model on an uncorrelated scale-free network with a power-law node degree distribution:

$$P(k) = c_{\lambda} k^{-\lambda}, \tag{3.2}$$

here, $c_{\lambda}$ is a normalization constant and $\lambda$ is the exponent of decay. The absence of correlations within the given link distribution means that the probability to create a link between two nodes is linearly proportional to their node degrees. Furthermore, one may consider the case of an annealed network, when the network configuration is fluctuating under the constraint of a given node degree distribution, see, e.g., [27, 28]. Alternatively, the links between the nodes may be randomly distributed but remain fixed

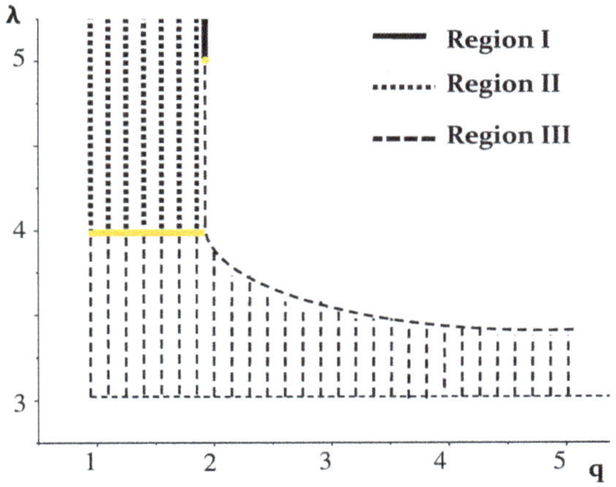


**Figure 1.** (Color online) Phase diagram of the Potts model on an uncorrelated scale-free network. The area of the second order phase transition is shown. Three regions correspond to three different universality classes. Logarithmic corrections to scaling appear for the values of $q$ and $\lambda$ shown by a light line and a light disc (brown online). See the text for a more detailed discussion.

in a given configuration, the so-called configurational model, see [5, 29]. The latter situation corresponds to the quenched case and is usually more complicated for analytical treatment.

The critical behaviour of the Potts model on an uncorrelated scale-free network has been considered in references [6–8]. It was found that the phase diagram of the model is uniquely defined by two parameters: the number of Potts states $q$ and the node degree distribution exponent $\lambda$. Here, we will complete the calculations of [6] where a comprehensive list of critical exponents governing the behaviour of thermodynamic functions in the second order phase transition regime was found. The results of [6] are exact for an annealed network and correspond to the mean field treatment of the quenched case. Our starting point will be the expression for the Helmholtz free energy obtained in reference [6] for different $q$. For non-integer $\lambda > 3$, the free energy reads:

$$F(\tau, M) = a_1 \tau M^2 + a_2 M^{\lambda-1} + \sum_{i=3}^{[\lambda-1]} a_i M^i + O(M^{[\lambda]}), \tag{3.3}$$

here, $M$ is magnetization, $a_i$ are non-universal coefficients, their explicit form is given in [6] and $[\lambda]$ is the integer part of $\lambda$. Note that the power law polynomial form (3.3) holds for the Helmholtz potential for non-integer $\lambda$ only. Logarithmic corrections appear in the case of integer values of $\lambda$. As we discuss below, this will lead to the changes in the critical behaviour at $\lambda = 4$ and $\lambda = 5$.

Figure 1 generalizes the information about the critical behavior of the Potts model on an uncorrelated scale-free network in the form of a phase diagram in the $q-\lambda$ plane [6, 8]. It has been shown that for $\lambda \leqslant 3$, the system remains in the ordered state for any finite temperature [6–8]. For $\lambda > 3$, the phase transition may be of the first or second order, depending on the specific values of $q$ and $\lambda$. The first order phase transition occurs at the values of $q$ and $\lambda$ that belong to the blank region above the dashed curve, $q > 2$, $\lambda > \lambda_c(q)$. Of the main interest for us will be the second order phase transition regime. This corresponds to three different regions shown in figure 1 that belong to three different universality classes. Region I, $\lambda > 5$, $q = 2$ (black solid line in figure 1) is governed by the Ising mean field critical exponents. Region II, $\lambda > 4$, $1 \leqslant q < 2$ (dotted area in the figure) is governed by the percolation mean field critical exponents. Region III ($3 < \lambda < 5$, $q = 2$; $3 < \lambda < 4$, $1 \leqslant q < 2$; $3 < \lambda \leqslant \lambda_c(q)$, $q > 2$) (dashed area in the figure) is characterized by the non-trivial $\lambda$-dependency of the critical exponents. Values of the critical exponents in all three regions are collected in table 1.

As it was mentioned above, logarithmic terms appear in the free energy at the integer values of $\lambda$. In turn, this leads to the appearance of logarithmic corrections [33–35] to the power-law scaling dependencies (2.1), (2.2) of thermodynamic functions at $\lambda = 5$ for the Ising model ($q = 2$) and at $\lambda = 4$ for $1 \leqslant q < 2$

**Table 1.** Critical indices of the Potts model on an uncorrelated scale-free network in three different regions of $q$ and $\lambda$ values, see figure 1. One recovers the results for the Ising model ($q = 2$ [30, 31]) and for percolation ($q = 1$) [32].

| | $\alpha$ | $\alpha_c$ | $\beta$ | $\delta$ | $\gamma$ | $\gamma_c$ | $\omega$ | $\omega_c$ |
|---|---|---|---|---|---|---|---|---|
| region I | 0 | 0 | 1/2 | 3 | 1 | 2/3 | 1/2 | 1/3 |
| region II | −1 | −1/2 | 1 | 2 | 1 | 1/2 | 0 | 0 |
| region III | $\frac{\lambda-5}{\lambda-3}$ | $\frac{\lambda-5}{\lambda-2}$ | $\frac{1}{\lambda-3}$ | $\lambda-2$ | 1 | $\frac{\lambda-3}{\lambda-2}$ | $\frac{\lambda-4}{\lambda-3}$ | $\frac{\lambda-4}{\lambda-2}$ |

[6]. Values of $q$ and $\lambda$ where the thermodynamic functions are governed by power-law singularities enhanced by the logarithmic corrections are shown in figure 1 by the light solid line and light disc (brown online).

In the forthcoming section we will be interested in the critical behaviour in the regions of the second order phase transition with the power law scaling. In particular, we will complete a quantitative description of three universality classes found in regions I, II and III (see figure 1) by calculating, in addition to the critical exponent, the scaling functions and amplitude ratios.

## 4. Critical amplitude ratios and scaling functions

The expression of the free energy of the Potts model on an uncorrelated scale-free network (3.3) will be a starting point for the analysis of the critical amplitude ratios and scaling functions. Passing to the dimensionless energy $f(m,\tau)$ and dimensionless magnetization $m$ and leaving the leading order contributions for small values of $m$, we can present (3.3) in three different regions of the phase diagram (figure 1) in the following form:

$$f(m,\tau) = \pm\frac{\tau}{2}m^2 + \frac{1}{4}m^4, \qquad \text{(Region I)}, \tag{4.1}$$

$$f(m,\tau) = \pm\frac{\tau}{2}m^2 + \frac{1}{4}m^3, \qquad \text{(Region II)}, \tag{4.2}$$

$$f(m,\tau) = \pm\frac{\tau}{2}m^2 + \frac{1}{4}m^{\lambda-1}, \qquad \text{(Region III)}, \tag{4.3}$$

the signs $\pm$ here and in what follows refer to the temperatures above and below the critical point $T_c$. Note that the positive sign of the second terms in (4.1)–(4.3) is due to the fact that coefficients $a_2$, $a_3$, and $a_4$ in (3.3) are positive definite in the regions III, II, and I, correspondingly. With the expressions for the free energy at hand it is straightforward to write down the equation of state and to derive the thermodynamic functions. The magnetic and entropic equations of state in the dimensionless variables $m$ and $\tau$ read:

$$h(m,\tau) = \left.\frac{\partial f(m,\tau)}{\partial m}\right|_\tau, \qquad s(m,\tau) = \mp\left.\frac{\partial f(m,\tau)}{\partial \tau}\right|_m. \tag{4.4}$$

Written explicitly in different regions of $q$, $\lambda$ the magnetic equation of state attains the following form:

$$h = m^3 \pm \tau m, \qquad \text{(Region I)}, \tag{4.5}$$

$$h = \frac{3}{4}m^2 \pm \tau m, \qquad \text{(Region II)}, \tag{4.6}$$

$$h = \frac{\lambda-1}{4}m^{\lambda-2} \pm \tau m, \qquad \text{(Region III)}. \tag{4.7}$$

The entropic equation of state is obtained by a temperature derivative at a constant magnetization $m$ while the explicit $\tau$-dependency is the same in all expressions (4.1)–(4.3). Therefore, the equation keeps the same form in all regions on $q$–$\lambda$ plane:

$$s = -m^2/2, \qquad \text{(Regions I–III)}. \tag{4.8}$$

**Table 2.** Scaling functions and critical amplitude ratios for the Potts model on an uncorrelated scale-free network.

| | Region I | Region II | Region III |
|---|---|---|---|
| $f_{\pm}(x)$ | $\pm\frac{x^2}{2}+\frac{x^4}{4}$ | $\pm\frac{x^2}{2}+\frac{x^3}{4}$ | $\pm\frac{x^2}{2}+\frac{x^{\lambda-1}}{4}$ |
| $H_{\pm}(x)$ | $x^3 \pm x$ | $\frac{3}{4}x^2 \pm x$ | $\frac{\lambda-1}{4}x^{\lambda-2} \pm x$ |
| $\mathscr{S}(x)$ | $-x^2/2$ | $-x^2/2$ | $-x^2/2$ |
| $\mathscr{C}_{\pm}(x)$ | $\frac{x^2}{3x^2\pm1}$ | $\frac{x^2}{3x/2\pm1}$ | $\frac{x^2}{(\lambda-1)(\lambda-2)x^{\lambda-3}/4\pm1}$ |
| $\chi_{\pm}(x)$ | $\frac{1}{3x^2\pm1}$ | $\frac{1}{3x/2\pm1}$ | $\frac{1}{(\lambda-1)(\lambda-2)x^{\lambda-3}/4\pm1}$ |
| $\mathscr{M}_{\pm}(x)$ | $\frac{x}{3x^2\pm1}$ | $\frac{x}{3x/2\pm1}$ | $\frac{x}{(\lambda-1)(\lambda-2)x^{\lambda-3}/4\pm1}$ |
| $A^+/A^-$ | 0 | 0 | 0 |
| $\Gamma^+/\Gamma^-$ | 2 | 1 | $\lambda-3$ |
| $R_{\chi}^+$ | 1 | 1 | 1 |
| $R_{\chi}^-$ | $\frac{1}{2}$ | 1 | $\frac{1}{\lambda-3}$ |
| $R_c^+$ | 0 | 0 | 0 |
| $R_c^-$ | $\frac{1}{4}$ | 1 | $\frac{1}{(\lambda-3)^2}$ |
| $R_A$ | $\frac{1}{3}$ | $\frac{1}{2}$ | $\frac{1}{\lambda-2}$ |

Thermodynamic functions $\chi_T$, $c_h$, and $m_T$ that characterize the response on an external action are directly obtained from the above equations of state. We do not present the explicit expressions here, being rather interested in the corresponding critical amplitude ratios. The latter are given in table 2. In the particular case $q=2$, by these expressions we recover the formerly obtained critical amplitude ratios for the Ising model on an uncorrelated scale-free network [23, 36], correcting at $3<\lambda<5$ the expression for $R_A$ given in [23][2]. Together with the previously derived set of critical exponents (see table 1), our results for the critical amplitude ratios quantify the universal features of critical behaviour of the Potts model in the second order phase transition regime.

Let us derive now the scaling functions for the free energy and other thermodynamic functions. Using the definition (2.4) and taking into account that the heat capacity and the order parameter critical exponents $\alpha$, $\beta$ take on different values in different regions of the phase diagram figure 1 (the formulas are given in table 1) we can recast Helmholtz potential $F(\tau, m)$ in terms of the scaling function $f_{\pm}(m/\tau^{\beta})$. The explicit expressions for the scaling function in all three regions of the phase diagram are given in table 2. Typical behaviour of the free energy scaling functions $f_+(x)$ and $f_-(x)$ is shown in figure 2 (a) and 2 (b), correspondingly.

At any value of $q$, the scaling functions share a common feature: their curvature gradually increases with an increase of $\lambda>3$. This happens up to some marginal value $\lambda=\lambda_c$. The marginal value $\lambda_c$ is $q$-dependent. For $\lambda>\lambda_c$ and $1\leqslant q\leqslant 2$, the scaling functions remain unchanged: their shape does not change with a further increase of $\lambda$. The logarithmic corrections to scaling appear at $\lambda=\lambda_c$ and the second order phase transition holds in this case for $\lambda>\lambda_c$ as well [6], see figure 1. Alternatively, for $\lambda>\lambda_c$ and $q>2$, the phase transition turns out to be of the first order. Curves I of figure 2 (plotted by solid lines) show the limiting behaviour of the scaling functions at $q=2$, $\lambda>5$ [note that $\lambda_c(q=2)=5$]: the functions remain unchanged for all $\lambda>5$. Similar behaviour holds for the case $1\leqslant q<2$, the value of $\lambda_c$, however, differs: $\lambda_c(1\leqslant q<2)=4$. This is shown by curves II in the figure, plotted by dashed lines. Finally, curves III (dotted lines) for $q=4$ are one of examples of the limiting behaviour of the scaling functions in the region $q>2$.

Entropy scaling function $\mathscr{S}(x)$ is defined by (2.6). Using expression (4.8) for the entropy and taking into account the values for the critical exponents $\alpha$ and $\beta$ given in table 1 we arrive at the entropy scaling function that remains unchanged in all regions on $q$–$\lambda$ plane: $\mathscr{S}(x)=-x^2/2$. There are different ways of representing the magnetic equation of state. In the Widom-Griffiths representation [37, 38] the magnetic

[2] In paper [23], using equation (2.3) to find $R_A$ the $\alpha$ exponent was substituted instead of $\alpha_c$ into the power of $D_c$.

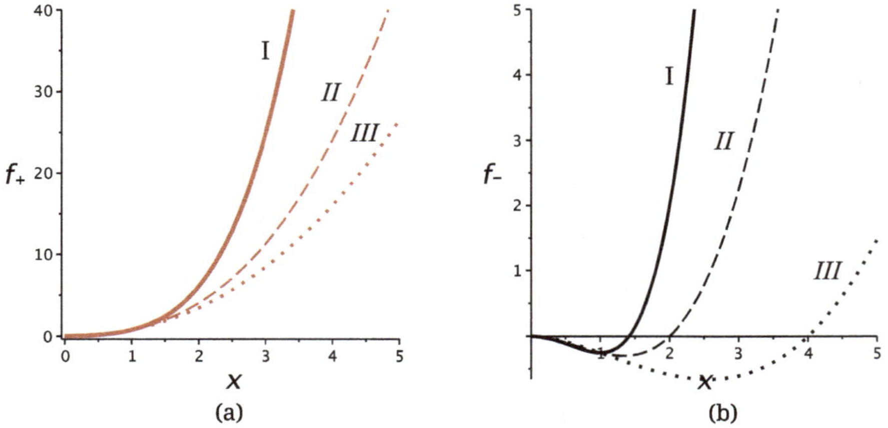


**Figure 2.** (Color online) Limiting behaviour of the free energy scaling functions $f_+$ [(a), $T > T_C$] and $f_-$ [(b), $T < T_C$]. The functions remain unchanged for $\lambda > 5$, $q = 2$ and $\lambda > 4$, $1 \leqslant q < 2$ (solid and dashed curves I and II, correspondingly). For $q > 2$ the phase transition turns out to be of the first order at $\lambda > \lambda_C(q)$. Dotted curves III: $q = 4$, $\lambda = \lambda_C(4) \simeq 3.5$. See the text for more details.

equation of state can be written in two equivalent forms:

$$h = m^{\delta} h_{\pm}\left(\tau / m^{1/\beta}\right), \qquad h = \tau^{\beta\delta} H_{\pm}\left(m / \tau^{\beta}\right), \tag{4.9}$$

with scaling functions $h_{\pm}(x)$ and $H_{\pm}(x)$. Alternatively, in Hankey-Stanley representation, the magnetization is written as [25]:

$$m = \tau^{\beta} \mu_{\pm}\left(h / \tau^{\beta\delta}\right) \tag{4.10}$$

with the scaling function $\mu_{\pm}(x)$.

Starting from the magnetic equation of state given in regions I–III by equations (4.5)–(4.7), it is straightforward to arrive at the scaling functions $H_{\pm}(x)$. We give the appropriate expressions in table 2. Subsequently, one can easily rewrite these expressions to get appropriate $h_{\pm}$-functions. Behaviour of the scaling functions $\mu_{\pm}(x)$ for different values of $\lambda$ and $q$ is shown in figure 3. From the explicit form of the equation of state it is easy to evaluate the asymptotic behaviour of the scaling functions. For $q = 2$ and $\lambda > 5$, one gets $\mu_{\pm}(x) \sim x^{1/3}$, $x \to \infty$. The functions tend to turn to infinity faster with a decrease of $\lambda$:

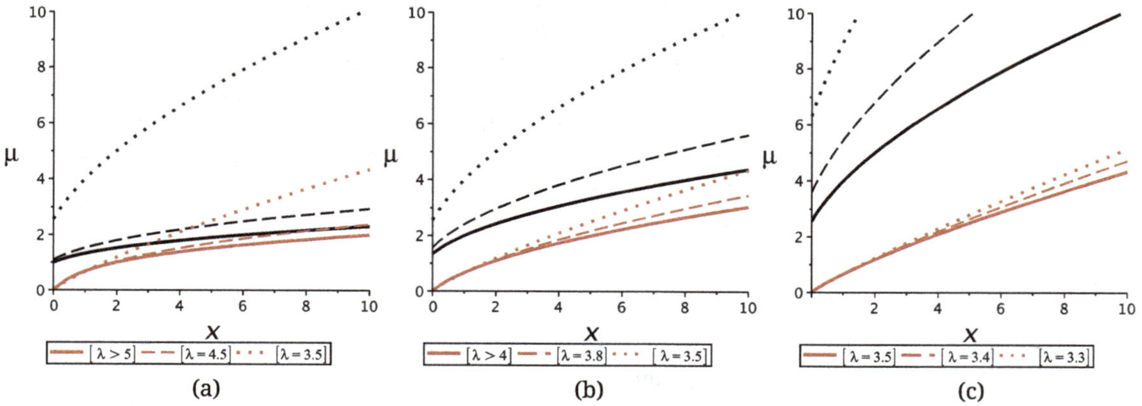


**Figure 3.** (Color online) Behaviour of the order parameter scaling functions $\mu_+(x)$ (light curves, brown online) and $\mu_-(x)$ (black curves) for different values of $\lambda$ and $q$. (a): $q = 2$, (b): $1 \leqslant q < 2$, (c): $q = 4$. Values of $\lambda$ are shown in the figures.

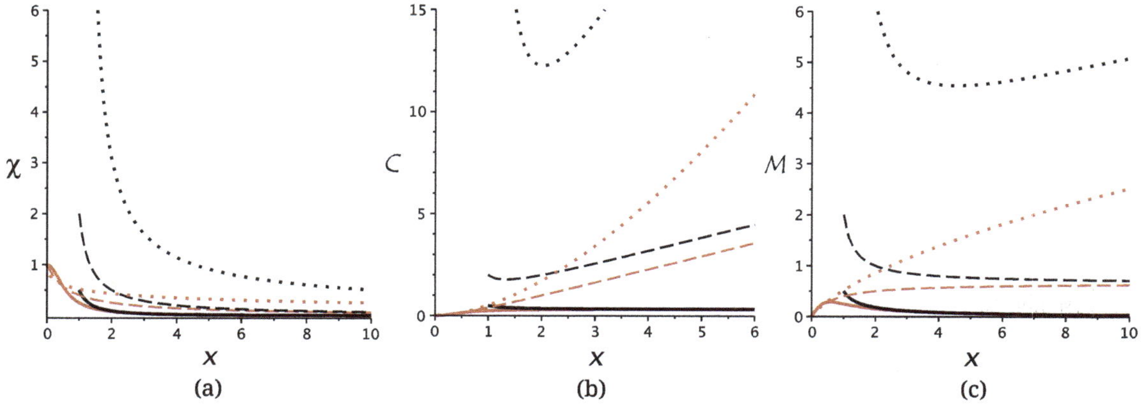


**Figure 4.** (Color online) Scaling functions for the isothermal susceptibility (a), heat capacity (b), and magnetocaloric coefficient (c). Light curves, brown online, $T > T_{\rm C}$: $\chi_+(x)$, $\mathscr{C}_+(x)$, $\mathscr{M}_+(x)$. Black curves, $T < T_{\rm C}$: $\chi_-(x)$, $\mathscr{C}_-(x)$, $\mathscr{M}_-(x)$. Solid, dashed and dotted curves correspond to the values of $q$ and $\lambda$ of the free energy scaling functions of figure 2.

$\mu_\pm(x) \sim x^{1/(\lambda-2)}$, $x \to \infty$ for $3 < \lambda < 5$. A similar feature is observed for the other values of $q$. At $1 \leqslant q < 2$, $\lambda > 4$ one gets $\mu_\pm(x) \sim \sqrt{x}$, $x \to \infty$ and $\mu_\pm(x) \sim x^{1/(\lambda-2)}$, $x \to \infty$ for $3 < \lambda < 4$. The last asymptotic behaviour also holds for $q > 2$ and $\lambda \leqslant \lambda_{\rm c}(q)$. Note that all light curves (brown online) in figure 3 start form the origin: this corresponds to the absence of spontaneous magnetization at $T > T_{\rm C}$. Correspondingly, the value of the scaling function $\mu_-(x)$ at $x = 0$ gives the spontaneous magnetization critical amplitude $B_-$, equation (2.2). As one can see in figure 3, the latter increases with a decrease of $\lambda$.

In figures 4 we show the behaviour of the scaling functions for thermodynamic observables that characterize the response of a system to an external action, i.e., the isothermal susceptibility [figure 4 (a)], heat capacity [figure 4 (b)], and magnetocaloric coefficient [figure 4 (c)]. The values of $q$ and $\lambda$, for which the curves are plotted are the same as those for the free energy scaling functions in figure 2: they reflect the limiting behaviour at some marginal value $\lambda_{\rm c}(q)$. At $1 \leqslant q \leqslant 2$ and $\lambda > \lambda_{\rm c}(q)$, the phase transition remains the second order but the critical exponents do not depend on $\lambda$ any more, the scaling function does not depend on $\lambda$ either. However, for $q > 2$, $\lambda > \lambda_{\rm c}(q)$, the phase transition turns to the first order and the scaling regime does not hold any more. In turn, in the region below $\lambda_{\rm c}$, the exponents acquire $\lambda$-dependency, so do the scaling functions, as is plotted in the figures.

## 5. Conclusions

In this paper we were interested in an analysis of the critical behaviour of the Potts model on uncorrelated scale-free network. In our previous work the list of critical exponents for the Potts model in the second order phase transition regime was obtained [6]. Here, we complete quantitative characteristics of the universal features by calculation of the amplitude ratios and scaling functions. Our results are exact for the annealed scale-free network and correspond to the mean field treatment of the quenched case.

We obtain general scaling functions for the order parameter, entropy, the constant-field heat capacity, magnetic susceptibility and the isothermal magnetocaloric coefficient near the critical point. The comprehensive list of scaling functions and critical amplitude ratios was obtained in different regions of $q$ and $\lambda$. It was shown that the critical amplitude ratios are $\lambda$-dependent similar to the critical exponents, so $\lambda$ plays the role of a global parameter of the system.

## Acknowledgements

This work was supported in part by FP7 EU IRSES projects No. 269139 'Dynamics and Cooperative Phenomena in Complex Physical and Biological Media', No. 295302 'Statistical Physics in Diverse Realizations' and by the Collège Doctoral 02 – 07 Statistical Physics of complex systems. It is my big pleasure to thank Bertrand Berche and Yurij Holovatch for useful comments and discussions.

## References


1. Fisher M.E., Rep. Prog. Phys., 1967, **30**, 615; doi:10.1088/0034-4885/30/2/306.
2. Kadanoff L., Götze W., Hamblen D., Hecht R., Lewis E.A.S., Palciauskas V.V., Rayl M., Swift J., Aspnes D., Kane J., Rev. Mod. Phys., 1967, **39**, 395; doi:10.1103/RevModPhys.39.395.
3. Domb C., The critical point, Taylor & Francis, London, 1996.
4. Privman V., Hohenberg P.C., Aharony A., In: Phase Transitions and Critical Phenomena, Vol. 14, Domb C., Lebowitz J.L. (Eds.), Academic Press, New York, 1991.
5. Dorogovtsev S.N., Goltsev A.V., Mendes J.F.F., Rev. Mod. Phys., 2008, **80**, 1275; doi:10.1103/RevModPhys.80.1275.
6. Krasnytska M., Berche B., Holovatch Yu., Condens. Matter Phys., 2013, **16**, 23602; doi:10.5488/CMP.16.23602.
7. Dorogovtsev S., Goltsev A.V., Mendes J.F.F., Eur. Phys. J. B, 2004, **38**, 177; doi:10.1140/epjb/e2004-00019-y.
8. Iglói F., Turban L., Phys. Rev. E, 2002, **66**, 036140; doi:10.1103/PhysRevE.66.036140.
9. Kasteleyn P.W., Fortuin C.M., J. Phys. Soc. Jpn. (Suppl.), 1969, **26**, 11.
10. Giri M.R., Stephen M.J., Grest G.S., Phys. Rev. B, 1977, **16**, 4971; doi:10.1103/PhysRevB.16.4971.
11. Stephen M.J., Phys. Lett. A, 1976, **56**, 149; doi:10.1016/0375-9601(76)90625-3.
12. Fortiun C.M., Kasteleyn P.W., Physica, 1972, **57**, 536; doi:10.1016/0031-8914(72)90045-6.
13. Majumdar S.N., Dhar D., Physica A, 1992, **185**, 129; doi:10.1016/0378-4371(92)90447-X.
14. Aharony A., J. Phys. C: Solid State Phys., 1978, **11**, L457; doi:10.1088/0022-3719/11/11/004.
15. Aharony A., Pfeuty P., J. Phys. C: Solid State Phys., 1979, **12**, L125; doi:10.1088/0022-3719/12/3/008.
16. Lubensky T.C., Isaacson J., Phys. Rev. Lett., 1978, **41**, 12; doi:10.1103/PhysRevLett.41.829.
17. Mukamel D., Fisher M.E., Domany E., Phys. Rev. Lett., 1976, **37**, 10; doi:10.1103/PhysRevLett.37.565.
18. Alexander S., Phys. Lett. A, 1975, **54**, 353; doi:10.1016/0375-9601(75)90766-5.
19. Bretz M., Phys. Rev. Lett., 1977, **38**, 9; doi:10.1103/PhysRevLett.38.501.
20. Domany E., Schick M., Walker J.S., Phys. Rev. Lett., 1977, **38**, 1148; doi:10.1103/PhysRevLett.38.1148.
21. Turner S., Sherratt J.A., J. Theor. Biol., 2002, **216**, 85; doi:10.1006/jtbi.2001.2522.
22. Laanait L., Messager A., Miracle-Sole S., Ruiz J., Shlosman S., Commun. Math. Phys., 1991, **140**, 81; doi:10.1007/BF02099291.
23. Von Ferber C., Folk R., Holovatch Yu., Kenna R., Palchykov V., Phys. Rev. E, 2011, **83**, 061114; doi:10.1103/PhysRevE.83.061114.
24. Stanley H.E., Rev. Mod. Phys., 1999, **71**, S358; doi:10.1103/RevModPhys.71.S358.
25. Hankey A., Stanley H.E., Phys. Rev. B, 1972, **6**, 3515; doi:10.1103/PhysRevB.6.3515.
26. Wu F.Y., Rev. Mod. Phys., 1982, **54**, 235; doi:10.1103/RevModPhys.54.235.
27. Lee S.H., Ha M., Jeong H., Noh J.D., Park H., Phys. Rev. E, 2009, **80**, 051127; doi:10.1103/PhysRevE.80.051127.
28. Bianconi G., Phys. Rev. E, 2012, **85**, 061113; doi:10.1103/PhysRevE.85.061113.
29. Dorogovtsev S.N., Goltsev A.V., Mendes J.F.F., Phys. Rev. E, 2002, **66**, 016104; doi:10.1103/PhysRevE.66.016104.
30. Leone M., Vázquez A., Vespignani A., Zecchina R., Eur. Phys. J. B, 2002, **28**, 191; doi:10.1140/epjb/e2002-00220-0.
31. Goltsev A.V., Dorogovtsev S., Mendes J.F.F., Phys. Rev. E, 2003, **67**, 026123; doi:10.1103/PhysRevE.67.026123.
32. Cohen R., ben-Avraham D., Havlin S., Phys. Rev. E, 2002, **66**, 036113; doi:10.1103/PhysRevE.66.036113.
33. Kenna R., Johnston D.A., Janke W., Phys. Rev. Lett., 2006, **96**, 115701; doi:10.1103/PhysRevLett.96.115701.
34. Kenna R., Johnston D.A., Janke W., Phys. Rev. Lett., 2006, **97**, 155702; doi:10.1103/PhysRevLett.97.155702.
35. Berche B., Butera P., Shchur L., J. Phys. A: Math. Theor., 2013, **46**, 095001; doi:10.1088/1751-8113/46/9/095001.
36. Palchykov V., von Ferber C., Folk R., Holovatch Yu., Kenna R., Phys. Rev. E, 2010, **82**, 011145; doi:10.1103/PhysRevE.82.011145.
37. Griffiths R.B., Phys. Rev., 1967, **158**, 176; doi:10.1103/PhysRev.158.176.
38. Widom B., J. Chem. Phys., 1965, **43**, 3898; doi:10.1063/1.1696618.

# 8

# Solvent primitive model of an electric double layer in slit-like pores: microscopic structure, adsorption and capacitance from a density functional approach

O. Pizio[1]*, S. Sokołowski[2]

[1] Instituto de Química, Universidad Nacional Autonoma de México, Circuito Exterior, Ciudad Universitaria, 04510 México, D.F., México

[2] Department for the Modelling of Physico-Chemical Processes, Maria Curie-Sklodowska University, Gliniana 33, Lublin, Poland

We investigate the electric double layer formed between charged walls of a slit-like pore and a solvent primitive model (SPM) for electrolyte solution. The recently developed version of the weighted density functional approach for electrostatic interparticle interaction is applied to the study of the density profiles, adsorption and selectivity of adsorption of ions and solvent species. Our principal focus, however, is in the dependence of differential capacitance on the applied voltage, on the electrode and on the pore width. We discuss the properties of the model with respect to the behavior of a primitive model, i.e., in the absence of a hard-sphere solvent. We observed that the differential capacitance of the SPM on the applied electrostatic potential has the camel-like shape unless the ion fraction is high. Moreover, it is documented that the dependence of differential capacitance of the SPM on the pore width is oscillatory, which is in close similarity to the primitive model.

**Key words:** *solvent primitive model, density functional, electrolyte solutions, adsorption, differential capacitance*

## 1. Introduction

The most frequently applied microscopic modelling for the electric double layer (EDL) formed at an interface between a charged solid surface involves the primitive model (PM) of the fluid ionic subsystem. Namely, it is assumed that ions are charged hard spheres immersed into a dielectric continuum having a certain dielectric constant. This very simplified model, compared to real systems in laboratory, has been used for the development and testing of theoretical approaches, as well as to explain experimental observations.

In the theoretical approaches, the dielectric discontinuity at the electrode-electrolyte interface is usually neglected. Another simplification commonly used in the problem of adsorption of PM electrolyte solutions into slit-like pores is to assume that the dielectric constant of the bulk fluid and inside a pore is the same. These comments just illustrate how far the present theoretical modelling is from an entirely satisfactory description of the EDL problems.

One step forward can be made by considering the solvent primitive model (SPM) rather than the PM in the EDL problems. The essence of the SPM is to take into account the effects of excluded volume, due to the presence of solvent molecules (most frequently considered as hard spheres) that are neglected in the PM. First attempts to investigate the SPM at a charged surface have been undertaken in references [1, 2]. More comprehensive efforts to explore the properties of the SPM at charged surfaces have been carried out using a density functional theory [3, 4] and Monte Carlo computer simulations [5, 6]. For

*E-mail: oapizio@gmail.com

the purposes of our study, it is worth mentioning that Tang et al. [3] used Tarazona's weighted density method to describe the hard sphere interaction, while the electrostatic contribution to the free energy functional was modelled assuming that the residual part of the direct correlation functions of nonuniform fluid is the same as in a bulk ionic system. On the other hand, in their recent investigation, Oleksy and Hansen [4] used a version of the density functional approach in which the electrostatic correlation contribution was neglected. In the same context, quite recently the SPM has been used to describe certain aspects of partitioning of electrolyte solutions through semipermeable membranes [7–9]. The importance of such sophistication of modelling, in spite of intrinsic impossibility to describe dielectric properties of the solvent medium, has been documented.

The present state of knowledge regarding the properties of the SPM electric double layer is still incomplete, in particular, concerning the problem of adsorption of electrolyte solutions in the slit-like pores, where an overlap of structures formed at two pore walls can cause some peculiarities of the density profiles, adsorption, dependence of the accumulated charge on the applied voltage and differential capacitance. The overlap of double layers formed at each wall has been involved in the interpretation of recent experimental observations of the dependence of the capacitance of an electrolyte solution on the pore width [10, 11], exhibiting a maximum for a particular very narrow pore of the width slightly larger than the value of the diameter of ions. Computer simulations performed for primitive type models, though with sophistication of the internal structure of ions in some cases, have confirmed the experimental results and provided a certain explanation of the peculiarities of the behavior of the differential capacitance in narrow pores [12–16].

The study of the effect of the differential capacitance of the SPM electric double layer on the value of electrostatic potential, on the pore walls and on the pore width is the principal issue of the present communication. To investigate this model, we use the recent successful weighted density functional approach proposed for a restricted primitive model of electrolyte solutions in contact with charged solid surface [17, 18]. Here, this approach is extended to a mixture of positive and negative ions and hard spheres confined in slit-like pores. In doing this, we use the recent developments dealing with the study of a similar problem, although at the level of the PM for electrolyte solutions in slit-like pores [19–22].

## 2. The model and theory

The SPM under consideration consists of three species, i.e., positive and negative ions (+, −) and solvent molecules mimicked as hard spheres (hs). For the sake of simplicity, in this work we assume that the diameters of all species are the same, $\sigma_+ = \sigma_- = \sigma_{\rm hs} = \sigma$. The valencies of cations and anions are the same $Z^{(+)} = |Z^{(-)}| = Z$. Moreover, we restrict to univalent ions in what follows, i.e., $Z = 1$. The interactions between species are as follows:

$$u^{(\alpha\gamma)}(r) = \begin{cases} \infty, & r < \sigma, \\ \frac{e^2 Z^{(\alpha)} Z^{(\gamma)}}{4\pi\epsilon\epsilon_0}\frac{1}{r}, & r > \sigma, \end{cases} \tag{1}$$

where $\alpha, \gamma = +, -, {\rm hs}$; $e$ denotes the magnitude of elementary charge, $\epsilon$ is the relative permittivity and $\epsilon_0$ is the permittivity of the vacuum. Also $Z^{(\rm hs)} = 0$, thus the solvent is just the fluid of hard spheres.

The mixture of three components is confined in a slit-like pore of the width $H$. The interaction of ions with the pore walls is described by the potential $v^{(\alpha)}(z) = v'^{(\alpha)}(z) = v'^{(\alpha)}(H - z)$ $(\alpha = +, -)$,

$$v'^{(\alpha)}(z) = v_{\rm hw}(z) + v_{\rm el}^{(\alpha)}(z), \tag{2}$$

where $v_{\rm hw}(z)$ is the hard-wall potential

$$v_{\rm hw}(z) = \begin{cases} \infty, & \text{for } z < \sigma/2 \text{ and } z > H - \sigma/2, \\ 0, & \text{otherwise}, \end{cases} \tag{3}$$

and

$$\beta v_{\rm el}^{(\alpha)}(z) = -2\pi l_{\rm B} Q Z^{(\alpha)} z \tag{4}$$

is the Coulomb potential. In the above $\beta = 1/kT$, $Qe$ is the surface charge density of the wall, $l_{\rm B} = e^2/(4\pi kT\epsilon\epsilon_0)$ denotes the Bjerrum length. Energetic aspects of interactions between ions for the model in hand are given in terms of reduced temperature $T^*_{\rm el} = \sigma / l_{\rm B}$. We assume that the interaction of solvent species with the pore walls, $v^{\rm (hs)}(z) = v'^{\rm (hs)}(z) = v'^{\rm (hs)}(H-z)$, is given in the form of Yukawa potential,

$$v'^{\rm (hs)}(z) = \begin{cases} \infty, & \text{for } z < \sigma/2 \text{ and } z > H - \sigma/2, \\ \varepsilon_{\rm gs} \exp[-\lambda_{\rm gs}(z - \sigma/2)]/z, & \text{otherwise.} \end{cases} \tag{5}$$

The confined mixture is in equilibrium with the bulk mixture composed of the same components. The bulk dimensionless densities of the species $\alpha = +, -, {\rm hs}$ are $\rho^*_\alpha = \rho_\alpha \sigma^3$ ($\rho^*_{\rm ion} = \rho^*_+ + \rho^*_-$).

We use the density functional approach, described more in detail in our recent works, see e.g., references [19, 20, 23]. In essence, we construct a thermodynamic potential for the system and then the equilibrium density profiles are obtained by minimizing the thermodynamic potential,

$$\Omega = F + \sum_{\alpha=+,-,{\rm hs}} \int {\rm d}\mathbf{r} \left[ v^{(\alpha)}(z) \rho^{(\alpha)}(z) - \mu_\alpha \right] + \int {\rm d}\mathbf{r} q(z) \Psi(z). \tag{6}$$

In the above $\rho^{(\alpha)}(z)$ and $\mu_\alpha$ are the local density and the chemical potential of the species $\alpha$, respectively, $F$ is the free energy functional and $q(z)$ is the charge density,

$$q(z)/e = \sum_{\alpha=+,-} Z^{(\alpha)} \rho^{(\alpha)}(z). \tag{7}$$

The electrostatic $\Psi(z)$ satisfies the Poisson equation,

$$\nabla^2 \Psi(z) = -\frac{4\pi}{\epsilon \epsilon_0} q(z). \tag{8}$$

The solution of differential equation (8) for the slit-pore geometry with walls of equal charge is perfomed similarly to reference [24], where the model is different, however. Moreover, the method of solving the Poisson equation for a set of interconnected slit-like pores with permeable walls was explained and analysed in every detail in the recent work by Kovacs et al. [25]. For the model of a single slit defined by equations (2)–(5) in the present study, the solution requires the choice of the boundary condition, namely of the value of the electrostatic potential at a wall, $V_0 = \Psi(z=0) = \Psi(z=H)$.

From the electro-neutrality condition of the system it follows that

$$Q + \int {\rm d}z q(z) = 0, \tag{9}$$

where $Qe$ is the surface charge density of the wall as we have already mentioned above.

The free energy of the system, $F$, is the sum of the ideal, $F_{\rm id}$, hard sphere, $F_{\rm hs}$ and residual electrostatic excess contribution, $F_{\rm el}$, arising from the coupling between electrostatic and hard-sphere interactions. The ideal part of the free energy, $F_{\rm id}$, is known exactly,

$$F_{\rm id}/kT = \sum_{\alpha=a,c,{\rm hs}} \int {\rm d}\mathbf{r} \rho^{(\alpha)}(z) \left[ \ln\left(\rho^{(\alpha)}(z)\right) - 1 \right]. \tag{10}$$

The excess free energy due to hard sphere interactions between the species $+, -$, and hs, $F_{\rm hs}$, is taken from the fundamental measure theory [26–30]. The details of the White Bear version of the fundamental measure theory are given in references [26–28] Finally, the residual electrostatic contribution $F_{\rm el}$ is described by using the so-called "weighted correlation approach", WCA-$k^2$ approximation, developed for nonuniform RPM ionic fluids by Wang et al. [17, 18] based on the analytic solution of the mean spherical approximation, cf. also reference [19]. The expressions used in the present study are given by equations (11)–(14) of our recent work [19]. They are omitted to avoid the unnecessary repetition.

At equilibrium, the density profiles minimize the thermodynamic potential $\Omega$, i.e.,

$$\frac{\delta \Omega}{\delta \rho^{(\alpha)}(\mathbf{r})} = 0, \qquad \alpha = +, -, {\rm hs}. \tag{11}$$

The resulting density profile equations can be straightforwardly derived by modifying those given in references [4, 17–19]

As we have mentioned above, the adsorption system is in equilibrium with the bulk SPM mixture, this equilibrium being determined by the equality of chemical potentials of each species in the bulk phase and in the pore. The bulk densities of ionic species satisfy the electro-neutrality condition $Z^{(+)}\rho_+ + Z^{(-)}\rho_- = 0$.

## 3. Results

Let us now specify a set of parameters of the model we study below. As already mentioned in the introduction, we restrict our attention to the model of equal diameters of all the species involved. Actually, we performed calculations for the model with a larger diameter of solvent species compared to the diameter of ions, but qualitatively the trends observed are very similar to those discussed below. The distance from the wall and pore width are given in reduced units, $z^* = z/\sigma$ and $H^* = H/\sigma$, respectively. Also, the electrostatic potential at the wall is considered in reduced units, $V^* = eV_0/kT$. Another introductory comment concerns the interaction between solvent hard sphere species and pore walls. It has been written in the form of Yukawa interaction. However, in the present study we just consider weakly adsorbing walls, $kT/\varepsilon_{gs} = 1$ and $\lambda_{gs} = 3$. This interaction has been introduced having in mind a possible extension of the SPM model in order to take into account the attractive interaction between solute particles in the spirit of works by Oleksy and Hansen [4, 31–33]. On the other hand, our interest is in a dense fluid with high fraction of solvent species and low ion content. Thus, in the majority of numerical calculations, the solvent bulk density is taken to be $\rho^*_{hs} = 0.5$.

### 3.1. Density profiles and adsorption

We begin the discussion by considering the microscopic structure and the resulting thermodynamic properties. The evolution of the density profile of a hard sphere solvent of the SPM with an increasing electrostatic potential on the wall is shown in figure 1. It can be seen that the contact value of the profile $\rho_{hs}(z)$ decreases with an increasing $V^*$, showing that hard spheres are expelled from the vicinity of the wall. The density of the second layer at $z^* = 2.5$ increases with an increasing $V^*$, reaches a maximum value at $V^* = 30.8$ and then slightly decreases with a further increase of the electrostatic potential. This proves that hard sphere solvent particles are again slightly expelled from the second layer at the expense of a weakly increasing density closer to the pore center. These trends are due to accumulation and simultaneous separation of ion species close to the charged surface of the pore, figure 2.

The density of counter-ions substantially increases close to the pore walls while the density of co-ions decreases at the contact and in the pore walls vicinity with an increasing $V^*$. However, structural changes also occur in the second layer around $z^* = 2.5$. In this layer, the co-ion density increases while opposite trends are seen for the counter-ions. It seems, however, that the presence of hard sphere solvent

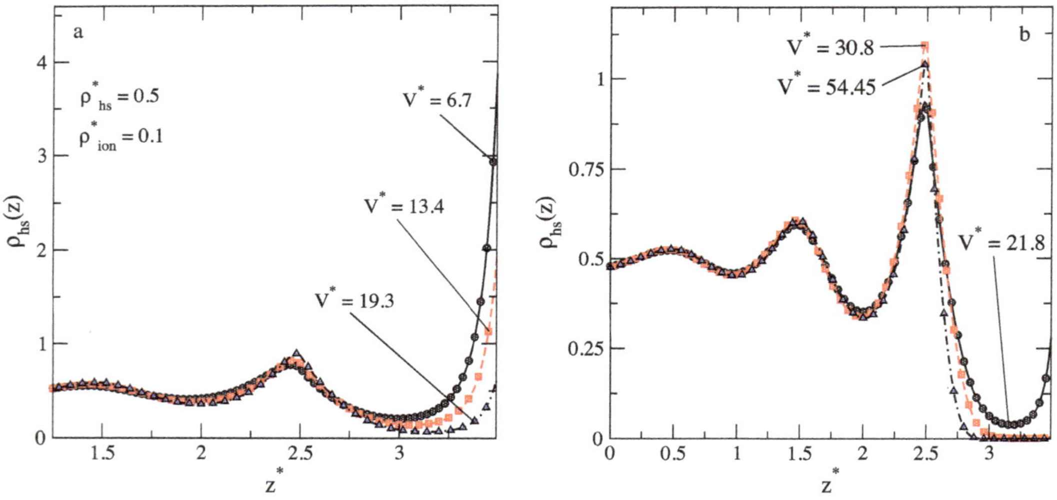


**Figure 1.** (Color online) Evolution of the density profiles of hard sphere species, $\rho_{hs}(z)$, of the SPM, with the applied electrostatic potential on the wall, $V^*$, at bulk density, $\rho^*_{hs} = 0.5$, $\rho^*_{ion} = 0.1$, in the slit-like pore of the width $H^* = 8$. The energetic parameters of the SPM are $T^*_{el} = 0.15$, $kT/\varepsilon_{gs} = 1.0$ here and in all the subsequent figures.

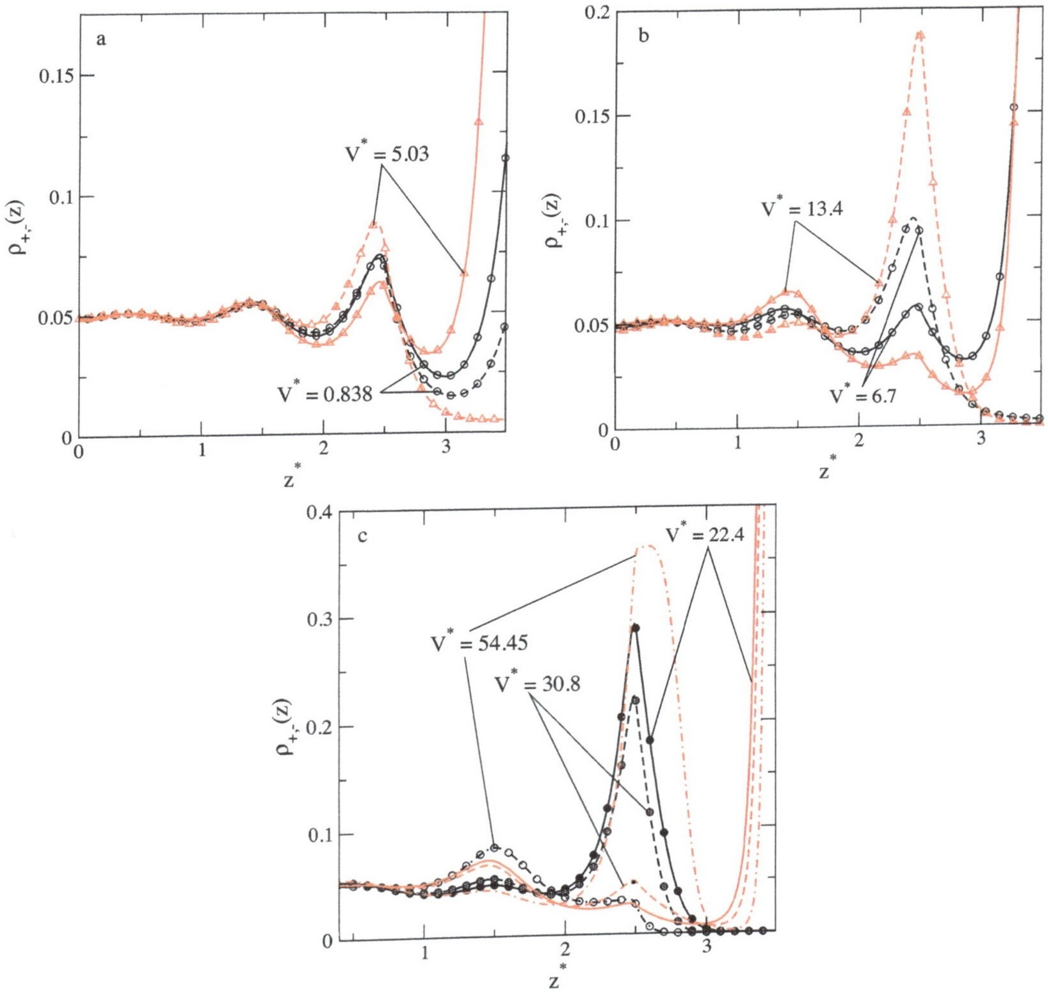


**Figure 2.** (Color online) Evolution of the density profiles of each ion species, $\rho_{+,-}(z)$, of the SPM, with the electrostatic potential applied to the wall, $V^*$. The system is the same as in figure 1.

species in the pore center promotes the separation of ions of the opposite charge close to the wall, thus playing a role of supporting the external field effects.

The trends of behavior of the average density of species in the pore with an increasing electrostatic potential are illustrated in figure 3. Excess adsorption of the species is defined as common:

$$A_\alpha^{\mathrm{ex}} = \int \mathrm{d}z \left[\rho_\alpha(z) - \rho_\alpha\right] \tag{12}$$

and the average density of the species is as follows:

$$\langle \rho_\alpha \rangle_H = \frac{1}{H} \int \mathrm{d}z \rho_\alpha(z). \tag{13}$$

From the panel (a) of this figure we learn that the excess adsorption of a hard sphere solvent substantially decreases with an increasing electrostatic potential and is negative in almost entire range of $V^*$. In all three cases considered, we kept constant the total density of the bulk solution at $\rho_{\mathrm{hs}} + \rho_{\mathrm{ion}} = 0.6$ and changed its composition by decreasing the ion density in the systems 1, 2, and 3. The curves behave differently at low, intermediate and high $V^*$. In a narrow region of rather small $V^*$ and at high values of $V^*$, the lowest excess adsorption is observed for the system 3 that has the lowest fraction of ions in the

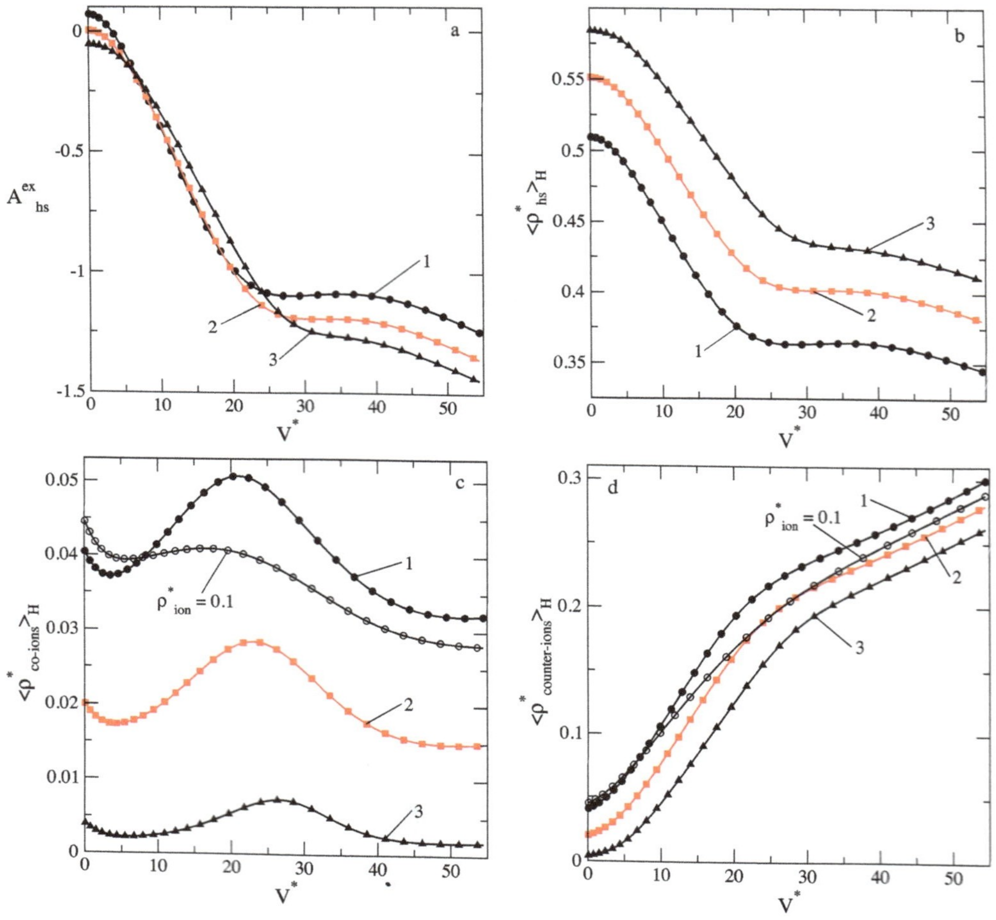


**Figure 3.** (Color online) Excess adsorption and average density of hard sphere species of the SPM in the slit like pore $H^* = 8$, panels (a) and (b), respectively. Average density of co-ions [panel (c)] and of counter-ions [panel (d)] in this pore. The nomenclature of systems is the following: 1 — $\rho^*_{hs} = 0.5$, $\rho^*_{ion} = 0.1$; 2 — $\rho^*_{hs} = 0.55$, $\rho^*_{ion} = 0.05$; 3 — $\rho^*_{hs} = 0.59$, $\rho^*_{ion} = 0.01$.

bulk phase. The average density of hard sphere species [panel (b)] decreases with an increasing $V^*$, its dependence on $V^*$ being non-monotonous, however.

This behavior of the excess adsorption and of the average density of solvent species is due to the changes of the average density (and distribution of ions) in the pore under the effect of external electric field. In particular, the behavior of the average density of co-ions [panel (c)] with an increasing $V^*$ having a maximum in the interval between 20 and 25 can be traced back to the corresponding density profiles showing how the co-ions are expelled from the vicinity of the wall and how they form a relatively dense second layer. Again, changes of the structure discussed in terms of the curves in figure 2 are manifest due to the different rate of growth at low and high $V^*$ of the average density of counter-ions [panel (d) of figure 3]. Most important, changes of the density of ion species and changes of distribution of solvent species upon increasing the electrostatic potential cause the changes of the dependence of the charge in the pore and consequently the changes in the shape of the differential capacitance.

## 3.2. Differential capacitance

The differential capacitance,

$$C_D = \left(\frac{\partial Q}{\partial V_0}\right)_{H,T,\mu_\alpha}, \tag{14}$$

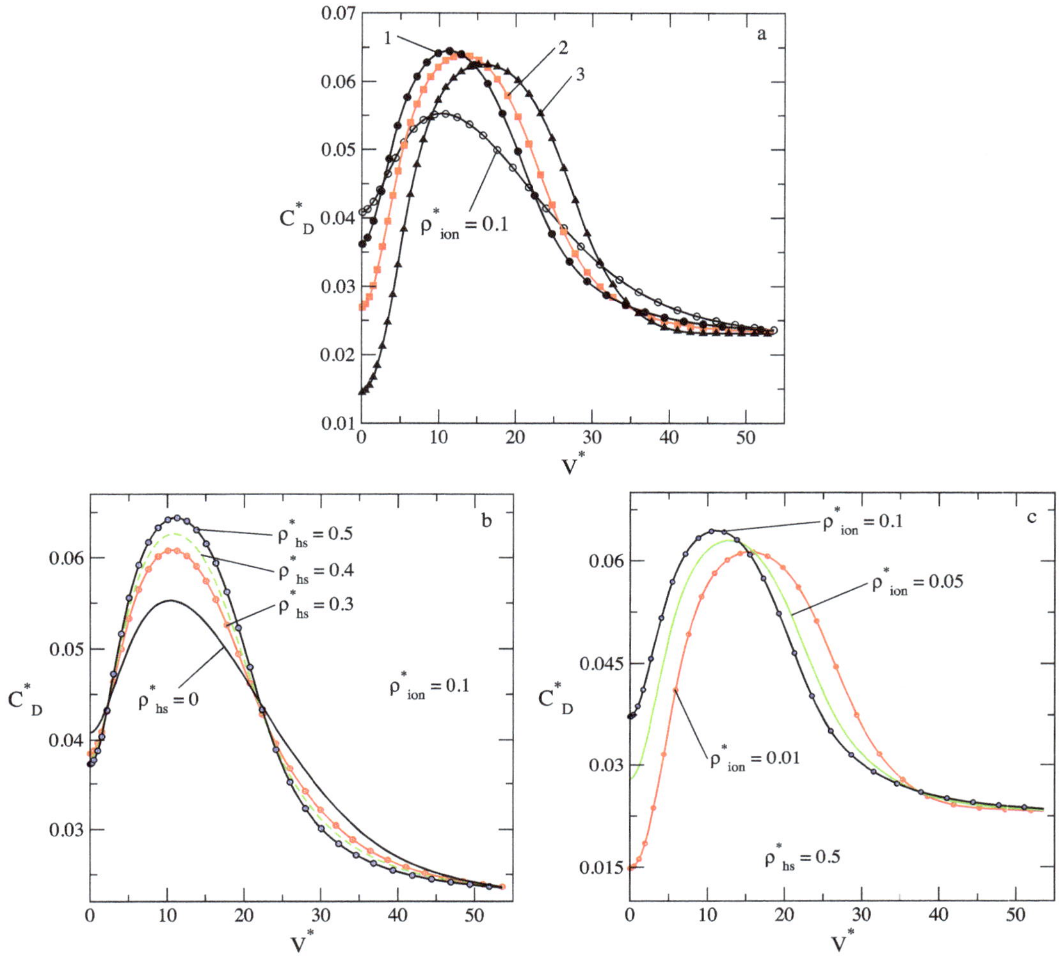


**Figure 4.** (Color online) The dependences of the differential capacitance, $C^*_D$, on the applied voltage, $V^*$ at a different bulk fluid density and a different composition shown in each panel. The nomenclature of systems 1, 2, and 3 [panel (a)] is given in the text.

is obtained by taking a derivative of the charge by electrostatic potential on the wall and is plotted as a function of electrostatic potential in figure 4. In all the cases studied we observe the camel-like shape of the differential capacitance. Considering the fixed total density as in figure 3, we see now that the highest maximum value of the capacitance is reached when the ion fraction is the highest, namely for the system 1 compared to 2 and 3. However, the value of the maximum is less sensitive to the ion fraction compared to the trough at a very small $V^*$ [panel (a) of figure 4]. At a very high $V^*$ the curves for three systems tend to almost equal value. If we compare the system 1 and its PM counterpart at the same ion density ($\rho^*_{ion} = 0.1$), then it appears that the differential capacitance curves behave qualitatively similarly. However, in the SPM case, the $C^*_D$ maximum is much higher compared to PM. Thus, it seems that the presence of solvent species enhances the separation of ions of the opposite charge by "putting" them slightly closer to the pore walls, where the electric field makes its job. In order to obtain higher values of the differential capacitance at maximum, one can either take a denser solvent (at a fixed ion density) like it is shown in the panel (b) of figure 4 or may increase the ion fraction at a fixed solvent density, like in the panel (c) of figure 4. To summarize, the presence of solvent species in the SPM permits to alter the values of differential capacitance in different regions of $V^*$, in comparison to PM. However, the overall shape remains qualitatively similar unless the ion fraction becomes high (in real systems one needs in fact to take into account the solubility limit).

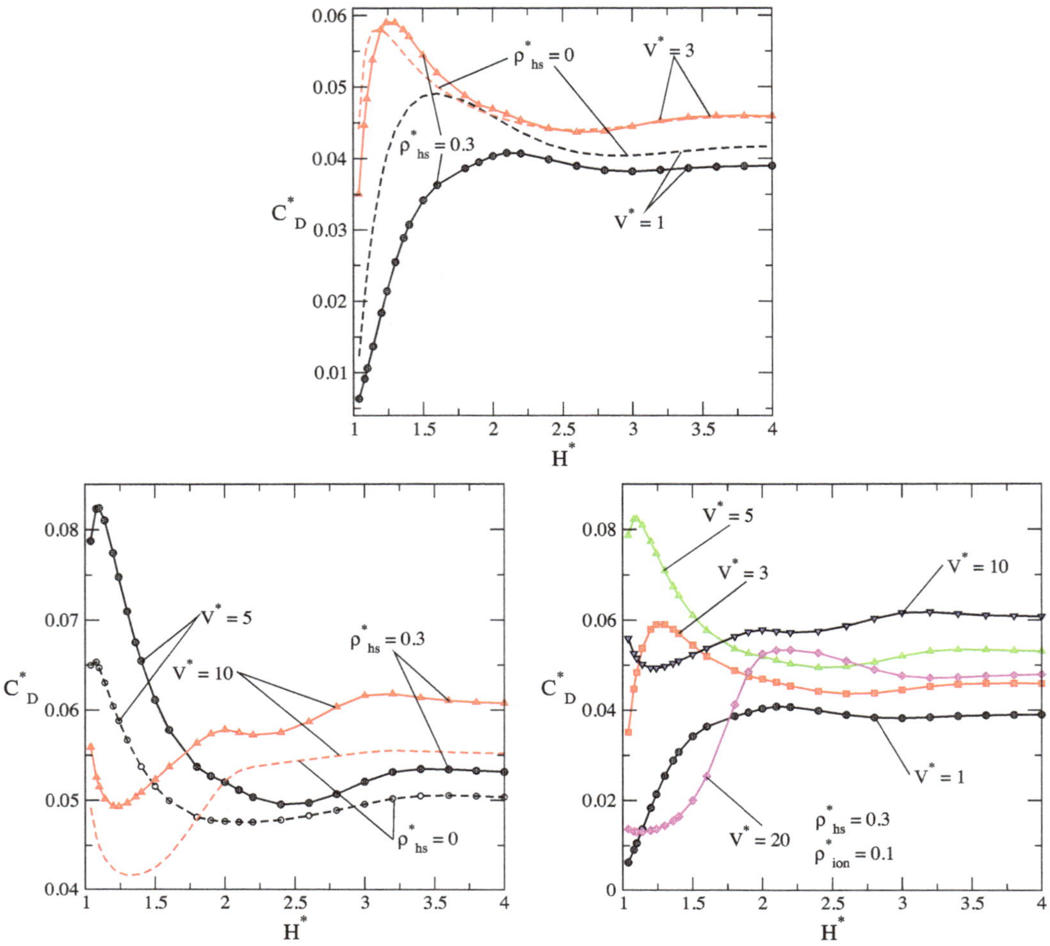


**Figure 5.** (Color online) The dependences of the differential capacitance, $C^*_{\mathrm{D}}$, on the pore width, $H^*$ at a different voltage, $V^*$. Panels (a) and (b) show a comparison of the results for the SPM and PM at the same conditions. Panel (c) contains the results for $C^*_{\mathrm{D}}(H^*)$ solely for the SPM at a different fixed voltage $V^*$.

The final issue we would like to discuss is the dependence of the differential capacitance on the pore width. This problem for a restricted primitive model of electrolyte solutions confined in slit-like pores was quite comprehensively discussed in the recent work from this laboratory [19]. In the panel (a) of figure 5 we compare the SPM and PM curves for $C^*_{\mathrm{D}}(H)$ at rather low values of the electrostatic potential, namely at $V^* = 1$ and $V^* = 3$. The curves for two models are of similar shape. However, the solvent affects the values for $C^*_{\mathrm{D}}$, especially in narrow pores. The first maximum of $C^*_{\mathrm{D}}$ can be either supressed (at $V^* = 1$) or enhanced (at $V^* = 3$) due to the solvent presence [figure 5 (a)]. The curves for SPM and PM eventually tend to zero if $H^*$ tends to its minimum value.

At higher values of $V^*$, $V^* = 5$ and $V^* = 10$ [panel (b) of figure 5], the qualitative features of the shape of functions in question are again similar for SPM and PM. Nevertheless, in the SPM case, we observe more pronounced oscillations of the differential capacitance on the pore width. In other words, the phase of overlap of the density profiles of ions formed at each wall (discussed in detail in [19]) is altered, due to the presence of solvent species. In close similarity to the PM system, the shape of the dependence of $C^*_{\mathrm{D}}(H^*)$ in the present SPM case alters depending on the value of the electrostatic potential $V^*$. The differential capacitance can either grow or drop in the region of very narrow pores depending on the choice of the voltage. Still, the oscillatory behavior (showing well pronounced and less pronounced maxima and several troughs) is observed for the confined SPM.

It is interesting to mention that Oleksy and Hansen observed the oscillatory curve for the solvation force between charged plates with the SPM-like solution in between. However, their calculations were performed under the condition of a constant charge on the plates rather than at a constant potential carried out in the present study. It seems that to establish the relation between the oscillatory curve for the differential capacitance and the dependence of the solvation force on the charged plates separation is of utmost importance in future research. In addition, we would like to emphasize that the model of this study permits several extensions. One of the promising extensions is the possibility to improve the model by introducing the concepts of chemical association in order to deal with the adsorption of either chain molecules or the network-forming solvent. Theoretical background is rather straightforward to be developed along the lines presented in e.g., [34].

## Acknowledgements

O.P. is grateful to David Vazquez for technical assistance at the Institute of Chemistry of the UNAM.

## References

1. Grimson M.J., Rickayzen G., Chem. Phys. Lett., 1982, **86**, 71; doi:10.1016/0009-2614(82)83119-9.
2. Groot R.D., Phys. Rev. A, 1985, **37**, 3456; doi:10.1103/PhysRevA.37.3456.
3. Tang Z., Scriven L.E., Davis H.T., J. Chem. Phys., 1992, **97**, 494; doi:10.1063/1.463595.
4. Oleksy A., Hansen J.-P., Molec. Phys., 2006, **104**, 2871; doi:10.1080/00268970600864491.
5. Boda D., Henderson D., J. Chem. Phys., 2000, **112**, 8934; doi:10.1063/1.481507.
6. Lamperski S., Zydor A., Electrochim. Acta, 2007, **52**, 2429; doi:10.1016/j.electacta.2006.08.045.
7. Boda D., Henderson D., Patrykiejew A., Sokolowski S., J. Chem. Phys., 2000, **113**, 802; doi:10.1063/1.481855.
8. Boda D., Henderson D., Patrykiejew A., Sokolowski S., J. Colloid Interface Sci., 2001, **239**, 432; doi:10.1006/jcis.2001.7560.
9. Henderson D., Bryk P., Sokolowski S., Wasan D.T., Phys. Rev. E, 2000, **61**, 3896; doi:10.1103/PhysRevE.61.3896.
10. Chmiola J., Yushin G., Gogotsi Y., Portet C., Simon P., Taberna P.L., Science, 2006, **313**, 1760; doi:10.1126/science.1132195.
11. Largeot C., Portet C., Chmiola J., Taberna P.L., Gogotsi Y., Simon P., J. Am. Chem. Soc., 2008, **130**, 2730; doi:10.1021/ja7106178.
12. Merlet C., Rotenberg B., Madden P.A., Tabrna P.L., Simon P., Gogotsi Y., Salanne M., Nat. Mater., 2012, **11**, 306; doi:10.1038/nmat3260.
13. Georgi N., Kornyshev A., Fedorov M., J. Electroanal. Chem., 2010, **649**, 261; doi:10.1016/j.jelechem.2010.07.004.
14. Kondrat S., Georgi N., Fedorov M., Kornyshev A., Phys. Chem. Chem. Phys., 2011, **13**, 11359; doi:10.1039/c1cp20798a.
15. Wu P., Huang J., Mernier V., Sumpter B.G., Qiao R., J. Phys. Chem. Lett., 2012, **3**, 1732; doi:10.1021/jz300506j.
16. Feng G., Cummings P.T., J. Phys. Chem. Lett., 2011, **2**, 2859; doi:10.1021/jz201312e.
17. Wang Z., Liu L., Neretnieks I., J. Phys.: Condens. Matter, 2011, **23**, 175002; doi:10.1088/0953-8984/23/17/175002.
18. Wang Z., Liu L., Neretnieks I., J. Chem. Phys., 2011, **135**, 244107; doi:10.1063/1.3672001.
19. Pizio O., Sokołowski S., Sokołowska Z., J. Chem. Phys., 2012, **137**, 234705; doi:10.1063/1.4771919.
20. Pizio O., Sokołowski S., J. Chem. Phys., 2013, **138**, 204715; doi:10.1063/1.4807777.
21. Jiang D., Jin Z., Wu J., Nano Lett., 2011, **11**, 5372; doi:10.1021/nl202952d.
22. Henderson D., J. Colloid Interface Sci., 2012, **374**, 345; doi:10.1016/j.jcis.2012.01.050.
23. Bryk P., Sokołowski S., J. Chem. Phys., 2004, **121**, 11314; doi:10.1063/1.1814075.
24. Henderson D., Bryk P., Sokołowski S., Wasan D.T., Phys. Rev. E, 2000, **61**, 3896; doi:10.1103/PhysRevE.61.3896.
25. Kovacs R., Valisko M., Boda D., Condens. Matter Phys., 2012, **15**, 23803; doi:10.5488/CMP.15.23803.
26. Yu Y.-X., Wu J., J. Chem. Phys., 2002, **117**, 2368; doi:10.1063/1.1491240.
27. Yu Y.-X., Wu J., J. Chem. Phys., 2002, **117**, 10156; doi:10.1063/1.1520530.
28. Yu Y.-X., Wu J., J. Chem. Phys., 2003, **118**, 3835; doi:10.1063/1.1539840.
29. Rosenfeld Y., Phys. Rev. Lett., 1989, **63**, 980; doi:10.1103/PhysRevLett.63.980.
30. Rosenfeld Y., Phys. Rev. A, 1990, **42**, 5978; doi:10.1103/PhysRevA.42.5978.
31. Oleksy A., Hansen J.-P., Molec. Phys., 2009, **107**, 2609; doi:10.1080/00268970903469022.
32. Oleksy A., Hansen J.-P., J. Chem. Phys., 2010, **132**, 204702; doi:10.1063/1.3428704.
33. Oleksy A., Hansen J.-P., Molec. Phys., 2011, **109**, 1275; doi:10.1080/00268976.2011.554903.
34. Bryk P., Sokołowski S., Pizio O., J. Chem. Phys., 2006, **125**, 024909; doi:10.1063/1.2212944.

# 9

# Self-consistent field theoretic simulations of amphiphilic triblock copolymer solutions: Polymer concentration and chain length effects

X.-G. Han[1,2]*, Y.-H. Ma[1,2]

[1] Inner Mongolia Key Laboratory for Utilization of Bayan Obo Multi-Metallic Resources: Elected State Key Laboratory, Inmongolia Science and Technology University, Baotou 014010, China

[2] School of Mathematics, Physics and Biology, Inmongolia Science and Technology University, Baotou 014010, China

Using the self-consistent field lattice model, polymer concentration $\bar{\phi}_P$ and chain length $N$ (keeping the length ratio of hydrophobic to hydrophilic blocks constant) the effects on temperature-dependent behavior of micelles are studied, in amphiphilic symmetric ABA triblock copolymer solutions. When chain length is increased, at fixed $\bar{\phi}_P$, micelles occur at higher temperature. The variations of average volume fraction of stickers $\bar{\phi}^{\mathrm{s}}_{\mathrm{co}}$ and the lattice site numbers $N^{\mathrm{ls}}_{\mathrm{co}}$ at the micellar cores with temperature are dependent on $N$ and $\bar{\phi}_P$, which demonstrates that the aggregation of micelles depends on $N$ and $\bar{\phi}_P$. Moreover, when $\bar{\phi}_P$ is increased, firstly a peak appears on the curve of specific heat $C_V$ for unimer-micelle transition, and then in addition a primary peak, the secondary peak, which results from the remicellization, is observed on the curve of $C_V$. For a long chain, in intermediate and high concentration regimes, the shape of specific heat peak markedly changes, and the peak tends to be a more broad peak. Finally, the aggregation behavior of micelles is explained by the aggregation way of amphiphilic triblock copolymer. The obtained results are helpful in understanding the micellar aggregation process.

**Key words:** *micelle, self-consistent field, amphiphilic copolymer*

## 1. Introduction

Polymeric micelles constitute a unique class of nanomaterials having a typical core-shell morphology. They are formed from amphiphilic block- or graft-copolymers in a selective solvent, where the non-soluble parts self-assemble to form the core of the micelles and the soluble parts form the solvated shell. The properties of micelle can be changed by the solution conditions such as concentration, temperature, and chain architecture. Such self-assembly phenomena of amphiphilic molecules are of principal importance in many biological and industrial processes. Recently, self-assembled bolaamphiphile nanotubes have been used as templates to produce metal-coated nanowires [1]. A detailed understanding of the aggregation process is crucial to understand and eventually control their formation for the related applications of micelles.

The triblock copolymers, made up of poly(ethyleneoxide) (PEO) and poly(propyleneoxide) (PPO) blocks, which are experimentally studied as amphiphilic molecules, have been the subject of intense research over the last two decades due to their unique solution behavior [2, 3]. Furthermore, the arrangement of the PPO and PEO blocks in the chain is the key factor affecting self-aggregation and phase behavior of these copolymers, which are well documented in literature [3–8]. The temperature induced aggregation behavior of triblock copolymers in aqueous solutions has received great attention during the

*xghan0@163.com

recent decades due to their fundamental and practical importance [3, 5–9]. Compared with experimental studies, however, related theoretical studies are few, especially to account for the effect of chain architecture. Han et al. [10] investigated the effects of the length of each hydrophobic end block and polymer concentration on micellar aggregation in amphiphilic symmetric ABA triblock copolymer solutions. It is found that the broadness of transition is affected by the length of hydrophobic end blocks (i.e., chain length). However, in associative polymers [11], when the total length of hydrophilic blocks is decreased, keeping the chain length constant, the broadness of transition concerned micelles increases. It is an obvious conclusion drawn that the broadness of the transition changes due to the length ratio of hydrophobic to hydrophilic blocks. The chain length is an important parameter to understand the thermodynamics of block copolymers in a selective solvent. In amphiphilic triblock copolymer solutions, however, the effect of chain length on micellar aggregation behavior has not been clarified so far, the length ratio of hydrophobic to hydrophilic blocks remaining constant.

As a mesoscopic polymer theory, the self-consistent field theory (SCFT) has its origin from the field theoretical approach by Edwards [12] and was explicitly adopted to deal with block copolymer structures by Helfand [13]. In recent years, Matsen and Schick proposed a powerful numerical spectral method that could be used to deal with complex microphases [14, 15]. This method is accurate enough but requests a prior knowledge of the symmetry of an ordered structure, which has hindered its application in predicting microphases of complex copolymer structures. Subsequently, Drolet and Fredrickson suggested a new combinatorial screening method [16, 17], which involves a direct implementation of SCFT in real space in an adaptive arbitrary cell. This method proves to be very successful and can be applied to complex copolymer melts. It has also been extended to predict the nanostructures of polymer-grafted nanoparticles [18], which have potential applications in the design and synthesis of hierarchical materials. In addition, SCFT allows us to investigate the aggregate morphology of amphiphilic block copolymers and their blends in a dilute solution [19–22]. Recently, Matsen extends SCFT to treat diblock copolymers with nongaussian chain of low molecular weight [23].

A lattice model is introduced to self-consistent mean-field theory to treat microphase separation for rod-coil block copolymers [24–26]. In our previous papers [10, 11, 27, 28], we have used the SCFT lattice model to study the phase behavior of physically associating polymer solutions. It is found that chain architecture and polymer concentration are important factors which affect the property of temperature-dependent aggregation behavior. Now, in amphiphilic ABA symmetric triblock copolymer solutions, we study chain length and polymer concentration effects on aggregation behavior. It is found that although the length ratio of hydrophobic to hydrophilic blocks remains constant, the increase in the aggregation degree of micelles is also dependent on the chain length, and it is explained by the way of aggregation of amphiphilic triblock copolymer.

## 2. Theory

This section briefly describes the self-consistent field theory (SCFT) lattice model for $n_P$ amphiphilic symmetric ABA triblock copolymers which are assumed to be incompressible. Each block molecule consists of $N_{\mathrm{ns}}$ nonsticker segments forming the middle B block and $N_{\mathrm{st}}$ sticker segments forming each end A block, distributed over a lattice. At the same time, $n_h$ solvent molecules are placed on the vacant lattice sites. Polymer monomers and solvent molecules have the same size and each occupies one lattice site. The total number of lattice sites is $N_L = n_h + n_P N$. The transfer matrix $\lambda$ is used to describe the polymer chain, which depends only on the chain model used. We assume that

$$\lambda_{r_s - r'_{s-1}}^{\alpha_s - \alpha_{s-1}} = \begin{cases} 0, & \alpha_s = \alpha_{s-1}, \\ 1/(z-1), & \text{otherwise}. \end{cases} \tag{2.1}$$

Here, $r'$ denotes the nearest neighboring site of $r$. $r_s$ and $\alpha_s$ denote the position and bond orientation of the $s$-th segment of the copolymer, respectively. $\alpha$ can be any of the allowed bond orientations depending on the lattice model used. $z$ is the coordination number of the lattice. This means that the chain is described as a random walk without a possibility of direct backfolding. Although self-intersections of a chain are not permitted, the excluded volume effect is sufficiently taken into account [29]. $G^{\alpha_s}(r, s|1)$ is the end segment distribution function of the $s$-th segment of the chain. Following the scheme of Schentiens and Leermakers [30], it is evaluated from the following recursive relation:

$$G^{\alpha_s}(r,s|1) = G(r,s)\sum_{r'_{s-1}}\sum_{\alpha_{s-1}}\lambda^{\alpha_s-\alpha_{s-1}}_{r_s-r'_{s-1}}G^{\alpha_{s-1}}(r',s-1|1), \tag{2.2}$$

where $G(r,s)$ is the free segment weighting factor and is expressed as

$$G(r,s) = \begin{cases} \exp[-\omega_{\mathrm{st}}(r_s)], & s\in \mathrm{st}, \\ \exp[-\omega_{\mathrm{ns}}(r_s)], & s\in \mathrm{ns}. \end{cases}$$

The initial condition is $G^{\alpha_1}(r,1|1) = G(r,1)$ for all the values of $\alpha_1$. $\sum_{r'_{s-1}}\sum_{\alpha_{s-1}}$ means the summation over all the possible positions and orientations of the $(s-1)$-th segment of the chain. Another end segment distribution function $G^{\alpha_s}(r,s|N)$ is evaluated from the following recursive relation:

$$G^{\alpha_s}(r,s|N) = G(r,s)\sum_{r'_{s+1}}\sum_{\alpha_{s+1}}\lambda^{\alpha_{s+1}-\alpha_s}_{r'_{s+1}-r_s}G^{\alpha_{s+1}}(r',s+1|N), \tag{2.3}$$

with the initial condition $G^{\alpha_N}(r,N|N) = G(r,N)$ for all the values of $\alpha_N$.

In this simulation, the free energy in the canonical ensemble $F$ is defined as

$$\frac{F[\omega_+,\omega_-]}{k_B T} = \sum_r\left\{\frac{1}{4\chi}\omega_-^2(r) - \omega_+(r)\right\} - n_P\ln Q_P[\omega_{\mathrm{st}},\omega_{\mathrm{ns}}] - n_h\ln Q_h[\omega_h], \tag{2.4}$$

where $\chi$ is the Flory-Huggins interaction parameter in the solutions, which equals $\frac{z}{2k_B T}\epsilon$, $z$ is the coordination number of the lattice used. $Q_h$ is the partition function of a solvent molecule subjected to the field $\omega_h(r) = \omega_+(r)$, which is defined as $Q_h = \frac{1}{n_h}\sum_r \exp[-\omega_h(r)]$. $Q_P$ is the partition function of a non-interaction polymer chain subjected to the fields $\omega_{\mathrm{st}}(r) = \omega_+(r) - \omega_-(r)$ and $\omega_{\mathrm{ns}}(r) = \omega_+(r)$, which act on sticker and nonsticky segments, respectively. $Q_P$ is expressed as $Q_P = \frac{1}{N_L}\frac{1}{z}\sum_{r_N}\sum_{\alpha_N}G^{\alpha_N}(r,N|1)$, where $r_N$ and $\alpha_N$ denote the position and orientation of the $N$-th segment of the chain, respectively. $\sum_{r_N}\sum_{\alpha_N}$ means the summation over all the possible positions and orientations of the $N$-th segment of the chain. Minimization of the free energy function $F$ with $\omega_-(r)$ and $\omega_+(r)$ leads to the following saddle point equations:

$$\omega_-(r) = 2\chi\phi_{\mathrm{st}}(r), \tag{2.5}$$

$$\phi_{\mathrm{st}}(r) + \phi_{\mathrm{ns}}(r) + \phi_h(r) = 1, \tag{2.6}$$

where

$$\phi_{\mathrm{st}}(r) = \frac{1}{N_L}\frac{1}{z}\frac{n_P}{Q_P}\sum_{s\in\mathrm{st}}\sum_{\alpha_s}\frac{G^{\alpha_s}(r,s|1)G^{\alpha_s}(r,s|N)}{G(r,s)} \tag{2.7}$$

and

$$\phi_{\mathrm{ns}}(r) = \frac{1}{N_L}\frac{1}{z}\frac{n_P}{Q_P}\sum_{s\in\mathrm{ns}}\sum_{\alpha_s}\frac{G^{\alpha_s}(r,s|1)G^{\alpha_s}(r,s|N)}{G(r,s)} \tag{2.8}$$

are the average numbers of sticker and nonsticky segments at $r$, respectively, and $\phi_h(r) = (1/N_L)(n_h/Q_h) \times \exp[-\omega_h(r)]$ is the average numbers of solvent molecules at $r$.

In our calculations, real space method is implemented to solve the SCFT equations in a cubic lattice with periodic boundary conditions, which is similar to our previous paper [27]. The configuration from SCFT equations is taken as a saddle point configuration. By comparing the free energies of the observed states from different initial fields, a relative stability of the observed morphologies can be obtained.

## 3. Result and discussion

In our studies, the property of symmetric ABA triblock copolymers is characterized by three tunable molecular parameters: $\chi$ (The Flory-Huggins interaction parameter), $N$ (The chain length of copolymer) and $N_{\mathrm{st}}/N_{\mathrm{ns}}$ (the length ratio of each hydrophobic end block to hydrophilic middle block). In this paper,

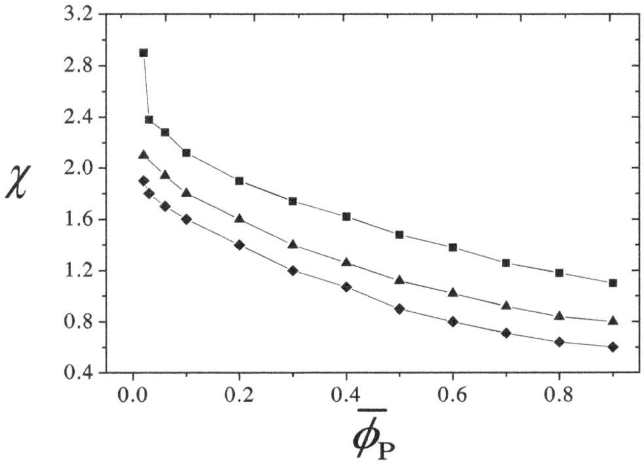


**Figure 1.** The phase diagram for the amphiphilic symmetric ABA tribolck copolymers with different chain length $N$. The boundary between homogenous solutions and micelle morphology is obtained. The squares, triangles and diamonds correspond to the boundaries for $N = 17, 26, 34$, respectively.

when the chain length is changed, the value of $N_{st}/N_{ns}$ ($\simeq 0.23$) remains constant. The aggregation behavior of micelle morphologies is focused when the length of copolymer is changed. Figure 1 shows the phase diagram of the systems with different chain length $N$. When $\chi$ is increased, the unimer-micelle transition occurs. At fixed $N$, the $\chi$ value on micellar boundary increases with decreasing $\bar{\phi}_P$. When $N$ is increased, at fixed $\bar{\phi}_P$, the $\chi$ value on micellar boundary shifts to a small value. It is noted that although the length ratio of each hydrophobic end block to hydrophilic middle block remains constant, the increase in the chain length of copolymer is also favorable to the occurrence of micelles in the system.

In order to demonstrate the effects of the chain length $N$ and polymer concentration $\bar{\phi}_P$ on aggregation of micelles, the variations of the average volume fraction of stickers $\bar{\phi}^{s}_{co}$ and the lattice site numbers $N^{ls}_{co}$ at the micellar cores ($\phi^{s}_{co} \geqslant 0.5$) with $\chi_r$ (the $\chi$ deviation from micellar boundary) in various polymer concentrations, for $N = 26$ and $N = 34$, are presented in figure 2 (a) and figure 4 (a), respectively. For $N = 26$, at $\bar{\phi}_P = 0.1$, $\bar{\phi}^{s}_{co}$, as well as the corresponding $N^{ls}_{co}$, smoothly rises with $\chi_r$ and then remains constant. When $\chi_r \geqslant 0.5$, $N^{ls}_{co}$ does not change with $\chi_r$, and the aggregation degree of micelles strengthens from the increase in $\bar{\phi}^{s}_{co}$. When $\bar{\phi}_P$ is increased the change of $\bar{\phi}^{s}_{co}$ with $\chi_r$ is not monotonous. At $\bar{\phi}_P = 0.3$ and $\bar{\phi}_P = 0.5$, when $\chi_r$ is increased, $\bar{\phi}^{s}_{co}$ firstly rises, and then a $\bar{\phi}^{s}_{co}$-lower region occurs in the range of $\chi_r$, and $\bar{\phi}^{s}_{co}$ finally tends to be constant. The corresponding $N^{ls}_{co}$ firstly rises, and then a jump occurs at the onset of the above lower region. When $\bar{\phi}_P = 0.8$, $\bar{\phi}^{s}_{co}$ always goes up with $\chi_r$, going with the slight surge of $N^{ls}_{co}$. It is seen that, at intermediate concentrations, when $\chi_r \geqslant 1.1$, micelles dissolve and remicellize, which is demonstrated by a decrease of the average volume fraction of stickers at micellar core with increasing from $\chi_r = 1.3$ to $\chi_r = 1.4$ (see figure 3). This can accelerate the further aggregation of micelles. At high concentrations, the behavior of micellar dissolution and remicellization is restrained. Only a few micelles dissolve to strengthen the aggregation degree of micelles. It is shown that the way of an increase in aggregation degree of micelles depends on polymer concentration. It is noted that when $\chi_r \geqslant 0.6$, $\bar{\phi}^{s}_{co}$ at fixed $\chi_r$ decreases with an increase in $\bar{\phi}_P$ for $N = 26$.

For $N = 34$ (see figure 4 (a)), when polymer concentration ($\bar{\phi}_P = 0.1$ and $0.3$) is not high, the tendencies of $\bar{\phi}^{s}_{co}$ and $N^{ls}_{co}$ to $\chi_r$ are similar to those of $N = 26$. At intermediate and high polymer concentrations, they are different from those of $N = 26$. When $\bar{\phi}_P = 0.5$, $\bar{\phi}^{s}_{co}$ and $N^{ls}_{co}$ always smoothly increase with $\chi_r$. At $\bar{\phi}_P = 0.8$, $\bar{\phi}^{s}_{co}$ always smoothly increase with $\chi_r$, but $N^{ls}_{co}$ goes down slowly with $\chi_r$. It is demonstrated that micelles almost do not dissolve at $\bar{\phi}_P = 0.5$. Consequently, the micellar further aggregation is restrained in a way near the micellar boundary. $\bar{\phi}^{s}_{co}$ at $\bar{\phi}_p = 0.5$ is larger than that of $\bar{\phi}_p = 0.8$ until $\chi_r \geqslant 1.5$, which is larger from the case of $N = 26$. It is demonstrated that, at intermediate and high concentrations, the further aggregation of micelles is markedly affected by the increase in $N$.

The heat capacity is an important thermodynamic signature to test the occurrence of a phase transition in a system. The shape of specific heat peak may also be a characteristic of transition. [11, 31]. In

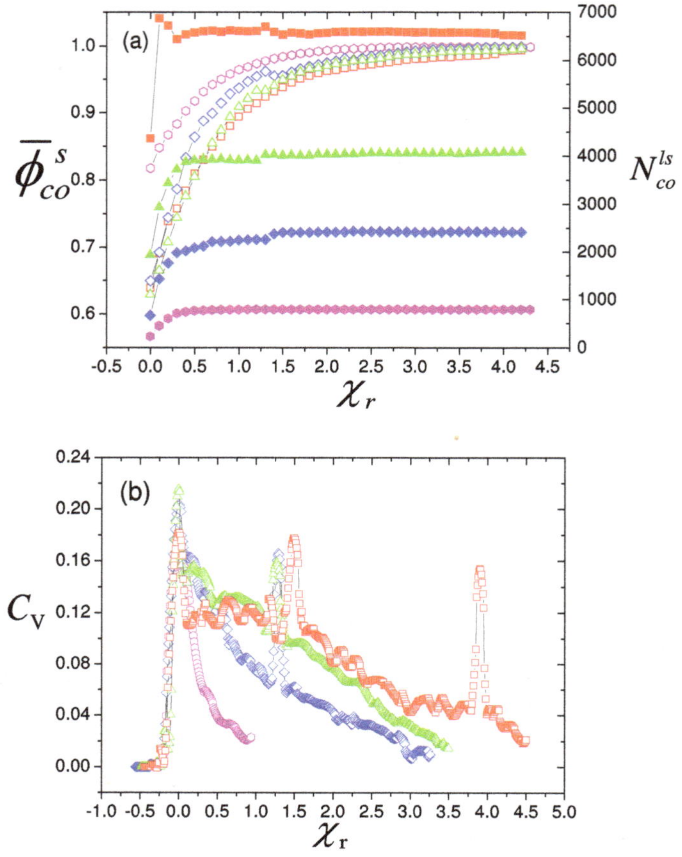


**Figure 2.** (Color online) The variations of average volume fractions of stickers $\bar{\phi}_{\mathrm{st}}$ and lattice site numbers $N_{\mathrm{co}}^{\mathrm{ls}}$ at the micellar cores in different amphiphilic ABA tribolck copolymers with the $\chi$ deviation from micellar boundary $\chi_r$, for various $\bar{\phi}_P$ at $N = 26$ is presented in figure 2 (a). The open and solid, open and solid triangles, open and solid diamonds, and open and solid hexagons denote the $\bar{\phi}_{\mathrm{st}}$ and $N_{\mathrm{co}}^{\mathrm{ls}}$ for $\bar{\phi}_P = 0.8, 0.5, 0.3, 0.1$, respectively; The changes of heat capacity for different $\bar{\phi}_P$ in figure 2 (a) with $\chi_r$ are shown in figure 2 (b). The squares, triangles, diamonds and hexagons denote the case of $\bar{\phi}_P = 0.8, 0.5, 0.3, 0.1$, respectively.

this work, the heat capacity per site of amphiphilic symmetric ABA triblock copolymers is expressed as (in the unit of $k_{\mathrm{B}}$):

$$C_V = \left(\frac{\partial U}{\partial T}\right)_{N_L, n_P} = \frac{1}{N_L}\chi^2 \frac{\partial}{\partial \chi}\left(\sum_r \phi_{\mathrm{st}}^2(r)\right). \tag{3.1}$$

The $C_V(\chi_r)$ curves for the unimer-micelle transition in various $\bar{\phi}_P$ at $N = 26$ and $N = 34$ are shown in figure 2 (b) and figure 4 (b), respectively. For unimer-micelle transition, an asymmetric specific heat peak appears. For $N = 26$, when $\bar{\phi}_P = 0.1$, there is only a peak on $C_V(\chi_r)$ curves. when $\bar{\phi}_P$ is increased, a primary and a secondary peaks, are observed as shown in figure 2 (at intermediate and high concentrations). When $\bar{\phi}_P = 0.3$ and 0.5, the primary peak is higher than the corresponding secondary peak. When $\bar{\phi}_P = 0.8$, a primary and two secondary peaks occur, and one of them is nearly as high as the primary peak. The primary and secondary peaks tend to be similar. The occurrence of the secondary peak is according to the saltation of the $\chi_r$ curves of the average volume fraction of stickers $\bar{\phi}_{\mathrm{co}}^{\mathrm{s}}$ and the lattice site numbers $N_{\mathrm{co}}^{\mathrm{ls}}$ at the micellar cores. Ignoring the secondary peak, the specific heat peak becomes broad with increasing $\bar{\phi}_P$. For $N = 34$, when polymer concentration is low, the peak is narrow and similar to the

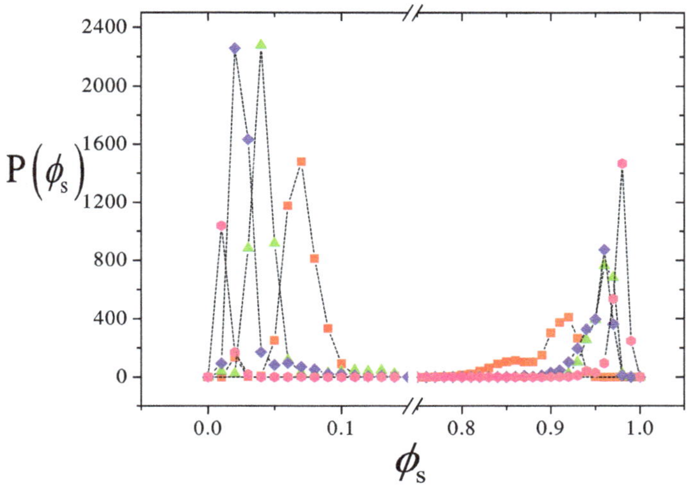


**Figure 3.** (Color online) The distributions of the numbers of micellar core sites with $\phi_{co}^{s}$ at $\bar{\phi}_P = 0.3$ and $N = 26$. The squares, triangles, diamonds and pentacles correspond to $\chi_r = 0.7, 1.3, 1.4, 2.4$, respectively.

corresponding case of $N = 26$. When $\bar{\phi}_P$ is increased, the peak shape changes and the peak also becomes broad. With increasing $\bar{\phi}_P$, the maximum of $C_V$ shifts to a big $\chi_r$ and the curves of $C_V(\chi_r)$ tend to be not smooth, thus the broad peak seems to be a primary peak. For a long chain, the peak shape changes markedly at intermediate and high concentrations.

When temperature drops to a certain extent, micelles appear, and with a further decrease in temperature the aggregation degree of micellar cores markedly strengthens. It is indicated above that the temperature-dependent aggregation behavior of micelles depends on polymer concentration and the chain length. The micellization of hydrophobic end blocks of triblock copolymer can be considered in the following ways. One is that both end blocks of each individual molecule could be incorporated into the same core, the other is that the two hydrophobic ends of triblock copolymer could be incorporated into two adjacent micelles. For a short chain, at low concentration, the first micellization way is dominant. When polymer concentration is increased, the possibility of the two hydrophobic ends of triblock copolymer to be incorporated into two adjacent micelles will rise markedly. At the same time, the aggregation degree of micelles on micellar boundary tends to decrease. Therefore, at intermediate concentrations, the further micellization of triblock copolymers is delayed for a while due to the correlations from chain connection among micelles. When temperature is decreased to some extent, many micelles will be dissolved, and then remicellize. On remicellization, the distribution width of the volume fraction of stickers at micellar cores increases which is different from the general variation of distribution for volume fraction of stickers at micellar cores with $\chi_r$. At the same time, the relationship among the micelles becomes stronger, as shown in figure 5. At high concentrations, the correlations among micelles strengthens compared with intermediate concentrations, only a few micelles dissolve and some new micelles form, which is demonstrated by keeping the tendency of $\bar{\phi}_{st}(\chi_r)$ to increase and the surge of $N_{ls}(\chi_r)$ with $\chi_r$. For a long chain, when concentration is low, the case is similar to that of a short chain. When the polymer concentration is increased, the chain length effect emerges. The two hydrophobic ends of triblock copolymer will be almost absolutely incorporated into two adjacent micelles or small aggregates ($\phi_{co}^{s} < 0.5$ in figure 3). When $\bar{\phi}_P = 0.5$, quite a few small aggregates form. The increase in aggregation degree of micelles is caused by dissolution of small aggregates, and thus the corresponding $N_{co}^{ls}$ does not surge. At high concentrations, micelles form easily, and the quantity of small aggregates decreases notably. Thus, the further aggregation of micelles results from the dissolution of a few micelles. In summary, at intermediate concentrations for a short chain, the increase in aggregation degree of micelles is by its dissolution and remicellization. When polymer concentration or chain length is increased, the way of dissolution and remicellization is restrained, and thus the temperature-dependent aggregation behavior of micelles is changed.

In the end of paper, the validity of self-consistent field theory for the above results should be illuminated. Capture can be an essential feature for the accounted effect of polymer concentration $\bar{\phi}_P$ and

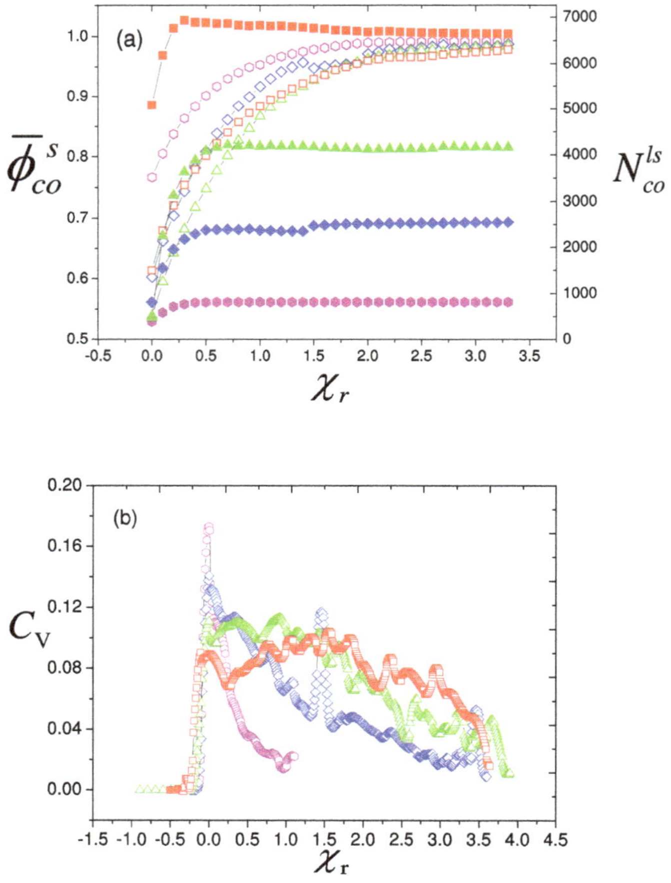


**Figure 4.** (Color online) The variations of the average volume fractions of stickers $\bar{\phi}_{st}$ and lattice site numbers $N_{co}^{ls}$ at the micellar cores in different amphiphilic symmetric ABA tribolck copolymers with the $\chi$ deviation from micellar boundary $\chi_r$, for various $\bar{\phi}_P$ at $N = 34$ is presented in figure 4 (a). The open and solid squares, open and solid triangles, open and solid diamonds, and open and solid hexagons denote the $\bar{\phi}_{st}$ and $N_{co}^{ls}$ for $\bar{\phi}_P = 0.8, 0.5, 0.3, 0.1$, respectively; The changes of heat capacity for different $\bar{\phi}_P$ in figure 4 (a) with $\chi_r$ is shown in figure 4 (b). The squares, triangles, diamonds and hexagons denote the case of $\bar{\phi}_P = 0.8, 0.5, 0.3, 0.1$, respectively.

chain length $N$. The chain accounted in the work should belong to gaussian chain. Using a self-consistent field lattice model, the phase diagram of coil-coil diblock copolymers for $N = 20$ in the three dimension space [24] is consistent with the Matsen-Schick phase diagram. [14], and the results for the solution of homopolymer length $N = 30$ in a two dimensional square lattice is also in reasonable agreement with the theoretical prediction [27]. Furthermore, the effect of relative chain length is also accounted for polymer blends [15]. The above effects from chain length $N = 26$ and $N = 34$ should be reasonable. SCFT is extensively applied to the study of the phase behavior of dilute amphiphilic block copolymer solutions, and the obtained results at $\bar{\phi}_P = 0.1$ have been proved by experimental observations [22]. The specific heat peak for the transition concerned with micelles is also calculated in physically associating polymer solutions [27], and the effect of concentration on specific heat peak (not shown) is in reasonable agreement with that of the related system [32]. Therefore, the concentration effect accounted for in the work by self-consistent field theory is reasonably qualitative.

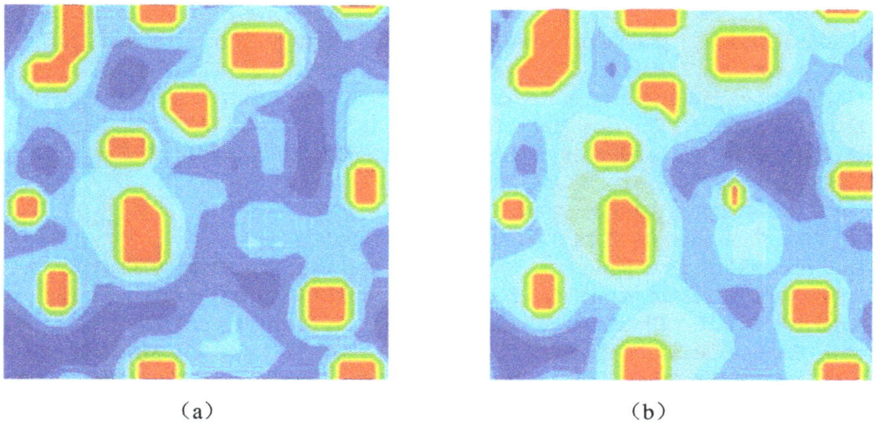


**Figure 5.** (Color online) The cross sections of the system are presented in figure 5. Figure 5 (a) and (b), which demonstrate the changes concerned with remicelliztion, correspond to the cases of $\chi_r = 1.3$ and $\chi_r = 1.4$, respectively, for $\bar{\phi}_P = 0.3$ and $N = 26$.

## 4. Conclusion and summary

Using the self-consistent field lattice model, polymer concentration $\bar{\phi}_P$ and the chain length $N$ (the length ratio of hydrophobic to hydrophilic blocks remains constant), the effects on the aggregation behavior of micelles are studied in amphiphilic symmetric ABA triblock copolymer solutions. When $N$ is increased, at fixed $\bar{\phi}_P$, micelles occur at a higher temperature. The variations of the average volume fraction of stickers $\bar{\phi}^{\mathrm{s}}_{\mathrm{co}}$ and the lattice site numbers $N^{\mathrm{ls}}_{\mathrm{co}}$ at the micellar cores with temperature depend on $N$ and $\bar{\phi}_P$, which is demonstrated by the change of the specific heat peak. For a short chain, when $\bar{\phi}_P$ is increased, firstly a peak appears on the curve of $C_V$ for the micellar appearance, and then, in addition to a primary peak, the secondary peak is observed. For a long chain, in intermediate and high concentration regimes, the shape of a specific heat peak changes markedly, and it tends to a broader primary peak, which is explained by the aggregation way of amphiphilic triblock copolymer. For a short chain, at intermediate concentrations, the way of two hydrophobic ends of triblock copolymer to be incorporated into two adjacent micelles is dominant. Therefore, the aggregation degree of the micelles increases by its dissolution and remicellization. When polymer concentration or chain length is increased, the way of dissolution and remicellization is restrained, and thus the temperature-dependent aggregation behavior of the micelles changes.

## Acknowledgements

This research is financially supported by the National Nature Science Foundations of China (11147132) and the Inner Mongolia municipality (2012MS0112), and the Innovative Foundation of Inner Mongolia University of Science and Technology (2011NCL018).

## References

1. Matsui H., Pan S., Gologan B., Jonas S.H., J. Phys. Chem. B, 2000, **104**, 9576; doi:10.1021/jp000762g.
2. Riess G., Prog. Polym. Sci., 2003, **28**, 1107; doi:10.1016/S0079-6700(03)00015-7.
3. Trong L.C., Djabourov M., Ponton A., J. Colloid Interface Sci., 2008, **328**, 278; doi:10.1016/j.jcis.2008.09.029.
4. Zhou Z., Chu B., Macromolecules, 1994, **27**, 2025; doi:10.1021/ma00086a008.
5. Chu B., Langmuir, 1996, **11**, 414; doi:10.1021/la00002a009.
6. Wu J., Xu Y., Dabros T., Hamza H., Colloids Surf. A: Physicochem. Eng. Asp., 2005, **252**, 79; doi:10.1016/j.colsurfa.2004.09.034.

7. Errico G.D., Paduan L., Khan A., J. Colloid Interface Sci., 2004, **279**, 379; doi:10.1016/j.jcis.2004.06.063.
8. Patel T., Bahadur P., Mata J., J. Colloid Interface Sci., 2010, **345**, 346, doi:10.1016/j.jcis.2010.01.079.
9. Hugouvieux V., Axelos M.A.V., Kol M., Soft Matter, 2011, 7, 2580; doi:10.1039/C0SM01018A.
10. Han X.G., Ma Y.H., Ouyang S.L., Condens. Matter Phys., 2013, **16**, 33601; doi:10.5488/CMP.16.33601.
11. Han X.G., Zhang X.F., Ma Y.H., Condens. Matter Phys., 2012, **15**, 3, 33602; doi:10.5488/CMP.15.33602.
12. Edwards S.F., Proc. Phys. Soc. London, 1965, **85**, 613; doi:10.1088/0370-1328/85/4/301.
13. Helfand E., J. Chem. Phys., 1975, **62**, 999; doi:10.1063/1.430517.
14. Matsen M.W., Schick M., Phys. Rev. Lett., 1994, **72**, 2660; doi:org/10.1103.
15. Matsen M.W., Macromolecules, 1995, **28**, 5765; doi:10.1021/ma00121a011.
16. Drolet F., Fredrickson G.H., Phys. Rev. Lett., 1999, **83**, 4317; doi:10.1103/PhysRevLett.83.4317.
17. Drolet F., Fredrickson G.H., Macromolecules, 2001, **34**, 5317; doi:10.1021/ma0100753.
18. Reister E., Fredrickson G.H., J. Chem. Phys., 2005, **123**, 214903; doi:/10.1063/1.2117008.
19. He X., Liang H., Huang L., Pan C., J. Phys. Chem. B, 2004, **108**, No. 5, 1731; doi:10.1021/jp0359337.
20. Jiang Y., Chen T., Ye F., Liang H., Shi A.-C., Macromolecules, 2005, **38**, 6710; doi:10.1021/ma050424j.
21. Wang R., Tang P., Qiu F., Yang Y., J. Phys. Chem. B, 2005, **109**, 17120; doi:10.1021/jp053248p.
22. Wang L.Q., Lin J.P., Soft Matter, 2011, 7, 3383; doi:10.1039/C0SM01079K.
23. Matsen M.W., Macromolecules, 2012, **45**, 8502; doi:10.1021/ma301788q.
24. Chen J.Z., Zhang C.X., Sun Z.Y., Zheng Y.S., An L.J., J. Chem. Phys., 2006, **124**, 104907; doi:10.1063/1.2176619.
25. Chen J.Z., Sun Z.Y., Zhang C.X., An L.J., Tong Z., J. Chem. Phys., 2007, **127**, 024105; doi:10.1063/1.2750337.
26. Chen J.Z., Sun Z.Y., Zhang C.X., An L.J., Tong Z., J. Chem. Phys., 2008, **128**, 074904; doi:10.1063/1.2831802.
27. Han X.G., Zhang C.X., J. Chem. Phys., 2010, **132**, 164905; doi:10.1063/1.3400648.
28. Han X.G., Zhang X.F., Ma Y.H., Zhang C.X., Guan Y.B., Condens. Matter Phys., 2011, **14**, 43601; doi:10.5488/CMP.14.43601.
29. Medvedevskikh Y.G., Condens. Matter Phys., 2001, **4**, 209; doi:10.5488/CMP.4.2.209.
30. Leermakers F.A.M., Scheutjens J.M.H.M., J. Chem. Phys., 1988, **89**, 3264; doi:10.1063/1.454931.
31. Douglas J.F., Dudowicz J., Freed K.F., J. Chem. Phys., 2006, **125**, 114907; doi:10.1063/1.2356863.
32. Dudowicz J., Freed K.F., Douglas J.F., J. Chem. Phys., 1999, **111**, 7116; doi:10.1063/1.480004.

# 10

# Domain-walls formation in binary nanoscopic finite systems

A. Patrykiejew, S. Sokołowski

Department for the Modelling of Physico-Chemical Processes, Faculty of Chemistry,
Maria Curie-Sklodowska University, 20031 Lublin, Poland

Using a simple one-dimensional Frenkel-Kontorowa type model, we have demonstrated that finite commensurate chains may undergo the commensurate-incommensurate (C-IC) transition when the chain is contaminated by isolated impurities attached to the chain ends. Monte Carlo (MC) simulation has shown that the same phenomenon appears in two-dimensional systems with impurities located at the peripheries of finite commensurate clusters.

**Key words:** *binary mixture, commensurate-incommensurate transitions, Monte Carlo simulation, finite systems, Frenkel-Kontorova model*

## 1. Introduction

In modern nanotechnologies one often deals with very small systems of countable numbers of atoms or molecules. In such cases, the finite size and boundary effects are large and bound to significantly affect the properties of the system with respect to its bulk counterpart [1]. Another important problem is the purity of small systems. While tiny amounts of impurities may be unimportant in macro-scale, the behavior of nanoscopic systems is considerably influenced even by a small number of impurity atoms [2]. Among the systems in which the presence of impurities may be of importance are those exhibiting the C-IC transition.

The C-IC transitions have been experimentally observed in a variety of systems including adsorbed films, [3–6] intercalated compounds [7, 8] composite crystals [9] and magnetically ordered structures of rare-earth compounds [10]. Theoretical studies of C-IC transitions have focused on the domain wall description of incommensurate phases [11–14]. According to the domain wall formalism, the IC phase is a collection of C domains separated by domain walls. The density within the domain walls may be lower or higher than the density of commensurate domains. In the former case, the walls are light and superlight, while in the latter the walls are heavy and superheavy [13].

The simplest theoretical approach which predicts the formation of domain walls is the one-dimensional Frenkel-Kontorova model [15]. The original FK model assumes that an infinite chain of atoms interacting via harmonic potential at zero temperature is subjected to a periodic (sinusoidal) external field. Depending on the misfit between the equilibrium distance of the harmonic potential, the period and amplitude of the external field, the FK model is capable of describing the C-IC transition. The FK model has been extended to two-dimensional systems [16, 17] to mixtures [18], systems with disorder [19] and has also been used to study finite chains [20–22].

The NPT Monte Carlo simulation has demonstrated [23, 24] that finite one-dimensional chains, either uniform or subjected to periodic field, exhibit structures that cannot appear in infinite chains. In particular, it has been shown that the chain experiences very large density fluctuations. In the case of chains on a periodic substrate, a number of different structures (registered, free floating, domain-wall incommensurate and resulting from the chain fragmentation) have been found to appear during a single run.

In one of our recent papers [25], we have shown that finite two-dimensional clusters of Kr adsorbed on graphite undergo the C-IC phase transition when contaminated by small amounts of Ar atoms. Computer simulation has demonstrated that the transition occurs already when the boundaries of a finite krypton island are covered with a single layer of argon atoms.

In this paper we address the issue of the influence of impurities on the behavior of finite one- and two-dimensional (1D and 2D) systems. We are interested in the effects of impurities located at the peripheries of finite 1D chains and 2D clusters of atoms subjected to the periodic external field. The field is assumed to be strong enough to enforce the formation of commensurate structures in pure systems and we consider the possibility of the C-IC transition driven by the presence of impurities. The paper is organized as follows. In the next section we discuss the behavior of one-dimensional systems in the framework of a modified Frenkel-Kontorova model. Then, in the third section we consider two-dimensional finite systems studied by Monte Carlo simulation. In the final section we summarize our findings.

## 2. One-dimensional Frenkel-Kontorova model

At first, we have considered a simple 1D finite chain of atoms at zero temperature and used the Frenkel-Kontorova (FK) model [15]. The energy of a finite chain consisting of $N$ atoms subjected to a periodic external potential and containing impurities can be written as follows:

$$E = \frac{1}{2}\left\{\sum_{i=1}^{N-1} K_i[x_{i+1} - x_i - b_i]^2 + \sum_{i=1}^{N} v_i[1-\cos(2\pi x_i/a)]\right\}, \tag{2.1}$$

where $K_i$ and $b_i$ are the elastic constant and the equilibrium distance for the pair $(i, i+1)$ and $v_i$ is the amplitude of the external field for the $i$-th particle. Having introduced the displacements $u_i = x_i/a - pi$ $(i = 1, 2, \ldots N)$ with $p$ being a positive integer (in this work we set $p = 2$), the energy given by eqn.(2.1) can be rewritten in units of $K_0 a^2/2$ ($K_0$ being the elastic constant for a pure chain) as follows:

$$E = \sum_{i=1}^{N-1} \hat{K}_i[u_{i+1} - u_i - m_i]^2 + \sum_{i=1}^{N} \hat{v}_i[1-\cos(2\pi u_i)]. \tag{2.2}$$

In general, the elastic constants $\hat{K}_i$ (the misfits $m_i$) can assume one of three possible values $k_{0,0} \equiv 1$, $k_{0,\mathrm{im}}$ or $k_{\mathrm{im,im}}$ ($m_{0,0}$, $m_{0,\mathrm{im}}$ or $m_{\mathrm{im,im}}$) depending on the composition of the pair $(i, i+1)$, and the amplitude $\hat{v}_i = v_i/a^2 K_0$ is equal either to $v_0$ or to $v_{\mathrm{im}}$.

In order to find the equilibrium configuration of the chain, the energy should attain its minimum value, specified by the condition stating that the forces $f_i = -\partial E/\partial u_i = 0$ for all $i$. We consider the systems in which a single impurity atom is located at one end of the chain (class I) and the systems with two impurity atoms located at both ends of the chain (class II) and put $k_{0,\mathrm{im}} = k_{\mathrm{im}}$ and $m_{0,\mathrm{im}} = m_{\mathrm{im}}$. The behavior of pure chains depends on $m_0$ and $v_0$. For the assumed value of $m_0 = -0.1$, pure chains are commensurate when $v_0$ exceeds the critical value $v_{0,c} \approx 0.004950$. The calculations have been done for $N$ between 21 and 401 and $v_0 = 0.006$.

The first series of calculations have been carried out for $k_{\mathrm{im}} = 1.0$, while $v_{\mathrm{im}}$ and $m_{\mathrm{im}}$ have been varied. Figure 1 shows the example of the results obtained for the chains with $N = 41$, $m_{\mathrm{im}} = -0.25$ and different values of $v_{\mathrm{im}}$. The lower and upper panels of figure 1 show the results for the systems of class I and II, respectively. It is evident that the systems belonging to both classes exhibit a qualitatively similar behavior. Of course, the systems of the class I lack the symmetry of atomic displacements with respect to the central atom. For $v_{\mathrm{im}}$ lower (higher) than $v_{\mathrm{im,l}}$ ($v_{\mathrm{im,u}}$), the chain remains commensurate, while for intermediate values of $v_{\mathrm{im}}$, there appear incommensurate structures with domain walls. When $v_{\mathrm{im}}$ is lower than $v_{\mathrm{im,l}}$, the energy cost to put the impurity out of registry position is low and the impurity can exhibit large displacements from the commensurate position, while the rest of the chain assumes a commensurate structure due to the domination of the surface over the elastic interaction. On the other hand, when $v_{\mathrm{im}} > v_{\mathrm{im,u}}$, the impurity is strongly pinned by the surface potential and the chain retains

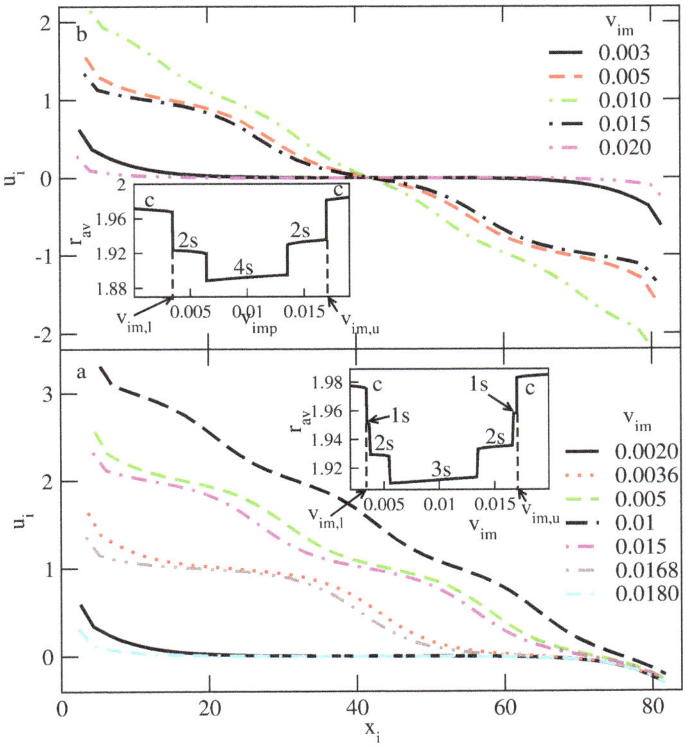


**Figure 1.** (Color online) Atomic displacements vs. atomic positions for the systems of class I (part a) and II (part b) and different values of $v_{im}$ (given in the figure). The calculations have been done $N = 41$, $m_{im} = -0.25$ and $k_{im} = 1.0$. The insets show the changes of the average nearest-neighbor distance vs. $v_{im}$ The regions marked by $C$ and $ks$ ($k = 1,2,3,4$) correspond to the commensurate structure to the incommensurate structures with $k$ domain walls, respectively.

the commensurate structure. Figure 1 shows that when $v_{im} > v_{im,u}$, the displacements of impurities are considerably lower than when $v_{im} < v_{im,l}$.

For intermediate values of $v_{im}$, the gain in elastic energy due to transition into the incommensurate structure is larger than the loss of surface energy. The insets to figure 1 show the changes of the average nearest-neighbor distance with $v_{im}$. In both classes of systems, we have found a series of transitions characterized by a different number of domain walls. The calculations for chains with $N$ up to 401 have shown that in longer chains a larger number of structures appear, and it seems that in the limit of $N \to \infty$ the transitions form a harmless staircase [11].

The values of $v_{im,l}$ and $v_{im,u}$ change with $m_{im}$ and there is a critical value of $m_{im,c}$ for which the difference $\Delta v_{im} = v_{im,u} - v_{im,l}$ goes to zero (figure 2). In the particular examples considered here ($N = 41$ and $k_{im} = 1.0$), $m_{im,c} \approx -0.1856$ for the systems of class I and II. The inset to figure 2 shows that $\Delta v_{im}$ scales with $m_{im,c} - m_{im}$ as follows:

$$\Delta v_{im} \propto (m_{im,c} - m_{im})^{1/2}, \qquad k = 1,2. \tag{2.3}$$

The same scaling appears for the transitions between different incommensurate structures, but the value of $m_{im,c}$ for each transition is different (figure 2).

We have then investigated the effects due to changes in the magnitude of $k_{im}$. Figure 3 gives the example of results for the systems of class I. We see that the values of $v_{im}$ at the transitions $C-1s$, $1s-2s$ and $2s-3s$ are nearly independent of $k_{im}$, while those corresponding to transitions $3s-2s$, $2s-1s$ and $1s-C$ (leading to the recovery of commensurability), exhibit a logarithmic dependence on $k_{im}$.

The logarithmic dependence of $v_{im}$ on $k_{im}$ at transitions $3s-2s$, $2s-1s$ and $1s-C$ has the same origin as the C-IC transition in a pure FK model [11]. On the other hand, the mechanism of transitions leading to commensurability when $v_{im}$ becomes very low is different. For sufficiently low $v_{im}$, the energy cost to put

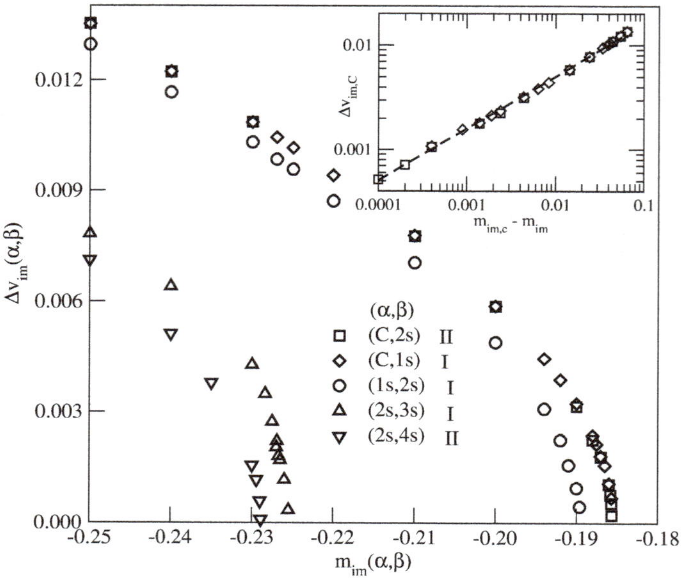


**Figure 2.** The phase diagrams showing the dependence of the difference between the upper and lower values of $v_{im}$ at the transition points ($\Delta v_{im}(\alpha,\beta)$) between different structures $(\alpha,\beta)$ and the impurity misfit $m_{im}(\alpha,\beta)$. The calculations have been done for $N = 41$ and $k_{im} = 1.0$ and for the classes I and II. The inset shows the scaling plot for the C-IC transition.

the impurity out of registry position and to restore the commensurate positions in the rest of the chain is low. Consequently, the impurity can exhibit a large displacement from the commensurate position, while the rest of the chain assumes a commensurate structure due to the domination of surface energy over the elastic energy. This is illustrated in the inset to figure 3, which shows atomic displacements for the

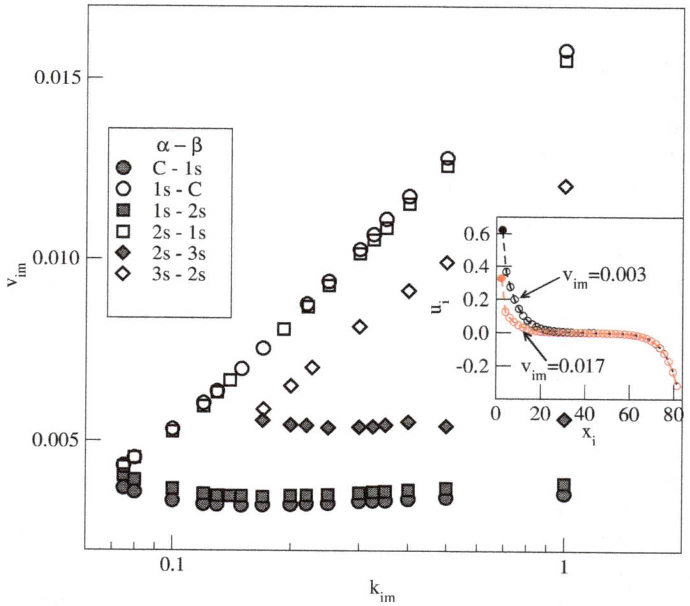


**Figure 3.** (Color online) The phase diagrams showing the dependence of $v_{im}$ on $k_{im}$ at the transition points between different structures. The calculations have been done for $N = 41$ and $m_{im} = -0.24$. The inset shows the atomic displacements versus actual atomic positions in the commensurate phase for $v_{im} = 0.003$ and $v_{im} = 0.017$ when $k_{im} = 1.0$. Filled symbols mark the impurity atom.

systems with $k_{\rm im} = 1.0$ and $v_{\rm im} = 0.003$ and $v_{\rm im} = 0.017$, while the rest of the parameters have been kept the same as in the main figure. Of course, there is an asymmetry of displacements in the chain but it is very small. In the commensurate phase at high values of $v_{\rm im} = 0.017$, the displacements at both ends are nearly the same.

Another question is whether a single impurity can drive the incommensurate system into commensurability? The answer is no. On the other hand, two impurities located at both ends of the chains do lead to the recovery of commensurability when $v_{\rm im}$ is sufficiently high. This is just the same as with the rope pinned either to one or two walls. In the first case, the rope hangs down freely. In the latter, the rope pinned to the opposite walls can be expanded to some extent. In the harmonic approximation, the chain always retains integrity, although when the interaction potential allows for dissociation, the chain may rupture [23, 24, 26] rather than restore the C structure.

## 3. Two-dimensional finite clusters

The phenomenon of reentrant commensurability is not restricted to the above discussed simple 1D model. It also appears in more realistic 2D systems in the ground state and at finite temperatures. To demonstrate this, we have performed MC simulation in the canonical ensemble for finite clusters decorated with impurities at the boundaries. The particles of $A$ and $B$ (impurity) have been placed in the substrate field

$$v_i(x,y) = -V_i\{\cos(\mathbf{q}_1\mathbf{r}) + \cos(\mathbf{q}_2\mathbf{r}) + \cos[(\mathbf{q}_1 - \mathbf{q}_2)\mathbf{r}]\} \tag{3.1}$$

corresponding to the graphite basal plane, where $V_i$ is the amplitude of the potential for the $i$th component and $\mathbf{q}_1$ and $\mathbf{q}_2$ are the reciprocal lattice vectors of the graphite basal plane. The initial configurations have been a hexagon of a perfectly ordered $(\sqrt{3}\times\sqrt{3})R30°$ structure consisting of $N_A z$ atoms of $A$ and $N_B$ atoms of $B$ placed along the one edge (case I), the adjacent three edges (case II) or along all six edges (case III) of the cluster. The interaction between the particles has been represented by the LJ(12,6) potential with the fixed $\sigma_{AA} = 1.46a$ ($a = 2.46$ Å is the graphite lattice constant taken as the unit of length) and $\varepsilon_{AA}/k_{\rm B} = 170$ K (assumed to be a unit of energy) and $V_A^* = V_A/\varepsilon_{AA} = 0.12$. In order to reduce the tendency of $B$ atoms towards aggregation, we have put $\varepsilon_{BB} = 0.5\varepsilon_{AA}$, while $\varepsilon_{AB}$ has been assumed to be equal to $\varepsilon_{AA}$. The parameters $\sigma_{BB} = \sigma_{AB}$ and $V_B$ have been varied.

The simulations have been carried out for the systems with $N_A = 271$ and $N_B = 11$ (case I), $N_B = 33$ (case II) and $N_B = 66$ (case III). Besides, we have also performed some runs for a larger system corresponding to case III, with $N_A = 811$ and $N_B = 114$ as well as with $N_A = 1189$ and $N_B = 138$. The simulation has been performed at reduced temperatures $T^* = k_{\rm B}T/\varepsilon_{AA}$ between 0.005 and 0.3. The formation of domain walls in the system has been monitored using the order parameter [25, 27]

$$\phi(\mathbf{r}) = \cos(\mathbf{q}_1\mathbf{r}) + \cos(\mathbf{q}_2\mathbf{r}) + \cos[(\mathbf{q}_1 - \mathbf{q}_2)\mathbf{r}], \tag{3.2}$$

and assuming that the atom is commensurate (incommensurate) when $\phi > 0$ ($\phi \leqslant 0$). We have calculated the average numbers of incommensurate atoms $A$ and $B$ neglecting the atoms located at the patch boundaries, i.e., those with less than 5 nearest neighbors.

It has been found that for any value of $\sigma_{BB}^* = \sigma_{BB}/a$ between 1.34 and 1.37 and for sufficiently low and sufficiently high values of $V_B^*$, the systems remain commensurate, while for the intermediate values of $V_B^*$, there appear networks of heavy walls separating commensurate domains.

In general, the results at low temperatures are qualitatively very similar to those obtained for 1D FK model. In particular, we have observed the formation of IC structures in the cases I, II and III, though the structure of the domain-wall networks is different in all cases. In general, the walls tend to assume the orientations perpendicular to the edges decorated with the impurity atoms. Therefore, in case III, we have found regular networks like that given in the leftmost part of figure 4. On the other hand, the systems with only one or three adjacent edges decorated by impurities form irregular networks of domain walls (see the middle and rightmost panels to figure 4).

The region of $V_B^*$ over which the IC phase is stable at low temperatures gradually decreases when the size of impurity atoms increases, so that for sufficiently large impurity atoms, only the C phase occurs independently of the magnitude of $V_B^*$ (see figure 5). The calculations for the larger system with $N_A = 811$

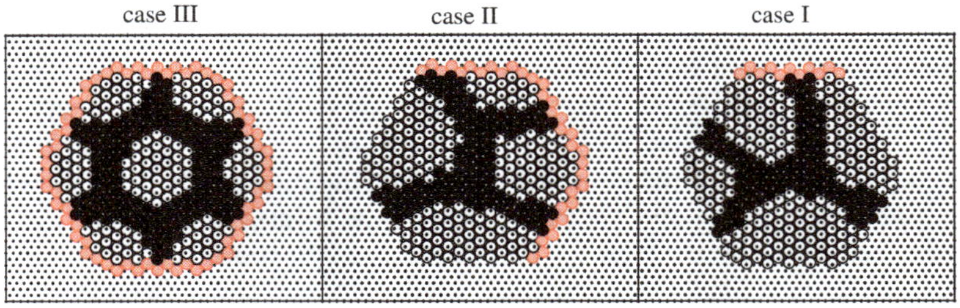


**Figure 4.** (Color online) The examples of snapshots for the system with $\sigma^*_{BB} = 1.36$ and $V^*_B = 0.12$ at $T^* = 0.02$ corresponding to the three cases considered. Shaded circles represent $B$ atoms. Open (filled) circles correspond to the commensurate (incommensurate) $A$ atoms.

and $N_B = 114$ (case III) have shown that maximum value of $\sigma^*_{AB}$ above which the C-IC transition does not appear slightly increases with the system size. Moreover, the simulations performed for larger systems have also demonstrated that at low temperatures the size of C domains does not change and only their number increases. This result is quite similar to that obtained for 1D FK model for the chains of different length. This is demonstrated by the snapshots given in the leftmost part of figure 4 and in figure 6.

The impurity induced changes of the surface stress are responsible for the C-IC transition, which leads to the compression of the finite island. When the amplitude of the impurity surface potential, $V^*_B$, is very low, the presence of impurities at the cluster boundaries does not appreciably affect the strain of the atoms in the cluster. Therefore, the surface stress of the impurity decorated island does not match differently from that of the pure system. For sufficiently large values of $V^*_B$, the impurities are strongly pinned over the surface potential minima as well as do not enlarge the strain in the island. Therefore, for sufficiently low and high $V^*_B$, the system retains the C structure. For intermediate values of $V^*_B$, the competing atom-atom and atom-external field interactions lead to a sufficiently large increase of a surface stress in order to trigger the C-IC transition. One should note a similarity of the herein reported impurity driven C-IC transition to the behavior of metal nanowires [28] and to the formation of IC phases in the subjected to uniaxial compression self-assembled monolayer gold nanoparticles supported on a fluid [29].

The results presented here describe only the low temperature behavior of the systems. The simulation has demonstrated that upon an increase of temperature, the IC structure gradually changes and the C structure is restored at temperatures still below the melting point of the cluster. The effects of thermal excitations on the domain-wall networks will be discussed elsewhere.

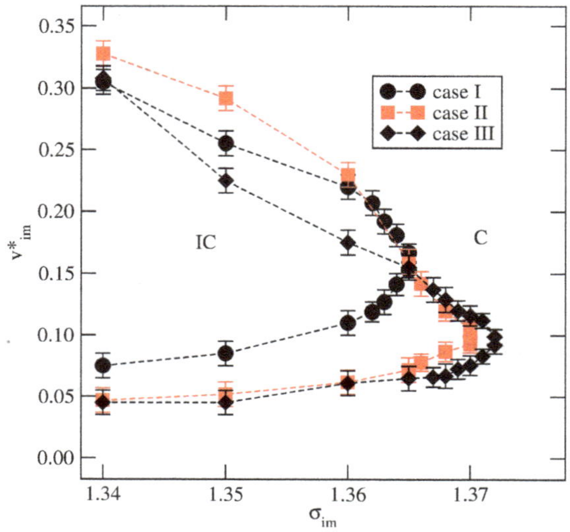


**Figure 5.** (Color online) The phase diagram showing the locations ($V^*_B$ versus $\sigma^*_{AB}$) of C-IC transition in the limit of $T^* \rightarrow 0$ in the systems with $N_A = 271$ and different concentration of impurity atoms.

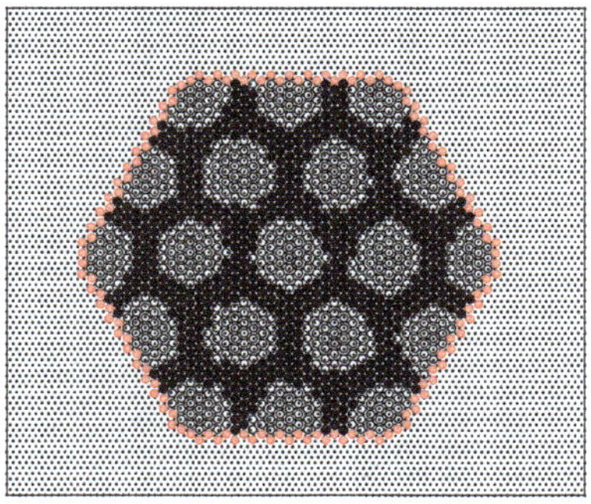

**Figure 6.** (Color online) The example of snapshot for the system with $N_A = 1189$ and $N_B = 138$, $\sigma_{AB} = 1.365$ and $V_B^* = 0.08$ recorded at $T^* = 0.01$ K. Shaded circles represent $B$ atoms. Open and filled circles correspond to the commensurate and incommensurate $A$ atoms.

## 4. Summary

In this work we have studied the impurity driven commensurate-incommensurate transitions in one- and two-dimensional finite systems at zero temperature. In the case of one-dimensional finite chains, we have considered the situations in which the impurity is located at one and two ends of the chain. It has been shown that in both situations, the C-IC transition occurs when the amplitude of the external field experienced by the impurity atoms falls into the region between the lower and upper threshold values. These limiting values of $v_{\mathrm{im}}$ depend upon the parameters characterizing the interaction between the atoms in the main chain, the amplitude of external field acting on the main chain atoms and the interaction between the main chain and the impurities. The number of solitons (domain walls) in the IC structure is different for the chains with one and two ends decorated with the impurity atoms and also depends upon the $v_{\mathrm{im}}$. This behavior is a consequence of different symmetry properties of the system.

In the case of two-dimensional finite clusters, a very similar C-IC transition has been found. In particular, we have observed that the transition occurs when only a part of the cluster boundary is covered with a single row of impurity atoms. Also, by analogy to the results of one-dimensional FK model, the transition occurs only between the lower and upper threshold values of $v_{\mathrm{im}}$.

It should be noted that the C-IC transition observed in two-dimensional systems occurs only when the parameters entering the potential describing the interactions between $AA$, $AB$, and $BB$ atoms are suitably chosen. First of all, the components $A$ and $B$ should not exhibit the tendency towards mixing. Otherwise, the impurity atoms $B$ are likely to penetrate the patch and different scenarios are possible. One, is the formation of a mixed commensurate phase. This has been found in the case of Ar-Kr finite patches at sufficiently high temperatures [25], when the $AB$ interaction potential parameters were obtained using the standard Lorentz-Berthelot mixing rules. The same mixture exhibits a different behavior at low temperatures, and does undergo the C-IC transition, although with the domain walls preferentially made of Ar atoms. Another requirement is that atoms $A$ should tend to order into the C phase, while atoms $B$ should order into the IC phase. This imposes certain restrictions on the choice of $\sigma^*_{AA}$, $\sigma_{BB}$ and the values of $V_A^*$ and $V_B^*$. In particular, $\sigma^*_{AA}$ and $V_A^*$ should assume the values ensuring that a pure $A$ patch is commensurate, but are likely to undergo the C-IC transition when the density exceeds the monolayer capacity or when subjected to an external force due to the presence of impurity atoms along the patch boundaries. Thus, $V_A^*$ cannot be too low or to high and the misfit between the sizes of surface lattice and adsorbate atoms is rather small. A good example of such a system is krypton adsorbed on graphite [5]. In fact, the parameters $\sigma^*_{AA}$ and $V_A^*$ used here are rather close to those describing the krypton adsorbed on the graphite basal plane [25]. On the other hand, the values of $\sigma_{BB}$ and $V_B^*$ should favor the formation of IC phase by pure component $B$. Thus, the misfit between the size of B atoms and the surface lattice should be sufficiently large (it can be positive as well as negative) and $V_B^*$ should be sufficiently small. However, there is still another requirement for the appearance of C-IC transition in finite patches. The $AB$ interaction should be sufficiently strong so that the layer of $B$ atoms along the patch boundary is stable. Otherwise, the atoms $B$ would prefer to form clusters rather than stay at the patch boundary. Besides, this condition is also necessary to exert a sufficiently large force upon the atoms $A$ inside the finite patch to trigger the C-IC transition. Usually, the C-IC transition occurs only when the film density exceeds the density of a fully filled commensurate phase. For the C-IC transition to occur in finite patches, the stability of the C phase should be low enough so that a rather small force exerted by a thin layer of impurity atoms along the patch boundary is able to drive the C-IC transition. From our earlier studies of adsorption of Ar-Kr, Ar-Xe and Kr-Xe mixtures [25, 30–32] on graphite it follows that only in the case of Ar-Kr mixture the C-IC transition occurs at submonolayer densities in finite patches of adsorbed phase.

One expects that at higher temperatures, thermal excitations may considerably influence the effects observed. This problem is currently under study and the results will be published elsewhere.

## Acknowledgements

This work was supported by ERA under the Grant PIRSES 268498.

## References

1. Esfarjani K., Mansoori G.A., In: Handbook of Theoretical and Computational Nanotechnology, Vol. 10, Rieth M., Schommers W. (Eds.), American Scientific Publishers, Los Angeles, 2005, 1–45.
2. Hwang I.-S., Fang C.-K., Chang S.-H., Phys. Rev. B, 2011, **83**, 134119. doi:10.1103/PhysRevB.83.134119.
3. Jaubert M., Glachant A., Bienfait M., Boato G., Phys. Rev. Lett., 1981, **46**, 1679; doi:10.1103/PhysRevLett.46.1679.
4. Krim J., Suzanne J., Shechter H., Wang R., Taub H., Surface Sci., 1985, **162**, 446; doi:10.1016/0039-6028(85)90933-1.
5. Stephens P.W., Heiney P.A., Birgeneau R.J., Horn P.M., Moncton D.E., Brown G.S., Phys. Rev. B, 1984, **29**, 3512; doi:10.1103/PhysRevB.29.3512.
6. Usachov D., Dobrotvorskii A.M., Varykhalov A., Rader O., Phys. Rev. B, 2008, **78**, 085403; doi:10.1103/PhysRevB.78.085403.
7. Clarke R., In: Ordering in Two Dimensions, Sinha S.K. (Ed.), North-Holland, Amsterdam, 1980, p. 53–58.
8. Li L.J., Lu W.J., Zhu X.D., Ling L.S., Qu Z., Sun Y.P., Europhys. Lett., 2012, **97**, 67005; doi:10.1209/0295-5075/97/67005.
9. Nuss J., Pfeiffer S., van Smallen S., Jansen M., Acta Crystalogr. B, 2010, **66**, 27; doi:10.1107/S0108768109053312.
10. Vokhmyanin A.P., Lee S., Jang K.-H., Podlesnyak A.A., Keller L., Prokeš K., Sikolenko V.V., Park J.-G., Skryabin Yu.N., Pirogov A.N., J. Magn. Magn. Mater., 2006, **300**, e411; doi:10.1016/j.jmmm.2005.10.179.
11. Bak P., Rep. Prog. Phys., 1982, **45**, 587; doi:10.1088/0034-4885/45/6/001.
12. Huse D.A., Fisher M., Phys. Rev. B, 1984, **29**, 239; doi:10.1103/PhysRevB.29.239.
13. Den Nijs M., In: Phase Transitions and Critical Phenomena, Vol. 12, Domb C., Lebowitz J.L. (Eds.), Acedemic Press, London, 1988, p. 219–333.
14. Patrykiejew A., Sokołowski S., Binder K., Surface Sci., 2002, **512**, 1; doi:10.1016/S0039-6028(02)01702-8.
15. Frenkel Y.I., Kontorova T., Zh. Eksp. Theor. Fiz., 1938, **8**, 1340 (in Russian).
16. Lomdahl P.S., Srolovitz D.J., Phys. Rev. Lett., 1986, **57**, 2702; doi:10.1103/PhysRevLett.57.2702.
17. Hamilton J.C., Phys. Rev. Lett., 2002, **88**, 126101; doi:10.1103/PhysRevLett.88.126101.
18. Daruka I., Hamilton J.C., J. Phys.: Condens. Matter, 2003, **15**, 1827; doi:10.1088/0953-8984/15/12/302.
19. Braun O.M., Kivshar J.S., Phys. Rep., 1998, **306**, 1; doi:10.1016/S0370-1573(98)00029-5.
20. Braiman Y., Baumgarten J., Jortner J., Klafter J., Phys. Rev. Lett., 1990, **65**, 2398; doi:10.1103/PhysRevLett.65.2398.
21. Braiman Y., Baumgarten J. Klafter J., Phys. Rev. B, 1993, **47**, 11159; doi:10.1103/PhysRevB.47.11159.
22. Vanossi A., Franchini A., Bortolani V., Surface Sci., 2002, **502-503**, 437; doi:10.1016/S0039-6028(01)01990-2.
23. Phillips J.M., Dash J.G., J. Stat. Phys., 2005, **120**, 721; doi:10.1007/s10955-005-5252-x.
24. Hartnett A.S., Phillips J.M., Phys. Rev. B, 2008, **77**, 035408; doi:10.1103/PhysRevB.77.035408.
25. Patrykiejew A., Rżysko W., Sokołowski S., J. Phys. Chem. C, 2011, **116**, 753; doi:10.1021/jp208323b.
26. Markov I., Trayanov A., J. Phys.: Condens. Matter, 1990, **2**, 6965; doi:10.1088/0953-8984/2/33/009.
27. Houlrik J. M., Landau D.P., Phys. Rev. B, 1991, **44**, 8962; doi:10.1103/PhysRevB.44.8962.
28. Diao J., Gall K., Dunn M.L., Nat. Mater., 2003, **2**, 656; doi:10.1038/nmat977.

29. Chua Y., Leahy B., Zhang M., You S., Lee K.Y.C., Coppersmith S.N., Lin B., PNAS, 2013, **110**, 824; doi:10.1073/pnas.1101630108.
30. Patrykiejew A., Condens. Matter Phys., 2012, **15**, 23601. doi:10.5488/CMP.15.23601.
31. Patrykiejew A., Sokołowski S., J. Chem. Phys., 2012, **136**, 144702; doi:10.1063/1.3699330.
32. Patrykiejew A., J. Phys.: Condens. Matter, 2013, **25**, 015001; doi:10.1088/0953-8984/25/1/015001.

# 11

# Quantum transport equations for Bose systems taking into account nonlinear hydrodynamic processes

P.A. Hlushak, M.V. Tokarchuk

Institute for condensed Matter Physics of the National Academy of Sciences of Ukraine
1 Svientsitskii St., 79011 Lviv, Ukraine

Using the method of nonequilibrium statistical operator by Zubarev, an approach is proposed for the description of kinetics which takes into account the nonlinear hydrodynamic fluctuations for a quantum Bose system. Non-equilibrium statistical operator is presented which consistently describes both the kinetic and nonlinear hydrodynamic processes. Both a kinetic equation for the nonequilibrium one-particle distribution function and a generalized Fokker-Planck equation for nonequilibrium distribution function of hydrodynamic variables (densities of momentum, energy and particle number) are obtained. A structure function of hydrodynamic fluctuations in cumulant representation is calculated, which makes it possible to analyse the generalized Fokker-Planck equation in Gaussian and higher approximations of the dynamic correlations of hydrodynamic variables which is important in describing the quantum turbulent processes.

**Key words:** *Bose system, helium, kinetics, hydrodynamics, correlation function, Fokker-Planck equation*

## 1. Introduction

The development of the nonequilibrium statistical theory which takes into account one-particle and collective physical processes is a difficult problem in modern physics. The separation of contributions from the kinetic and hydrodynamic fluctuations into time correlation functions, excitation spectrum, and transport coefficients allows one to obtain more information on physical processes at different time and spatial intervals that define the dynamic properties of a system. A considerable success was achieved in the papers [1–4] in which the approach of a consistent description of kinetics and hydrodynamics of classical dense gases and fluids is proposed based on the Zubarev method of nonequilibrium statistical operator [5–8]. It is appropriate to apply this approach for the study of the dynamics of quantum liquids such as liquid helium in the normal and superfluid states. The quantum system of Bose particles serves as a physical model in theoretical descriptions of both equilibrium and nonequilibrium properties of liquid helium and trapped weakly-interacting Bose gases [9, 10]. Many books [11–15] and articles are devoted to a theoretical description of this system.

In the studies by Morozov [16, 17] and Lebedev, Sukhorukov and Khalatnikov [18] the theoretical approaches were proposed to the description of nonlinear hydrodynamic fluctuations connected with the problem of calculating the dispersion for kinetic transport coefficients and the spectrum of collective modes in the low-frequency area of the superfluid Bose liquid. The microscopical derivation of hydrodynamic equations of a superfluid liquid taking dissipative processes into account was presented by Kovalevsky, Lavrinenko, Peletminsky and Sokolovsky [19], where kinetic coefficients are expressed in terms of time correlation functions of the corresponding flux operators. The generalized Fokker-Planck equation for nonequilibrium distribution function of slow variables for quantum systems was obtained by Morozov [20].

The problems of building a kinetic equation for Bose systems based on the microscopic approach were considered by Akhiezer and Peletminsky [21] and by Kirkpatrick and Dorfman [22–24]. The results of [22, 23] were extended and used to describe the trapped weakly-interacting Bose gases at finite temperatures [25, 26]. Lauck, Vasconcellos and Luzzi [27] developed a nonlinear quantum transport theory for a far from equilibrium many-body system, which is based on nonequilibrium statistical operator method. The hierarchy of generalized evolution equations of dissipative processes in a Bose fluid was derived by Madureira, Vasconcellos and Luzzi [28, 29]. The proposed approach may be suitable for a description of the transport coefficients in molecular hydrodynamics, where the coefficients are frequency and wavelength dependent. Molecular hydrodynamics of nondegenerate Bose gas [30, 31] and degenerate one [32, 33] was derived using the method of two-temporal Green functions [34, 35].

The nonequilibrium statistical operator of many-particle Bose system which consistently describes the kinetics and hydrodynamics, was derived in [36, 37]. The quantum nonequilibrium one-particle distribution function and the average value of density of interaction potential energy have been selected as parameters of a consistent description of the nonequilibrium state. Generalized transport equations were obtained for strongly and weakly nonequilibrium Bose systems with separate contributions from both the kinetic and potential energies of particle interaction.

The aim of the present paper is to construct transport equations for a quantum system that take into account nonlinear hydrodynamic processes. Large-scale fluctuations in a system, which are related to the nonlinear hydrodynamical processes, play an essential role in the transition from normal to superfluid state [38], in the transition from laminar to turbulent flow, and in the acoustic turbulence in superfluid helium [39, 40]. Similar problems arise while describing low-frequency anomalies in kinetic equations related to "long tails" of correlation functions [41–44]. To achieve this aim, a nonequilibrium statistical operator, which consistently describes both the kinetic and nonlinear hydrodynamical fluctuations in a quantum liquid, is derived. Then, a coupled set of kinetic equations are obtained for quantum one-particle distribution function and generalized Fokker-Plank equations for the functional of hydrodynamical variables: densities of particle number, momentum and energy. A structure function of hydrodynamic fluctuations is calculated using a cumulant representation. It provides a possibility to analyse the generalized Fokker-Planck equation in Gaussian and higher approximations of dynamic correlations of hydrodynamic variables which is important in describing the phase transitions and quantum turbulent processes.

## 2. Kinetic equation for nonequilibrium Wigner function and Fokker-Planck equation for distribution function of hydrodynamic variables

The observable average values of energy density $\langle\hat{\varepsilon}_{\mathbf{q}}\rangle^t$, momentum density $\langle\hat{P}_{\mathbf{q}}\rangle^t$, and particle numbers density $\langle\hat{n}_{\mathbf{q}}\rangle^t$ are the reduced description parameters of the hydrodynamical nonequilibrium state of a normal Bose liquid characterized by the energy, momentum and mass flow processes. Operators for these physical quantities are defined through the Klimontovich operator of the phase density of particle number $\hat{n}_{\mathbf{q}}(\mathbf{p}) = \hat{a}^+_{\mathbf{p}-\frac{\mathbf{q}}{2}}\hat{a}_{\mathbf{p}+\frac{\mathbf{q}}{2}}$

$$\hat{n}_{\mathbf{q}} = \frac{1}{\sqrt{N}}\sum_{\mathbf{p}}\hat{n}_{\mathbf{q}}(\mathbf{p}), \qquad \hat{\mathbf{P}}_{\mathbf{q}} = \frac{1}{\sqrt{N}}\sum_{\mathbf{p}}\mathbf{p}\,\hat{n}_{\mathbf{q}}(\mathbf{p}), \qquad \hat{\varepsilon}_{\mathbf{q}} = \hat{\varepsilon}^{\rm kin}_{\mathbf{q}} + \hat{\varepsilon}^{\rm int}_{\mathbf{q}}, \tag{2.1}$$

where $\hat{\varepsilon}^{\rm kin}_{\mathbf{q}}$ and $\hat{\varepsilon}^{\rm int}_{\mathbf{q}}$ are Fourier-components of the operators of kinetic and potential energy densities

$$\hat{\varepsilon}^{\rm kin}_{\mathbf{q}} = \frac{1}{\sqrt{N}}\sum_{\mathbf{p}}\left(\frac{p^2}{2m} - \frac{q^2}{8m}\right)\hat{n}_{\mathbf{q}}(\mathbf{p}), \qquad \hat{\varepsilon}^{\rm int}_{\mathbf{q}} = \frac{1}{2V\sqrt{N}}\sum_{\mathbf{k},\mathbf{p},\mathbf{p}'}\nu(k)\hat{a}^+_{\mathbf{p}+\frac{\mathbf{k}-\mathbf{q}}{2}}\hat{n}_{\mathbf{k}}(\mathbf{p}')\hat{a}_{\mathbf{p}-\frac{\mathbf{k}-\mathbf{q}}{2}}. \tag{2.2}$$

The average value of the phase density operator of particle number is equal to the nonequilibrium one-particle distribution function $f_1(\mathbf{q},\mathbf{p},t) = \langle\hat{n}_{\mathbf{q}}(\mathbf{p})\rangle^t$, which satisfies the kinetic equation for a quantum Bose system.

The agreement between the kinetics and hydrodynamics for dilute Bose gas does not cause any problems because in this case the density is a small parameter. Therefore, only the quantum one-particle distribution function $f_1(\mathbf{q},\mathbf{p};t)$ can be chosen as the parameter of a reduced description. At a transition to quantum Bose liquids, the contribution of collective correlations, which are described by average potential energy of interaction, is more important than one-particle correlations connected with $f_1(\mathbf{q},\mathbf{p};t)$. Hence, in order to consistently describe the kinetics and hydrodynamics of Bose liquid, the one-particle nonequilibrium distribution function along with the average potential energy of interaction should be chosen as the parameters of a reduced description [36, 37]. The nonequilibrium state of such a quantum system is completely described by a nonequilibrium statistical operator $\hat{\varrho}(t)$ which satisfies the quantum Liouville equation

$$\frac{\partial}{\partial t}\hat{\varrho}(t) + iL_N\hat{\varrho}(t) = -\varepsilon\left[\hat{\varrho}(t) - \hat{\varrho}_{\mathrm{q}}(t)\right]. \tag{2.3}$$

The infinitesimal source $\varepsilon$ in the right-hand side of this equation breaks the symmetry of the Liouville equation with respect to $t \to -t$ and selects retarded solutions ($\varepsilon \to +0$ after limiting thermodynamic transition). The quasi-equilibrium statistical operator $\hat{\varrho}_{\mathrm{q}}(t)$ is determined from the condition of the informational entropy extremum at the conservation of normalization condition $\mathrm{Sp}\,\hat{\varrho}_{\mathrm{q}}(t) = 1$ for fixed values of $\langle\hat{n}_{\mathbf{q}}(\mathbf{p})\rangle^t$ and $\langle\hat{\varepsilon}^{\mathrm{int}}_{\mathbf{q}}\rangle^t$ [36, 37]:

$$\hat{\varrho}_{\mathrm{q}}(t) = \exp\left\{-\Phi(t) - \sum_{\mathbf{q}}\beta_{-\mathbf{q}}(t)\hat{\varepsilon}^{\mathrm{int}}_{\mathbf{q}} - \sum_{\mathbf{q},\mathbf{p}}\gamma_{-\mathbf{q}}(\mathbf{p};t)\hat{n}_{\mathbf{q}}(\mathbf{p})\right\}, \tag{2.4}$$

where the Lagrangian multipliers $\beta_{-\mathbf{q}}(t)$, $\gamma_{-\mathbf{q}}(\mathbf{p};t)$ are determined from the self-consistent conditions:

$$\langle\hat{n}_{\mathbf{q}}(\mathbf{p})\rangle^t = \langle\hat{n}_{\mathbf{q}}(\mathbf{p})\rangle^t_{\mathrm{q}}, \qquad \langle\hat{\varepsilon}^{\mathrm{int}}_{\mathbf{q}}\rangle^t = \langle\hat{\varepsilon}^{\mathrm{int}}_{\mathbf{q}}\rangle^t_{\mathrm{q}}.$$

Here, $\langle(\ldots)\rangle^t = \mathrm{Sp}(\ldots)\hat{\varrho}(t)$ and $\langle(\ldots)\rangle^t_{\mathrm{q}} = \mathrm{Sp}(\ldots)\hat{\varrho}_{\mathrm{q}}(t)$. The Massieu-Plank functional

$$\Phi(t) = \ln\mathrm{Sp}\exp\left\{-\sum_{\mathbf{q}}\beta_{-\mathbf{q}}(t)\hat{\varepsilon}^{\mathrm{int}}_{\mathbf{q}} - \sum_{\mathbf{q},\mathbf{p}}\gamma_{-\mathbf{q}}(\mathbf{p};t)\hat{n}_{\mathbf{q}}(\mathbf{p})\right\} \tag{2.5}$$

is determined from the normalization condition $\mathrm{Sp}\,\hat{\varrho}_{\mathrm{q}}(t) = 1$.

The system of equations for the one-particle distribution function and the average density of potential energy are strongly nonlinear[36, 37]. The system can be used for a consistent description of the kinetics and hydrodynamics of both strongly and weakly nonequilibrium states of the Bose systems. Projecting the transport equations on the components of the vector $\Psi(\mathbf{p}) = \{1, \mathbf{p}, p^2/(2m) - q^2/(8m)\}$ yields the equations of nonlinear hydrodynamics, in which the transport processes of kinetic and potential parts of the energy are described by two interdependent equations. Obviously, such equations of nonlinear hydrodynamics provide more opportunities to describe in detail the mutual transformation of kinetic and potential energies during nonequilibrium processes in the system.

In this paper, as previously [36, 37], the nonequilibrium quantum distribution function $f_1(\mathbf{q},\mathbf{p};t) = \langle\hat{n}_{\mathbf{q}}(\mathbf{p})\rangle^t$ is chosen as a parameter to describe one-particle correlations. However, to describe the collective processes in a quantum system, we introduce the operator function

$$\hat{f}(\mathbf{a}) = \int d\mathbf{x}\, e^{i\mathbf{x}(\hat{\mathbf{a}}-\mathbf{a})}, \qquad d\mathbf{x} = \prod_{m=1}^{5}\prod_{\mathbf{k}}\frac{dx_{m\mathbf{k}}}{2\pi}, \tag{2.6}$$

where $\hat{\mathbf{a}} = \{\hat{a}_{1\mathbf{k}}, \hat{a}_{2\mathbf{k}}, \hat{a}_{3\mathbf{k}}\}$, $\hat{a}_{1\mathbf{k}} = \hat{n}_{\mathbf{k}}$, $\hat{a}_{2\mathbf{k}} = \hat{\mathbf{P}}_{\mathbf{k}}$, $\hat{a}_{3\mathbf{k}} = \hat{\varepsilon}_{\mathbf{k}} = \hat{\varepsilon}^{\mathrm{kin}}_{\mathbf{k}} + \hat{\varepsilon}^{\mathrm{int}}_{\mathbf{k}}$ are the Fourier-components of the operators of particle number, momentum and energy densities (2.1). The scalar values $a_{m\mathbf{k}} = \{n_{\mathbf{k}}, \mathbf{P}_{\mathbf{k}}, \varepsilon_{\mathbf{k}}\}$ are the corresponding collective variables. The average values of the operator function (2.6) represent a microscopic distribution function of hydrodynamic variables obtained in accordance with Weyl correspondence rule from the classical distribution function [20]

$$f(\mathbf{a}) = \delta(\mathbf{A} - \mathbf{a}) = \prod_{m=1}^{N}\prod_{\mathbf{k}}\delta(A_{m\mathbf{k}} - a_{m\mathbf{k}}), \tag{2.7}$$

where $\mathbf{A} = \{A_{1\mathbf{k}}\ldots, A_{N\mathbf{k}}\}$ are the classical dynamical variables.

The average values $f_1(\mathbf{q},\mathbf{p};t) = \langle\hat{n}_{\mathbf{k}}(\mathbf{p})\rangle^t$, $f(\mathbf{a};t) = \langle\hat{f}(\mathbf{a})\rangle^t$ are calculated using the nonequilibrium statistical operator $\hat{\varrho}(t)$, which satisfies the Liouville equation. In line with the idea of a reduced description of the nonequilibrium state, the statistical operator $\hat{\varrho}(t)$ should functionally depend on the quantum one-particle distribution function and on the distribution functions of hydrodynamic variables

$$\hat{\varrho}(t) = \hat{\varrho}\left[f_1(\mathbf{q},\mathbf{p};t), f(\mathbf{a};t)\right].$$

Thus, the task is to find a solution of the Liouville equation for $\hat{\varrho}(t)$ which has the above form. To this end, we use the method of Zubarev nonequilibrium statistical operator [5–8]. We consider the Liouville equation (2.3) with infinitely small source. The source correctly selects retarded solutions in accordance with the reduced description of nonequilibrium state of a system. The quasi-equilibrium statistical operator $\hat{\varrho}_{\mathrm{q}}(t)$ is determined in a usual way, from the condition of the maximum informational entropy functional:

$$S[\hat{\varrho}'] = -\mathrm{Sp}\left\{\hat{\varrho}'\ln\hat{\varrho}'\right\} - \sum_{\mathbf{p}}\gamma_{-\mathbf{q}}(\mathbf{p};t)\mathrm{Sp}\left\{\hat{\varrho}'\hat{n}_{\mathbf{q}}(\mathbf{p})\right\} - \int \mathrm{d}a F(\mathbf{a};t)\mathrm{Sp}\left\{\hat{\varrho}'\hat{f}(\mathbf{a})\right\}.$$

Then, the quasi-equilibrium statistical operator can be written as

$$\hat{\varrho}_{\mathrm{q}}(t) = \exp\left\{-\Phi(t) - \sum_{\mathbf{q},\mathbf{p}}\gamma_{-\mathbf{q}}(\mathbf{p};t)\hat{n}_{\mathbf{q}}(\mathbf{p}) - \int \mathrm{d}\mathbf{a} F(\mathbf{a};t)\hat{f}(\mathbf{a})\right\}, \tag{2.8}$$

where $\mathrm{d}\mathbf{a} \to \{\mathrm{d}n_{\mathbf{k}}, \mathrm{d}\mathbf{P}_{\mathbf{k}}, \mathrm{d}\varepsilon_{\mathbf{k}}\}$. The Massieu-Plank functional $\Phi(t)$ is determined from the normalization condition $\mathrm{Sp}\,\hat{\varrho}_{\mathrm{q}}(t) = 1$:

$$\Phi(t) = \ln\mathrm{Sp}\left\{\exp\left[-\sum_{\mathbf{q},\mathbf{p}}\gamma_{-\mathbf{q}}(\mathbf{p};t)\hat{n}_{\mathbf{q}}(\mathbf{p}) - \int \mathrm{d}\mathbf{a} F(\mathbf{a};t)\hat{f}(\mathbf{a})\right]\right\}.$$

The functions $\gamma_{-\mathbf{q}}(\mathbf{p};t)$ and $F(\mathbf{a},t)$ are Lagrange multipliers and can be defined from self-consistent conditions

$$f_1(\mathbf{q},\mathbf{p};t) = \langle\hat{n}_{\mathbf{q}}(\mathbf{p})\rangle^t = \langle\hat{n}_{\mathbf{q}}(\mathbf{p})\rangle^t_{\mathrm{q}}, \qquad f(\mathbf{a};t) = \langle\hat{f}(\mathbf{a})\rangle^t = \langle\hat{f}(\mathbf{a})\rangle^t_{\mathrm{q}}. \tag{2.9}$$

The generalized solution of equation (2.3) in nonequilibrium statistical operator method by Zubarev can be presented in the following form:

$$\hat{\varrho}(t) = \hat{\varrho}_{\mathrm{q}}(t) - \int_{-\infty}^{t} \mathrm{d}t'\,\mathrm{e}^{\varepsilon(t'-t)}\hat{T}_{\mathrm{q}}(t;t')\left[1 - P_{\mathrm{q}}(t')\right]\mathrm{i}L_N\hat{\varrho}_{\mathrm{q}}(t'), \tag{2.10}$$

where

$$\hat{T}_{\mathrm{q}}(t;t') = \exp_+\left\{-\int_{t'}^{t}\mathrm{d}t'\left[1 - P_{\mathrm{q}}(t')\right]\mathrm{i}L_N\right\} \tag{2.11}$$

is the generalized time evolution operator that takes the projection into account. To obtain the solution (2.10) we used the Kawasaki-Gunton projection operator [7, 45] which in our case acts on the arbitrary statistical operator according to the rule:

$$\begin{aligned} P_{\mathrm{q}}(t)\hat{\varrho}' &= \hat{\varrho}_{\mathrm{q}}(t)\mathrm{Sp}\hat{\varrho}' + \sum_{\mathbf{q},\mathbf{p}}\frac{\partial\hat{\varrho}_{\mathrm{q}}(t)}{\partial\langle\hat{n}_{\mathbf{q}}(\mathbf{p})\rangle^t}\left\{\mathrm{Sp}\left[\hat{n}_{\mathbf{q}}(\mathbf{p})\hat{\varrho}'\right] - \langle\hat{n}_{\mathbf{q}}(\mathbf{p})\rangle^t\mathrm{Sp}\hat{\varrho}'\right\} \\ &+ \int d\mathbf{a}\frac{\partial\hat{\varrho}_{\mathrm{q}}(t)}{\partial f(\mathbf{a};t)}\left\{\mathrm{Sp}\left[\hat{f}(\mathbf{a})\hat{\varrho}'\right] - f(\mathbf{a};t)\mathrm{Sp}\hat{\varrho}'\right\}. \end{aligned} \tag{2.12}$$

We consider the action of the Liouville operator on the quasi-equilibrium operator (2.8):

$$\begin{aligned} \mathrm{i}L_N\hat{\varrho}_{\mathrm{q}}(t) &= -\sum_{\mathbf{q},\mathbf{p}}\gamma_{-\mathbf{q}}(\mathbf{p};t)\int_0^1 \mathrm{d}\tau[\hat{\varrho}_{\mathrm{q}}(t)]^{\tau}\dot{\hat{n}}_{\mathbf{q}}(\mathbf{p})[\hat{\varrho}_{\mathrm{q}}(t)]^{1-\tau} \\ &- \int \mathrm{d}\mathbf{a}F(\mathbf{a};t)\int_0^1 \mathrm{d}\tau[\hat{\varrho}_{\mathrm{q}}(t)]^{\tau}\mathrm{i}L_N\hat{f}(a)[\hat{\varrho}_{\mathrm{q}}(t)]^{1-\tau}, \end{aligned} \tag{2.13}$$

where $\dot{\hat{n}}_{\mathbf{q}}(\mathbf{p}) = \mathrm{i}L_N \hat{n}_{\mathbf{q}}(\mathbf{p})$. One introduces the operator function as in [20]

$$\hat{J}(\mathbf{a}) = \int \mathrm{d}\mathbf{x} \mathrm{e}^{\mathrm{i}\mathbf{x}(\hat{\mathbf{a}}-\mathbf{a})} \int_0^1 \mathrm{d}\tau \mathrm{e}^{-\mathrm{i}\tau\mathbf{x}\hat{\mathbf{a}}} \mathrm{i}L_N \hat{\mathbf{a}} \mathrm{e}^{\mathrm{i}\tau\mathbf{x}\hat{\mathbf{a}}}, \tag{2.14}$$

then,

$$\mathrm{i}\hat{L}_N \hat{f}(\mathbf{a}) = -\frac{\partial}{\partial \mathbf{a}} \hat{J}(\mathbf{a}) = -\sum_{m=1}^{5} \sum_{\mathbf{k}} \frac{\partial \hat{J}_{m\mathbf{k}}(\mathbf{a})}{\partial \mathbf{a}_{m\mathbf{k}}}. \tag{2.15}$$

The second term in the right-hand side of (2.13) can be represented as follows:

$$\int \mathrm{d}\mathbf{a} F(\mathbf{a};t) \int_0^1 \mathrm{d}\tau [\hat{\varrho}_{\mathrm{q}}(t)]^{\tau} \left(-\frac{\partial}{\partial \mathbf{a}} \hat{J}(\mathbf{a})\right) [\hat{\varrho}_{\mathrm{q}}(t)]^{1-\tau} = \int \mathrm{d}\mathbf{a} \left(\frac{\partial}{\partial \mathbf{a}} F(\mathbf{a};t)\right) \int_0^1 \mathrm{d}\tau [\hat{\varrho}_{\mathrm{q}}(t)]^{\tau} \hat{J}(\mathbf{a}) \left[\hat{\varrho}_{\mathrm{q}}(t)\right]^{1-\tau}.$$

Now the expression (2.13) reads

$$\mathrm{i}L_N \hat{\varrho}_{\mathrm{q}}(t) = -\sum_{\mathbf{q},\mathbf{p}} \gamma_{-\mathbf{q}}(\mathbf{p};t) \int_0^1 \mathrm{d}\tau \, \dot{\hat{n}}_{\mathbf{q}}(\mathbf{p};\tau) \hat{\varrho}_{\mathrm{q}}(t) + \int \mathrm{d}\mathbf{a} \left(\frac{\partial}{\partial \mathbf{a}} F(\mathbf{a};t)\right) \int_0^1 \mathrm{d}\tau \, \hat{J}(\mathbf{a};\tau) \hat{\varrho}_{\mathrm{q}}(t), \tag{2.16}$$

where

$$\dot{\hat{n}}_{\mathbf{q}}(\mathbf{p};\tau) = [\hat{\varrho}_{\mathrm{q}}(t)]^{\tau} \dot{\hat{n}}_{\mathbf{q}}(\mathbf{p}) [\hat{\varrho}_{\mathrm{q}}(t)]^{-\tau}, \qquad \hat{J}(\mathbf{a};\tau) = [\hat{\varrho}_{\mathrm{q}}(t)]^{\tau} \hat{J}(\mathbf{a}) [\hat{\varrho}_{\mathrm{q}}(t)]^{-\tau}. \tag{2.17}$$

Taking into account (2.16) we represent the nonequilibrium statistical operator (2.10) in the following form:

$$\begin{aligned}
\hat{\varrho}(t) &= \hat{\varrho}_{\mathrm{q}}(t) + \sum_{\mathbf{q},\mathbf{p}} \int_{-\infty}^{t} \mathrm{d}t' \mathrm{e}^{\varepsilon(t'-t)} T_{\mathrm{q}}(t,t') \left[1 - P_{\mathrm{q}}(t')\right] \int_0^1 \mathrm{d}\tau \, \dot{\hat{n}}_{\mathbf{q}}(\mathbf{p};\tau) \gamma_{-\mathbf{q}}(\mathbf{p};t') \hat{\varrho}_{\mathrm{q}}(t') \\
&+ \int \mathrm{d}\mathbf{a} \int_{-\infty}^{t} \mathrm{d}t' \mathrm{e}^{\varepsilon(t'-t)} T_{\mathrm{q}}(t,t') \left[1 - P_{\mathrm{q}}(t')\right] \int_0^1 \mathrm{d}\tau \, \hat{J}(\mathbf{a};\tau) \frac{\partial}{\partial \mathbf{a}} F(\mathbf{a};t') \varrho_{\mathrm{q}}(t') \\
&= \hat{\varrho}_{\mathrm{q}}(t) + \sum_{\mathbf{q},\mathbf{p}} \int_{-\infty}^{t} \mathrm{d}t' \mathrm{e}^{\varepsilon(t'-t)} T_{\mathrm{q}}(t,t') \left[1 - P_{\mathrm{q}}(t')\right] \int_0^1 \mathrm{d}\tau \, \dot{\hat{n}}_{\mathbf{q}}(\mathbf{p};\tau) \gamma_{-\mathbf{q}}(\mathbf{p};t') \hat{\varrho}_{\mathrm{q}}(t') \\
&+ \sum_{\mathbf{q}} \int \mathrm{d}\mathbf{a} \int_{-\infty}^{t} \mathrm{d}t' \mathrm{e}^{\varepsilon(t'-t)} T_{\mathrm{q}}(t,t') \left[1 - P_{\mathrm{q}}(t')\right] \\
&\times \int_0^1 \mathrm{d}\tau \left\{ \hat{J}_{n_{\mathbf{q}}}(\mathbf{a};\tau) \frac{\partial}{\partial n_{\mathbf{q}}} F(\mathbf{a};t') + \hat{\mathbf{J}}_{\mathbf{P}_{\mathbf{q}}}(\mathbf{a};\tau) \cdot \frac{\partial}{\partial \mathbf{P}_{\mathbf{q}}} F(\mathbf{a};t') + \hat{J}_{\varepsilon_{\mathbf{q}}}(\mathbf{a};\tau) \frac{\partial}{\partial \varepsilon_{\mathbf{q}}} F(\mathbf{a};t') \right\} \varrho_{\mathrm{q}}(t'),
\end{aligned} \tag{2.18}$$

which contains both nondissipative $\hat{\varrho}_{\mathrm{q}}(t)$ and dissipative parts that consistently describe non-Markovian kinetic and hydrodynamic processes with microscopic flows $\dot{\hat{n}}_{\mathbf{q}}(\mathbf{p};\tau)$, $\hat{J}(\mathbf{a};\tau)$. Moreover,

$$\begin{aligned}
\dot{\hat{n}}_{\mathbf{q}}(\mathbf{p}) &= -\mathrm{i}[\hat{n}_{\mathbf{q}}(\mathbf{p}), \hat{H}]_{-} = -\mathrm{i}\frac{(\mathbf{p}\cdot\mathbf{q})}{m} \hat{n}_{\mathbf{q}}(\mathbf{p}) \\
&-\mathrm{i}\frac{\sqrt{N}}{V} \sum_{\mathbf{k},\mathbf{p}'} \nu(k) \left(\delta_{\mathbf{p}',\mathbf{p}-\frac{\mathbf{k}}{2}} - \delta_{\mathbf{p}',\mathbf{p}+\frac{\mathbf{k}}{2}}\right) \hat{a}^{+}_{\mathbf{p}'+\frac{\mathbf{k}-\mathbf{q}}{2}} \hat{n}_{\mathbf{k}} \hat{a}_{\mathbf{p}'-\frac{\mathbf{k}-\mathbf{q}}{2}}
\end{aligned} \tag{2.19}$$

and one obtains the microscopic conservation law of particle density $\hat{n}_{\mathbf{q}}$:

$$\dot{\hat{n}}_{\mathbf{q}} = -\mathrm{i}[\hat{n}_{\mathbf{q}}, \hat{H}]_{-} = -\mathrm{i}(\mathbf{q}\cdot\hat{\mathbf{J}}_{\mathbf{q}}) = -\frac{\mathrm{i}}{m}(\mathbf{q}\cdot\hat{\mathbf{P}}_{\mathbf{q}}), \tag{2.20}$$

where $\hat{\mathbf{J}}_{\mathbf{q}}$ is the flow density operator of Bose particles. Respectively, the expressions $J(\hat{\mathbf{P}}_{\mathbf{q}};\tau)$, $J(\hat{\varepsilon}_{\mathbf{q}};\tau)$ contain the microscopic conservation law of the momentum density operator

$$\dot{\hat{P}}_{\mathbf{q}}^{\alpha} = -\frac{\mathrm{i}}{\sqrt{N}}\sum_{\mathbf{p}}\frac{p_\alpha}{m}(\mathbf{p}\cdot\mathbf{q})\hat{n}_{\mathbf{q}}(\mathbf{p}) - \frac{\mathrm{i}\sqrt{N}}{2V}\sum_{\mathbf{k}}[\nu(\mathbf{k})k_\alpha + \nu(\mathbf{k}+\mathbf{q})(-k_\alpha+q_\alpha)]\hat{n}_{\mathbf{k}}\hat{n}_{-\mathbf{k}+\mathbf{q}} \tag{2.21}$$

and the microscopic conservation law of the complete energy density operator

$$\begin{aligned}\dot{\hat{\varepsilon}}_{\mathbf{q}} &= -\frac{\mathrm{i}}{\sqrt{N}}\sum_{\mathbf{p}}\left(\frac{p^2}{2m}-\frac{q^2}{8m}\right)\frac{(\mathbf{p}\cdot\mathbf{q})}{m}\hat{n}_{\mathbf{q}}(\mathbf{p}) \\ &\quad -\frac{\mathrm{i}\sqrt{N}}{2V}\sum_{\mathbf{k},\alpha}\left\{\frac{\nu(\mathbf{k})-\nu(-\mathbf{k}+\mathbf{q})}{2}k_\alpha + \frac{\nu(\mathbf{k})+\nu(-\mathbf{k}+\mathbf{q})}{2}q_\alpha\right\} \\ &\quad \times\frac{1}{m}\left[\hat{n}_{\mathbf{k}},\hat{P}^{\alpha}_{-\mathbf{k}+\mathbf{q}}\right]_{+} + \frac{1}{2V}\sum_{\mathbf{k}}\nu(k)\dot{\hat{n}}_{\mathbf{q}}.\end{aligned} \tag{2.22}$$

The nonequilibrium statistical operator (2.18) is a functional of the reduced description parameters $f_1(\mathbf{q},\mathbf{p};t)$ and $f(a;t)$ contained in self-consistent conditions (2.9) of determination of Lagrange multiplier $\gamma_{\mathbf{q}}(\mathbf{p};t)$. The values $F(\mathbf{a};t)$ and $f_1(\mathbf{q},\mathbf{p};t)$ are necessary for a complete description of transport processes in a system. With this aim in mind, we use the conditions:

$$\frac{\partial}{\partial t}\langle\hat{n}_{\mathbf{q}}(\mathbf{p})\rangle^t = \frac{\partial}{\partial t}f_1(\mathbf{q},\mathbf{p};t) = \langle\dot{\hat{n}}_{\mathbf{q}}(\mathbf{p})\rangle^t, \qquad \frac{\partial}{\partial t}f(\mathbf{a};t) = \mathrm{Sp}\left(\hat{\varrho}(t)iL_N\hat{f}(\mathbf{a})\right).$$

Calculating the average values in the right-hand parts of these equations with nonequilibrium statistical operator (2.18) we obtain a system of the following transport equations:

$$\begin{aligned}\frac{\partial}{\partial t}f_1(\mathbf{q},\mathbf{p};t) &+ \mathrm{i}\frac{(\mathbf{q}\cdot\mathbf{p})}{m}f_1(\mathbf{q},\mathbf{p};t) = \frac{\mathrm{i}\sqrt{N}}{V}\sum_{\mathbf{k},\mathbf{p}'}\nu(k)\left(\delta_{\mathbf{p}',\mathbf{p}+\frac{\mathbf{k}}{2}} - \delta_{\mathbf{p}',\mathbf{p}-\frac{\mathbf{k}}{2}}\right)f_2(\mathbf{q},\mathbf{p},\mathbf{p}',\mathbf{k};t) \\ &+ \sum_{\mathbf{q}',\mathbf{p}'}\int_{-\infty}^{t}dt'e^{\varepsilon(t'-t)}\varphi_{nn}(\mathbf{q},\mathbf{q}',\mathbf{p},\mathbf{p}';t,t')\gamma_{-\mathbf{q}'}(\mathbf{p}';t') \\ &- \int d\mathbf{a}\int_{-\infty}^{t}dt'e^{\varepsilon(t'-t)}\varphi_{nJ}(\mathbf{q},\mathbf{p},;t,t')\frac{\partial}{\partial\mathbf{a}}F(\mathbf{a};t'),\end{aligned} \tag{2.23}$$

$$\begin{aligned}\frac{\partial}{\partial t}f(\mathbf{a};t) &+ \frac{\partial}{\partial\mathbf{a}}\langle\hat{J}(\mathbf{a})\rangle_{\mathrm{q}}^{t} = -\sum_{\mathbf{q}',\mathbf{p}'}\int_{-\infty}^{t}dt'e^{\varepsilon(t'-t)}\frac{\partial}{\partial\mathbf{a}}\varphi_{Jn}(\mathbf{a},\mathbf{q}',\mathbf{p}';t,t')\gamma_{-\mathbf{q}'}(\mathbf{p}';t') \\ &+ \int d\mathbf{a}'\int_{-\infty}^{t}dt'e^{\varepsilon(t'-t)}\frac{\partial}{\partial\mathbf{a}}\varphi_{JJ}(\mathbf{a},\mathbf{a}';t,t')\frac{\partial}{\partial\mathbf{a}'}F(\mathbf{a}';t'),\end{aligned} \tag{2.24}$$

where

$$f_2(\mathbf{q},\mathbf{p},\mathbf{p}',\mathbf{k};t) = \mathrm{Sp}\left\{\hat{a}^{+}_{\mathbf{p}'+\frac{\mathbf{k}-\mathbf{q}}{2}}\hat{n}_{\mathbf{q}}\hat{a}_{\mathbf{p}'-\frac{\mathbf{k}-\mathbf{q}}{2}}\hat{\varrho}_{\mathrm{q}}(t)\right\} \tag{2.25}$$

is the two-particle quasi-equilibrium distribution function of Bose particles,

$$\varphi_{nn}(\mathbf{q},\mathbf{q}',\mathbf{p},\mathbf{p}';t,t') = \langle\hat{I}_n(\mathbf{q},\mathbf{p};t)\,\hat{T}_{\mathrm{q}}(t,t')\int_0^1 d\tau\,\hat{I}_n(\mathbf{q}',\mathbf{p}';t',\tau)\rangle_{\mathrm{q}}^{t'}. \tag{2.26}$$

is the transport kernel (memory function) which describes the dissipation of kinetic processes. Quantities

$\varphi_{nJ}(\mathbf{q},\mathbf{p},\mathbf{a}';t,t')$, $\varphi_{Jn}(\mathbf{a},\mathbf{q},\mathbf{p};t,t')$ are the matrices with elements

$$\varphi_{nJ_l}(\mathbf{q},\mathbf{p},a';t,t') = \langle \hat{I}_n(\mathbf{q},\mathbf{p};t)\,\hat{T}_{\mathrm{q}}(t,t')\int_0^1 \mathrm{d}\tau\,\hat{I}_{J_l}(\mathbf{a}';t',\tau)\rangle_{\mathrm{q}}^{t'},$$

$$\varphi_{J_l n}(\mathbf{a},\mathbf{q},\mathbf{p};t,t') = \langle \hat{I}_{J_l}(\mathbf{a};t)\,\hat{T}_{\mathrm{q}}(t,t')\int_0^1 \mathrm{d}\tau\,\hat{I}_n(\mathbf{q},\mathbf{p};t',\tau)\rangle_{\mathrm{q}}^{t'}. \tag{2.27}$$

These elements are the transport kernels which describe the dissipation between kinetic and hydrodynamic processes. $\varphi_{JJ}(a,a';t,t')$ is the matrix with elements

$$\varphi_{J_lJ_f}(\mathbf{a},\mathbf{a}';t,t') = \langle \hat{I}_{J_l}(\mathbf{a};t)\,\hat{T}_{\mathrm{q}}(t,t')\int_0^1 \mathrm{d}\tau\,\hat{I}_{J_f}(\mathbf{a}';t',\tau)\rangle_{\mathrm{q}}^{t'}, \tag{2.28}$$

which describe the dissipation of hydrodynamic processes in a quantum Bose fluid. Transport kernels (2.26)–(2.28) are constructed on generalized flows

$$\hat{I}_n(\mathbf{q},\mathbf{p};t) = Q(t)\dot{\hat{n}}_{\mathbf{q}}(\mathbf{p}), \qquad \hat{I}_{J_l}(\mathbf{a};t) = Q(t)\dot{\hat{J}}_l(\mathbf{a}), \tag{2.29}$$

where $Q(t) = 1 - P(t)$. Operator $P(t)$ is the generalized projection Mori operator which acts on any operator $\hat{A}$ according to the rule

$$P(t)\hat{A} = \langle \hat{A}\rangle_{\mathrm{q}}^t + \sum_{\mathbf{q},\mathbf{p}} \frac{\partial\langle \hat{A}\rangle_{\mathrm{q}}^t}{\partial\langle \hat{n}_{\mathbf{q}}(\mathbf{p})\rangle^t}\left(\hat{n}_{\mathbf{q}}(\mathbf{p}) - \langle \hat{n}_{\mathbf{q}}(\mathbf{p})\rangle^t\right) + \int da\frac{\partial\langle \hat{A}\rangle_{\mathrm{q}}^t}{\partial\langle \hat{f}(\mathbf{a})\rangle^t}\left(\hat{f}(\mathbf{a}) - \langle \hat{f}(\mathbf{a})\rangle^t\right)$$

and corresponds to the structure of the projection Kawasaki-Gunton operator $P_{\mathrm{q}}(t)$ (2.12). It is important to note that the transport kernels contain both a contribution from quantum diffusion in the coordinate and momentum space and a contribution from the generalized function "force-force". One can readily derive this by substituting (2.19) into $\varphi_{nn}(\mathbf{q},\mathbf{q}',\mathbf{p},\mathbf{p}';t,t')$ and open the contribution from kinetic and potential parts of (2.19):

$$\begin{aligned}\varphi_{nn}(\mathbf{q},\mathbf{p},\mathbf{q}',\mathbf{p}';t,t') &= \frac{1}{m^2}\mathbf{q}\cdot D_{nn}(\mathbf{q},\mathbf{p},\mathbf{q}',\mathbf{p}';t,t')\cdot\mathbf{q}' - \frac{1}{m}\mathbf{q}\cdot D_{nF}(\mathbf{q},\mathbf{p},\mathbf{q}',\mathbf{p}';t,t')\\ &\quad - D_{Fn}(\mathbf{q},\mathbf{p},\mathbf{q}',\mathbf{p}';t,t')\frac{1}{m}\mathbf{q}' + D_{FF}(\mathbf{q},\mathbf{p},\mathbf{q}',\mathbf{p}';t,t'),\end{aligned} \tag{2.30}$$

where

$$D_{nn}(\mathbf{q},\mathbf{p},\mathbf{q}',\mathbf{p}';t,t') = \langle Q(t)\cdot\mathbf{p}\hat{n}_{\mathbf{q}}(\mathbf{p})\cdot\hat{T}_{\mathrm{q}}(t,t')Q(t')\cdot\mathbf{p}'\hat{n}_{\mathbf{q}}(\mathbf{p}')\rangle_{\mathrm{q}}^t \tag{2.31}$$

is the generalized diffusion coefficient of quantum particles in the space $\{\mathbf{q},\mathbf{p}\}$. Other components have the following structure:

$$\begin{aligned}D_{nF}(\mathbf{q},\mathbf{p},\mathbf{q}',\mathbf{p}';t,t') &= \langle Q(t)\cdot\mathbf{p}\hat{n}_{\mathbf{q}}(\mathbf{p})\cdot\hat{T}_{\mathrm{q}}(t,t')\,Q(t')\cdot F_{\mathbf{q}'}(\mathbf{p}')\rangle_{\mathrm{q}}^t,\\ D_{Fn}(\mathbf{q},\mathbf{p},\mathbf{q}',\mathbf{p}';t,t') &= \langle Q(t)\cdot F_{\mathbf{q}}(\mathbf{p})\cdot\hat{T}_{\mathrm{q}}(t,t')\,Q(t')\cdot\mathbf{p}'\hat{n}_{\mathbf{q}'}(\mathbf{p}')\rangle_{\mathrm{q}}^t,\\ D_{FF}(\mathbf{q},\mathbf{p},\mathbf{q}',\mathbf{p}';t,t') &= \langle Q(t)\cdot F_{\mathbf{q}}(\mathbf{p})\cdot\hat{T}_{\mathrm{q}}(t,t')\,Q(t')\cdot F_{\mathbf{q}'}(\mathbf{p}')\rangle_{\mathrm{q}}^t,\end{aligned} \tag{2.32}$$

where

$$F_{\mathbf{q}}(\mathbf{p}) = \frac{\mathrm{i}\sqrt{N}}{V}\sum_{\mathbf{k},\mathbf{p}'} \nu(k)\left(\delta_{\mathbf{p}',\mathbf{p}+\frac{\mathbf{k}}{2}} - \delta_{\mathbf{p}',\mathbf{p}-\frac{\mathbf{k}}{2}}\right)\hat{a}^+_{\mathbf{p}'+\frac{\mathbf{k}-\mathbf{q}}{2}}\hat{n}_{\mathbf{q}}(\mathbf{p})\hat{a}_{\mathbf{p}'-\frac{\mathbf{k}-\mathbf{q}}{2}}.$$

For a detailed study of the mutual effect of kinetic and hydrodynamic processes we will allocate the "kinetic" part in quasi-equilibrium statistical operator using the operator representation:

$$\hat{\varrho}_{\mathrm{q}}(t) = \hat{\varrho}_{\mathrm{q}}^{\mathrm{k}}(t) - \int \mathrm{d}aF(\mathbf{a};t)\int_0^1 \mathrm{d}\tau U(F|\tau)\hat{f}(\mathbf{a};\tau)\hat{\varrho}_{\mathrm{q}}^{\mathrm{k}}(t), \tag{2.33}$$

where

$$\hat{\varrho}^{\mathrm{k}}_{\mathrm{q}}(t) = \exp\left\{-\Phi^{\mathrm{k}}(t) - \sum_{\mathbf{q},\mathbf{p}} \gamma_{-\mathbf{q}}(\mathbf{p};t)\hat{n}_{\mathbf{q}}(\mathbf{p})\right\}, \qquad \Phi^{\mathrm{k}}(t) = \ln \mathrm{Sp} \exp\left\{-\sum_{\mathbf{q},\mathbf{p}} \gamma_{-\mathbf{q}}(\mathbf{p};t)\hat{n}_{\mathbf{q}}(\mathbf{p})\right\} \tag{2.34}$$

is the quasi-equilibrium statistical operator which is the basis of the kinetic level of description, and

$$\hat{f}(\mathbf{a};\tau) = [\hat{\varrho}^{\mathrm{k}}_{\mathrm{q}}(t)]^{\tau}\hat{f}(\mathbf{a})[\hat{\varrho}^{\mathrm{k}}_{\mathrm{q}}(t)]^{-\tau}. \tag{2.35}$$

The operator $U(F|\tau)$ satisfies the equation

$$U(F|\tau) = 1 - \int d\mathbf{a}F(\mathbf{a};t)\int_0^{\tau} d\tau' U(F|\tau')\hat{f}(\mathbf{a};\tau'). \tag{2.36}$$

We use the expression (2.33) for determining the Lagrange multiplier $F(a;t)$ from the self-consistent condition:

$$f(\mathbf{a};t) = \langle\hat{f}(\mathbf{a})\rangle^t_{\mathrm{q}} = \langle\hat{f}(\mathbf{a})\rangle^t_{\mathrm{k}} - \int d\mathbf{a}' W(\mathbf{a},\mathbf{a}';t,\tau)F(\mathbf{a}';t), \tag{2.37}$$

where

$$W(\mathbf{a},\mathbf{a}';t) = \int_0^1 d\tau \langle\hat{f}(\mathbf{a})U(F|\tau)\hat{f}(\mathbf{a}';\tau)\rangle^t_{\mathrm{k}} \tag{2.38}$$

is the structure function, in which the averaging is implemented with quasi-equilibrium statistical operator (2.34). From (2.37) we find $F(a;t)$:

$$F(\mathbf{a};t) = -\int d\mathbf{a}'\delta f(\mathbf{a}';t)W_{-1}(\mathbf{a},\mathbf{a}';t), \tag{2.39}$$

where

$$\delta f(\mathbf{a};t) = f(\mathbf{a};t) - \langle\hat{f}(\mathbf{a})\rangle^t_{\mathrm{k}} = \langle\hat{f}(\mathbf{a})\rangle^t - \langle\hat{f}(\mathbf{a})\rangle^t_{\mathrm{k}} \tag{2.40}$$

are the fluctuations of the distribution function of hydrodynamic variables determined as a difference between the complete distribution function and the one averaged with operator $\hat{\varrho}^{\mathrm{k}}_{\mathrm{q}}(t)$. Taking into account equation (2.39) the quasi-equilibrium statistical operator $\hat{\varrho}_{\mathrm{q}}(t)$ can be written as follows:

$$\hat{\varrho}_{\mathrm{q}}(t) = \hat{\varrho}^{\mathrm{k}}_{\mathrm{q}}(t) + \int d\mathbf{a}\int d\mathbf{a}'\int_0^1 d\tau U(F|\tau)W_{-1}(\mathbf{a}',\mathbf{a};t)\hat{f}(\mathbf{a};\tau)\delta f(\mathbf{a}';t)\hat{\varrho}^{\mathrm{k}}_{\mathrm{q}}(t). \tag{2.41}$$

The function $W_{-1}(\mathbf{a},\mathbf{a}';t)$ is the inverse to the structure function $W(\mathbf{a},\mathbf{a}';t)$ and is the solution of the integral equation:

$$\int d\mathbf{a}'' W(\mathbf{a},\mathbf{a}'';t)W_{-1}(\mathbf{a}'',\mathbf{a}';t) = \delta(\mathbf{a}-\mathbf{a}'). \tag{2.42}$$

Only the functions $W(\mathbf{a},\mathbf{a}';t)$ and $W_{-1}(\mathbf{a},\mathbf{a}';t)$ satisfy the equation (2.42) which have singular parts:

$$W(\mathbf{a},\mathbf{a}';t) = W(\mathbf{a};t)[\delta(\mathbf{a}-\mathbf{a}') + R(\mathbf{a},\mathbf{a}';t)], \quad W_{-1}(\mathbf{a},\mathbf{a}';t) = W_{-1}(\mathbf{a};t)[\delta(\mathbf{a}-\mathbf{a}') + r(\mathbf{a},\mathbf{a}';t)],$$

where $R(\mathbf{a},\mathbf{a}';t)$ and $r(\mathbf{a},\mathbf{a}';t)$ are the regular parts and

$$W(\mathbf{a};t) = \int d\mathbf{a}'\, W(\mathbf{a},\mathbf{a}';t), \qquad W_{-1}(\mathbf{a};t) = \int d\mathbf{a}'\, W_{-1}(\mathbf{a},\mathbf{a}';t).$$

An important point is that the expression (2.39) is the equation to determine $F(\mathbf{a};t)$, because the function $W_{-1}(\mathbf{a},\mathbf{a}';t)$ depends on $F(\mathbf{a};t)$ according to the structure function $W(\mathbf{a},\mathbf{a}';t)$, which in turn depends on $U(F|\tau)$ (2.36).

By means of (2.39) and (2.41) we rewrite the equation system (2.23), (2.24) in the following form:

$$\begin{aligned}\frac{\partial}{\partial t} f_1(\mathbf{q},\mathbf{p};t) &+ \mathrm{i}\frac{(\mathbf{q}\cdot\mathbf{p})}{m} f_1(\mathbf{q},\mathbf{p};t) = \frac{\mathrm{i}\sqrt{N}}{V}\sum_{\mathbf{k},\mathbf{p}'} \nu(k)\left(\delta_{\mathbf{p}',\mathbf{p}+\frac{\mathbf{k}}{2}} - \delta_{\mathbf{p}',\mathbf{p}-\frac{\mathbf{k}}{2}}\right) f_2(\mathbf{p},\mathbf{q},\mathbf{p}',\mathbf{k};t)\\ &+ \sum_{\mathbf{q}',\mathbf{p}'} \int_{-\infty}^{t} dt' e^{\varepsilon(t'-t)} \varphi_{nn}(\mathbf{q},\mathbf{q}',\mathbf{p},\mathbf{p}';t,t')\gamma_{-\mathbf{q}'}(\mathbf{p}';t')\\ &+ \int d\mathbf{a}' \int d\mathbf{a}'' \int_{-\infty}^{t} dt' e^{\varepsilon(t'-t)} \varphi_{nJ}(\mathbf{q},\mathbf{p},;t,t') \frac{\partial}{\partial \mathbf{a}'} W_{-1}(\mathbf{a}',\mathbf{a}'';t')\delta f(\mathbf{a};t'),\end{aligned} \tag{2.43}$$

$$\begin{aligned}\frac{\partial}{\partial t}\delta f(\mathbf{a};t) &- \sum_{\mathbf{q}',\mathbf{p}'} \Omega_{fn}(\mathbf{a},\mathbf{q}',\mathbf{p}';t)\gamma_{-\mathbf{q}'}(\mathbf{p}';t') + \frac{\partial}{\partial \mathbf{a}}\int d\mathbf{a}' \int d\mathbf{a}'' v(\mathbf{a},\mathbf{a}'';t) W_{-1}(\mathbf{a}'',\mathbf{a}';t)\delta f(\mathbf{a}';t)\\ = &- \sum_{\mathbf{q}',\mathbf{p}'} \int_{-\infty}^{t} dt' e^{\varepsilon(t'-t)} \frac{\partial}{\partial \mathbf{a}} \varphi_{Jn}(\mathbf{a},\mathbf{q}',\mathbf{p}';t,t')\gamma_{-\mathbf{q}'}(\mathbf{p}';t')\\ &- \int d\mathbf{a}' \int d\mathbf{a}'' \int_{-\infty}^{t} dt' e^{\varepsilon(t'-t)} \frac{\partial}{\partial \mathbf{a}} \varphi_{JJ}(\mathbf{a},\mathbf{a}';t,t') \frac{\partial}{\partial \mathbf{a}'} W_{-1}(\mathbf{a}',\mathbf{a}'';t')\delta f(\mathbf{a}'';t'),\end{aligned} \tag{2.44}$$

where the generalized hydrodynamic velocities

$$v(\mathbf{a},\mathbf{a}';t) = \int d\mathbf{a}'' \int_0^1 d\tau \mathrm{Sp}\left\{\hat{J}(\mathbf{a}) Q(F|\tau) \hat{f}(\mathbf{a}'';\tau)\hat{\varrho}_{\mathrm{q}}^{\mathrm{k}}(t)\right\} W_{-1}(\mathbf{a}'',\mathbf{a}';t). \tag{2.45}$$

were introduced. While deriving these equations, the following property was used:

$$\frac{\partial}{\partial \mathbf{a}}\langle \hat{J}(\mathbf{a})\rangle_{\mathrm{k}}^{t} = -\frac{\partial}{\partial t}\langle \hat{f}(\mathbf{a})\rangle_{\mathrm{k}}^{t} - \sum_{\mathbf{q}',\mathbf{p}'} \Omega_{fn}(a;\mathbf{q}',\mathbf{p}';t)\gamma_{-\mathbf{q}'}(\mathbf{p}';t'),$$

where

$$\Omega_{fn}(a;\mathbf{q}',\mathbf{p}';t) = \mathrm{Sp}\left\{ \hat{f}(\mathbf{a}) \int_0^1 d\tau [\hat{\varrho}_{\mathrm{q}}^{\mathrm{k}}(t)]^{\tau} I_n^{\mathrm{k}}(\mathbf{q}',\mathbf{p}';t)[\hat{\varrho}_{\mathrm{q}}^{\mathrm{k}}(t)]^{1-\tau}\right\}$$

is the time correlation function between $\hat{f}(\mathbf{a})$ and $I_n^{\mathrm{k}}(\mathbf{q}',\mathbf{p}';t)$. Here, the generalized kinetic flows $I_n^{\mathrm{k}}(\mathbf{q},\mathbf{p};t) = [1 - P^{\mathrm{k}}(t)]\dot{\hat{n}}_{\mathbf{q}}(\mathbf{p})$ and the kinetic generalized projection Mori operator

$$P^{\mathrm{k}}(t)\hat{A} = \langle \hat{A}\rangle_{\mathrm{k}}^{t} + \sum_{\mathbf{q},\mathbf{p}} \frac{\delta\langle \hat{A}\rangle_{\mathrm{k}}^{t}}{\delta\langle \hat{n}_{\mathbf{q}}(\mathbf{p})\rangle^{t}}\left(\hat{n}_{\mathbf{q}}(\mathbf{p}) - \langle \hat{n}_{\mathbf{q}}(\mathbf{p})\rangle^{t}\right)$$

were introduced.

The two limiting cases follow from the system of transport equations (2.43), (2.44). First, unless we consider the nonlinear hydrodynamic correlations we will obtain the kinetic equation for Wigner function of quantum Bose particles. Second, if we do not take into account the kinetic processes, then we will obtain a Fokker-Plank equation for distribution function $f(\mathbf{a};t)$, that corresponds to the results of the article [20]:

$$\begin{aligned}\frac{\partial}{\partial t} f(\mathbf{a};t) &+ \frac{\partial}{\partial \mathbf{a}}\int d\mathbf{a}' v(\mathbf{a},\mathbf{a}') f(\mathbf{a}';t) =\\ &- \int_{-\infty}^{t} dt' e^{\varepsilon(t'-t)} \frac{\partial}{\partial \mathbf{a}} \int d\mathbf{a}' K(\mathbf{a},\mathbf{a}';t-t') \frac{\partial}{\partial \mathbf{a}'}\int d\mathbf{a}'' W_{-1}(\mathbf{a}',\mathbf{a}'') f(\mathbf{a}'';t),\end{aligned} \tag{2.46}$$

where (2.45) transforms to $v(\mathbf{a},\mathbf{a}')$ from [20]

$$v(\mathbf{a},\mathbf{a}') = \int d\mathbf{a}''\mathrm{Sp}\{\hat{J}(\mathbf{a})\hat{f}(\mathbf{a}'')W_{-1}(\mathbf{a}',\mathbf{a}'')\} \tag{2.47}$$

and the structure function is reduced to

$$W(\mathbf{a},\mathbf{a}') = \mathrm{Sp}\{\hat{f}(\mathbf{a})\hat{f}(\mathbf{a}')\}. \tag{2.48}$$

Accordingly, $K(\mathbf{a},\mathbf{a}';t)$ is the matrix with elements

$$K_{lf}(\mathbf{a},\mathbf{a}';t) = \mathrm{Sp}\{\hat{I}_l(\mathbf{a})\hat{T}_{\mathrm{q}}(t,t')\hat{I}_f(\mathbf{a}')\}, \tag{2.49}$$

where $\hat{I}_l(\mathbf{a}) = (1-P)\hat{J}_l(\mathbf{a})$ are the dissipative flows with projection operator

$$P\hat{A} = \int d\mathbf{a}\int d\mathbf{a}'\hat{f}(\mathbf{a})W_{-1}(\mathbf{a},\mathbf{a}')\mathrm{Sp}\{\hat{A}\hat{f}(\mathbf{a}')\}.$$

If we neglect the memory effects on the hydrodynamical level in the system of transport equation (2.43), (2.44), then we will obtain the following system of equations:

$$\begin{aligned}\frac{\partial}{\partial t}f_1(\mathbf{q},\mathbf{p};t) &+ \mathrm{i}\frac{(\mathbf{q}\cdot\mathbf{p})}{m}f_1(\mathbf{q},\mathbf{p};t)\\ &= \frac{\mathrm{i}\sqrt{N}}{V}\sum_{\mathbf{k},\mathbf{p}'}\nu(k)\left(\delta_{\mathbf{p}',\mathbf{p}+\frac{\mathbf{k}}{2}}-\delta_{\mathbf{p}',\mathbf{p}-\frac{\mathbf{k}}{2}}\right)f_2(\mathbf{p},\mathbf{q},\mathbf{p}',\mathbf{k};t)\\ &+ \sum_{\mathbf{q}',\mathbf{p}'}\int_{-\infty}^{t}dt'e^{\varepsilon(t'-t)}\varphi_{nn}(\mathbf{q},\mathbf{q}',\mathbf{p},\mathbf{p}';t,t')\gamma_{-\mathbf{q}'}(\mathbf{p}';t')\\ &- \int d\mathbf{a}'\varphi_{nJ}(\mathbf{q},\mathbf{p},;\mathbf{a}')\frac{\partial}{\partial\mathbf{a}'}W_{-1}(\mathbf{a}';t)\delta f(\mathbf{a};t),\end{aligned} \tag{2.50}$$

$$\begin{aligned}\frac{\partial}{\partial t}\delta f(\mathbf{a};t) &+ \frac{\partial}{\partial a}v(\mathbf{a};t)\delta f(\mathbf{a};t)\\ &= -\sum_{\mathbf{q}'\mathbf{p}'}\frac{\partial}{\partial\mathbf{a}}\varphi_{Jn}(\mathbf{a},\mathbf{q}',\mathbf{p}')\gamma_{-\mathbf{q}'}(\mathbf{p}';t)-\frac{\partial}{\partial\mathbf{a}}\varphi_{JJ}(\mathbf{a})\frac{\partial}{\partial\mathbf{a}}W_{-1}(\mathbf{a};t)\delta f(\mathbf{a};t),\end{aligned} \tag{2.51}$$

in which the memory effects on the kinetic level are preserved. In this system, the following designations are used:

$$\varphi_{nJ_l}(\mathbf{q},\mathbf{p},\mathbf{a}) = \int_{-\infty}^{0} dte^{\varepsilon t}\langle\hat{I}_n(\mathbf{q},\mathbf{p})\hat{T}_{\mathrm{q}}(t)\hat{I}_{J_l}(\mathbf{a})\rangle_{\mathrm{k}}^{t}, \tag{2.52}$$

$$\varphi_{J_lJ_f}(\mathbf{a}) = \int d\mathbf{a}'\int_{-\infty}^{0} dte^{\varepsilon t}\langle\hat{I}_{J_l}(\mathbf{a})\hat{T}_{\mathrm{q}}(t)\hat{I}_{J_f}(\mathbf{a}')\rangle_{\mathrm{k}}^{t}, \tag{2.53}$$

where $v(\mathbf{a};t)$ is the contribution from a singular part of generalized velocity $v(\mathbf{a},\mathbf{a}';t) = v(\mathbf{a};t)\delta(\mathbf{a}-\mathbf{a}') + u(\mathbf{a},\mathbf{a}',t)$, during which $v(\mathbf{a};t) = \langle J(\mathbf{a};\tau)\rangle_{\mathrm{k}}^{t}$. Another point to emphasize is that a contribution only from a singular part of the structure function $W_{-1}(\mathbf{a},\mathbf{a}';t)$ is present in the equation system, namely $W_{-1}(\mathbf{a};t)$. Such local approximation can be used near the critical point when the values that strongly fluctuate are hydrodynamical variables and long-wave components of the order parameter.

A hard problem to examine the nonlinear fluctuations based on the equation system (2.43), (2.44) is to calculate the structure function $W(a,a';t)$ and generalized hydrodynamical velocities $v(\mathbf{a};t)$. To this end, we use the method of iterations. In the first approximation for the operator function $U(F|\tau)$, we take

$$U^{(1)}(F|\tau) = 1,$$

which follows from (2.36), and then by substiting into (2.38), we obtain the first approximation for the structure function:

$$W^{(1)}(\mathbf{a},\mathbf{a}';t)=\int\limits_0^1 \mathrm{d}\tau\langle\hat{f}(\mathbf{a})\hat{f}(\mathbf{a}';\tau)\rangle_{\mathrm{k}}^{t}. \tag{2.54}$$

In a similar manner using the following approximation

$$U^{(2)}(F|\tau)=-\int \mathrm{d}\mathbf{a}F(\mathbf{a};t)\int\limits_0^{\tau}\mathrm{d}\tau'\,\hat{f}(\mathbf{a};\tau')$$

for the second approximation of the structure function, one obtains

$$W^{(2)}(\mathbf{a},\mathbf{a}';t)=-\int \mathrm{d}\mathbf{a}''F(\mathbf{a}'',t)\int\limits_0^1\mathrm{d}\tau\int\limits_0^{\tau}\mathrm{d}\tau'\langle\hat{f}(\mathbf{a})\hat{f}(\mathbf{a}'';\tau)\hat{f}(\mathbf{a}';\tau)\rangle_{\mathrm{k}}^{t}. \tag{2.55}$$

In a such a manner, the structure function is equal to

$$W(\mathbf{a},\mathbf{a}';t)=W^{(1)}(\mathbf{a},\mathbf{a}';t)+W^{(2)}(\mathbf{a},\mathbf{a}';t)+\ldots. \tag{2.56}$$

To calculate the first approximation of the structure function, we present it as follows:

$$W^{(1)}(\mathbf{a},\mathbf{a}';t)=W^{(1)}(\mathbf{a};t)\left[\delta(\mathbf{a}-\mathbf{a}')+R^{(1)}(\mathbf{a},\mathbf{a}';t)\right], \tag{2.57}$$

where

$$W^{(1)}(\mathbf{a};t)=\int \mathrm{d}\mathbf{a}'\,W^{(1)}(\mathbf{a},\mathbf{a}';t)=\langle\hat{f}(a)\rangle_{\mathrm{k}}^{t}.$$

After simple transformations, for the regular part of the structure function we can obtain [20]

$$R^{(1)}(\mathbf{a},\mathbf{a}';t)=\int\mathrm{d}\mathbf{x}\int\mathrm{d}\mathbf{x}'\mathrm{e}^{\mathrm{i}\mathbf{x}'\mathbf{a}'-\mathrm{i}\mathbf{x}\mathbf{a}}\int\limits_0^1 d\tau\langle \mathrm{e}^{\mathrm{i}\mathbf{x}\hat{\mathbf{a}}}\mathrm{e}^{-\mathrm{i}\mathbf{x}'\hat{\mathbf{a}}'(\tau)}-\mathrm{e}^{\mathrm{i}(\mathbf{x}-\mathbf{x}')\hat{\mathbf{a}}}\rangle_{\mathrm{k}}^{t}\frac{1}{W^{(1)}(\mathbf{a};t)}.$$

From this expression it follows that if the basis operators commute with one another, then $R(\mathbf{a},\mathbf{a}';t)$ vanishes. Thus, the presence of the regular part of $W(\mathbf{a},\mathbf{a}';t)$ (reciprocally in function $W_{-1}(\mathbf{a},\mathbf{a}';t)$) is characteristic only of a quantum system.

Now we calculate the function $W^{(1)}(\mathbf{a};t)$

$$W^{(1)}(\mathbf{a};t)=\int\mathrm{d}\mathbf{x}\mathrm{e}^{-\mathrm{i}\mathbf{x}\mathbf{a}}\,\mathrm{Sp}\left\{\mathrm{e}^{\mathrm{i}\mathbf{x}\hat{\mathbf{a}}}\,\varrho_{\mathrm{q}}^{\mathrm{k}}(t)\right\}. \tag{2.58}$$

Using the cumulant expansion, we write it in the following form:

$$W^{(1)}(\mathbf{a};t)=\int\mathrm{d}\mathbf{x}\exp\left\{-\mathrm{i}\sum_{\alpha}x_\alpha a_\alpha+\sum_{l=1}\frac{(\mathrm{i})^l}{l!}\sum_{\alpha_1\ldots\alpha_l}x_{\alpha_1}\ldots x_{\alpha_l}M_{\alpha_1\ldots\alpha_l}\right\}, \tag{2.59}$$

where $\alpha=\{m,\mathbf{k}\}$ and $\sum_\alpha(\ldots)=\sum_{m=1}^{5}\sum_{\mathbf{k}}(\ldots)$. To calculate cumulants $M_{\alpha_1\ldots\alpha_l}$, we write the average from the exponent as follows:

$$\begin{aligned}\langle \mathrm{e}^{\sum_\alpha \mathrm{i}x_\alpha\hat{a}_\alpha}\rangle_{\mathrm{k}}^{t} &= \sum_{l=0}^{\infty}\frac{\mathrm{i}^l}{l!}\sum_{\alpha_1\ldots\alpha_l}x_{\alpha_1}\ldots x_{\alpha_l}\langle a_{\alpha_1}\ldots a_{\alpha_l}\rangle_{\mathrm{k}}^{t}\\ &= \exp\left\{\sum_{l=1}^{\infty}\frac{\mathrm{i}^l}{l!}\sum_{\alpha_1\ldots\alpha_l}x_{\alpha_1}\ldots x_{\alpha_l}M_{\alpha_1\ldots\alpha_l}\right\}\end{aligned}$$

and expand the right-hand side in a series. We compare the coefficients at the same products of $x_\alpha$ and for the first three cumulants obtain:

$$\begin{aligned} M_\alpha &= \langle\hat{a}_\alpha\rangle^t_{\text{cum}} = \langle\hat{a}_\alpha\rangle^t_{\text{k}}, \\ M_{\alpha_1\alpha_2} &= \langle\hat{a}_{\alpha_1}\hat{a}_{\alpha_2}\rangle^t_{\text{cum}} = \langle\hat{a}_{\alpha_1}\hat{a}_{\alpha_2}\rangle^t_{\text{k}} - \langle\hat{a}_{\alpha_1}\rangle^t_{\text{k}}\langle\hat{a}_{\alpha_2}\rangle^t_{\text{k}}, \\ M_{\alpha_1\alpha_2\alpha_3} &= \langle\hat{a}_{\alpha_1}\hat{a}_{\alpha_2}\hat{a}_{\alpha_3}\rangle^t_{\text{cum}} = \langle\hat{a}_{\alpha_1}\hat{a}_{\alpha_2}\hat{a}_{\alpha_3}\rangle^t_{\text{k}} - \frac{3}{2}\langle\hat{a}_{\alpha_1}\rangle^t_{\text{k}}\langle\hat{a}_{\alpha_2}\hat{a}_{\alpha_3}\rangle^t_{\text{k}} \\ &\quad -\frac{3}{2}\langle\hat{a}_{\alpha_1}\hat{a}_{\alpha_2}\rangle^t_{\text{k}}\langle\hat{a}_{\alpha_3}\rangle^t_{\text{k}} + 2\langle\hat{a}_{\alpha_1}\rangle^t_{\text{k}}\langle\hat{a}_{\alpha_2}\rangle^t_{\text{k}}\langle\hat{a}_{\alpha_3}\rangle^t_{\text{k}}, \end{aligned} \tag{2.60}$$

which are averaged with quasi-equilibrium statistical operator $\hat{\varrho}^{\text{k}}_{\text{q}}(t)$.

Now, we separate the sum over $l$ in the exponent (2.59) in two parts: with $l \leqslant 2$ and $l \geqslant 3$. Thus, we select a Gaussian component and expand the rest in a series. As a result, we have

$$W^{(1)}(\mathbf{a};t) = \int \mathrm{d}\mathbf{x} \exp\left\{\mathrm{i}\sum_\alpha x_\alpha(M_\alpha - a_\alpha) - \frac{1}{2}\sum_{\alpha_1\alpha_2} x_{\alpha_1}x_{\alpha_2}M_{\alpha_1\alpha_2}\right\}\left\{1 + \Lambda + \frac{1}{2}\Lambda^2 + \frac{1}{3!}\Lambda^3 + \ldots\right\}, \tag{2.61}$$

where

$$\Lambda = \sum_{l\geqslant 3}\frac{\mathrm{i}^l}{l!}\sum_{\alpha_1\alpha_2}\ldots\sum_{\alpha_l} x_{\alpha_1}x_{\alpha_2}\ldots x_{\alpha_l}M_{\alpha_1\alpha_2\ldots\alpha_l}.$$

Let us consider the Gaussian approximation

$$W_G(\mathbf{a};t) = \int \mathrm{d}\mathbf{x} \exp\left\{\mathrm{i}\sum_\alpha x_\alpha(M_\alpha - a_\alpha) - \frac{1}{2}\sum_{\alpha_1\alpha_2} x_{\alpha_1}x_{\alpha_2}M_{\alpha_1\alpha_2}\right\} \tag{2.62}$$

and integrate it with respect to $\mathbf{x}$. To this end, we should transform the quadratic form $x_{\alpha_1}x_{\alpha_2}M_{\alpha_1\alpha_2}$ in the exponent into a diagonal form. We need to solve the equation for the eigenvalues and eigenvectors of the symmetric matrix $M_{\alpha_1\alpha_2}$

$$\sum_{\alpha_2} M_{\alpha_1\alpha_2}v_{\alpha\alpha_2} = z_\alpha v_{\alpha\alpha_1}, \qquad \det|M_{\alpha_1\alpha_2} - z_\alpha\mathbf{I}| = 0,$$

where $v_{\alpha\alpha_1}$, $z_\alpha$ are the eigenvectors and eigenvalues, $\mathbf{I}$ is the unit matrix. Then, out of the eigenvectors we construct the transition matrix $v_{\alpha_1\alpha_2}$ to the new variables $\mathbf{y}$ where the coefficient matrix is diagonal and consists of eigenvalues $\bar{Q}_\alpha = \bar{M}_{\alpha\alpha_1} = \delta_{\alpha\alpha_1}z_\alpha$. Old variables $\mathbf{x}$ are connected with the new ones by the relation $x_\alpha = \sum_{\alpha_1} v_{\alpha\alpha_1}y_{\alpha_1}$. Now, the exponent in (2.62) has the form of perfect squares $\sum_{\alpha_1\alpha_2} x_{\alpha_1}x_{\alpha_2}M_{\alpha_1\alpha_2} = \sum_\alpha \bar{Q}_\alpha y^2_\alpha$ and can be integrated with respect to the variables $\mathbf{y}$.

## 3. Concluding remarks

The nonequilibrium statistical operator was obtained for a consistent description of kinetic and nonlinear hydrodynamic fluctuations in a quantum Bose system. In order to consider the kinetic processes, the nonequilibrium one-particle Wigner function is used as a parameter in a reduced description. The distribution function of hydrodynamic variables is chosen for the study of nonlinear hydrodynamic fluctuations. While deriving a quasi-equilibrium statistical operator, the "kinetic" part $\hat{\varrho}^{\text{k}}_{\text{q}}(t)$ and the part connected with the superoperator $U(F|\tau)$, that leads to the approximation in terms of correlation functions for $\hat{f}(a)$, were examined in detail. The equation of the Fokker-Plank type was obtained, which is related to the kinetic equation. The nonequilibrium distribution function makes it possible to calculate the averaged values of $\langle\hat{a}_l\ldots\hat{a}_l\rangle^t = \int \mathrm{d}a\,(a_l\ldots a_l)\,f(a;t)$, and to obtain operators $\hat{\mathbf{J}}_{\mathbf{q}}$ for flux density and Reynolds-type chain of equations for values $\langle\hat{\mathbf{J}}_{\mathbf{q}}\rangle^t$, $\langle\hat{\mathbf{J}}_{\mathbf{q}}\hat{\mathbf{J}}_{\mathbf{q}'}\rangle^t$, $\langle\hat{\mathbf{J}}_{\mathbf{q}}\hat{\mathbf{J}}_{\mathbf{q}'}\hat{\mathbf{J}}_{\mathbf{q}''}\rangle^t$ as in classical case [46]. This result is important for the study of quantum turbulence phenomena [39].

We have considered the first two approximations for $U^{(1)}(F|\tau)$ and $U^{(2)}(F|\tau)$, which allowed us to obtain the structure of function $W(a,a';t)$ given by Eq. (2.56). We proposed a method to calculate the structure function in the first approximation for $W^{(1)}(a, a'; t)$ using the cumulant expansion (approximation with respect to correlations) with a Gaussian distribution for collective variables. Similar calculations can be used for hydrodynamic velocities (2.45) which is important for microscopic derivation of the generalized transport kernels (2.28). Such a calculation of the structure function enables us to consider the Fokker-Plank equation in Gauss approximations and higher, and to obtain a chain of Reynolds-type equations for time correlation functions $\langle \hat{a}_l \ldots \hat{a}_j \rangle^t$. The generalized hydrodynamical velocities $v(\mathbf{a}; t)$ will be calculated in the same manner as in the classical case [46–48] using the cumulant representations in Gauss approximations or higher. Then, the transport kernels in the Fokker-Plank equation in the mode-coupling-like form could be presented similarly to the classical case [49]. The case of quantum statistics requires a special consideration which will be presented in a subsequent paper.

## References


1. Zubarev D.N., Morozov V.G., Theor. Math. Phys., 1984, **60**, No. 2, 814; doi:10.1007/BF01018982.
2. Zubarev D.N., Morozov V.G., Omelyan I.P., Tokarchuk M.V., Theor. Math. Phys., 1991, **87**, No. 1, 412; doi:10.1007/BF01016582.
3. Zubarev D.N., Morozov V.G., Omelyan I.P., Tokarchuk M.V., Theor. Math. Phys., 1993, **96**, No. 3, 997; doi:10.1007/BF01019063.
4. Tokarchuk M.V., Omelyan I.P., Kobryn O.E., Condens. Matter Phys., 1998, **1**, 687; doi:10.5488/CMP.1.4.687.
5. Zubarev D.N., Non-equilibrium Statistical Thermodynamics, New York, Consultant Bureau, 1974.
6. Zubarev D.N., Morozov V.G., In: Collection of scientific works of Mathematical Institute of USSR Academy of Sciences, Vol. 191, Nauka, Moscow, 1989, p. 140 (in Russian).
7. Zubarev D.N., Morozov V.G., Röpke G., Statistical Mechanics of Non-equilibrium Processes, Vol. 1. Basic Concepts, Kinetic Theory, Akademie Verlag-Wiley VCN, Berlin, 1996.
8. Zubarev D.N., Morozov V.G., Röpke G., Statistical Mechanics of Non-equilibrium Processes, Vol. 2. Relaxation and Hydrodynamics Processes, Akademie Verlag-Wiley VCN, Berlin, 1996.
9. Pethick C.J., Smith H., Bose-Einstein Condensation in Dilute Gases, Cambridge University Press, Cambridge, 2008.
10. Griffin A., Nikuni T., Zaremba E., Bose Condensed Gases at Finite Temperatures, Cambridge University Press, Cambridge, 2009.
11. Khalatnikov I.M., Theory of Superfluidity, Nauka, Moscow, 1971 (in Russian).
12. Patterman S., Superfluid Hydrodynamics, North-Holland, Amsterdam, 1974.
13. Glyde H.R., Exitations in Liquid and Solid Helium, Clarendon Press, Oxford, 1994.
14. Griffin A., Exitations in a Bose-condensed Liquid. Cambridge University Press, Cambridge, 1993.
15. Kovalevsky M.Yu., Peletminskii S.V., Statistical Mechanics of Quantum Liquids and Crystals, Fizmatlit, Moscow, 2006 (in Russian).
16. Morozov V.G., Physica A, 1983, **117**, 511; doi:10.1016/0378-4371(83)90129-2.
17. Morozov V.G., Theor. Math. Phys., 1986, **67**, No. 1, 404; doi:10.1007/BF01028894.
18. Lebedev V.V., Sukhorukov A.I., Khalatnikov I.M., J. Exp. Theor. Phys., 1981, **53**, No. 4, 733.
19. Kovalevsky M.Yu., Lavrinenko N.M., Peletminsky S.V., Sokolovsky A.I., Theor. Math. Phys., 1982, **50**, No. 3, 296; doi:10.1007/BF01016462.
20. Morozov V.G., Theor. Math. Phys., 1981, **48**, No. 3, 807; doi:10.1007/BF01019317.
21. Akhiezer A.I., Peletminsky S.V., Methods of Statistical Physics, Pergamon Press, New York, 1981.
22. Kirkpatrick T.R., Dorfman I.R., J. Low Temp. Phys., 1985, **58**, No. 3/4, 301; doi:10.1007/BF00681309.
23. Kirkpatrick T.R., Dorfman I.R., J. Low Temp. Phys., 1985, **58**, No. 5/6, 399; doi:10.1007/BF00681133.
24. Kirkpatrick T.R., Dorfman I.R., J. Low Temp. Phys., 1985, **59**, No. 1/2, 1; doi:10.1007/BF00681501.
25. Zaremba E., Griffin A., Nikuni T., Phys. Rev. A, 1998, **57**, 4695; doi:10.1103/PhysRevA.57.4695.
26. Zaremba E., Nikuni T., Griffin A., J. Low Temp. Phys., 1999, **116**, No. 3/4, 277; doi:10.1023/A:1021846002995.
27. Lauck L., Vasconcellos Á.R., Luzzi R., Physica A, 1990, **168**, 789; doi:10.1016/0378-4371(90)90031-M.
28. Madureira J.R., Vasconcellos Á.R., Luzzi R., J. Chem. Phys., 1998, **108**, 7568; doi:10.1063/1.476191.
29. Madureira J.R., Vasconcellos Á.R., Luzzi R., J. Chem. Phys., 1998, **108**, 7580; doi:10.1063/1.476192.
30. Tserkovnikov Yu.A., Theor. Math. Phys., 1990, **85**, 1096; doi:10.1007/BF01017252.
31. Tserkovnikov Yu.A., Theor. Math. Phys., 1990, **85**, 1192; doi:10.1007/BF01086849.
32. Tserkovnikov Yu.A., Theor. Math. Phys., 1992, **93**, 1367; doi:10.1007/BF01016395.
33. Tserkovnikov Yu.A., Theor. Math. Phys. 1995, **105**, 1249; doi:10.1007/BF02067493.
34. Zubarev D.N., Sov. Phys. Usp., 1960, **3**, 320; doi:10.1070/PU1960v003n03ABEH003275.
35. Tserkovnikov Yu.A., Theor. Math. Phys., 1986, **69**, 1254; doi:10.1007/BF01017624.
36. Vakarchuk I.O., Hlushak P.A., Tokarchuk M.V., Ukr. Fiz. Zh., 1997, **42**, 1150 (in Ukrainian).

37. Hlushak P.A., Tokarchuk M.V., Condens. Matter Phys., 2004, **7**, No. 3(39), 639; doi:10.5488/CMP.7.3.639.
38. Onuki A., Phase Transition Dynamics, Cambridge University Press, Cambridge, 2004.
39. Progress in Low Temperature Physics: Quantum Turbulence, vol. XVI, Tsubota M., Halperin W.P. (Eds.), Elsevier, Amsterdam, 2006.
40. Nemirovskii S.K., Phys. Rep., 2013, **524**, 85; doi:10.1016/j.physrep.2012.10.005.
41. Résibois P., de Leener M., Classical Kinetic Theory of Fluids, John Willey & Sons, New York, 1977.
42. Kawasaki K., In: Phase Transition and Critical Phenomena, Vol. 5A, Domb C., Green M.S. (Eds.), Academic, New York, 1976, p. 165–411.
43. Zubarev D.N., Morozov V.G., Physica A, 1983, **120**, No. 3, 411; doi:10.1016/0378-4371(83)90062-6.
44. Peletmiskii S.V., Slusarenko Yu.V., Prob. At. Sci. Technol., 1992, **3**(24), 145 (in Russian).
45. Kawasaki K., Gunton J.D., Phys. Rev. A, 1973, **8**, 2048; doi:10.1103/PhysRevA.8.2048.
46. Zubarev D.N., Theor. Math. Phys., 1982, **59**, No. 1, 1004; doi:10.1007/BF01014797.
47. Idzyk I.M., Ighatyuk V.V., Tokarchuk M.V., Ukr. Fiz. Zh., 1996, **41**, No. 10, 1017 (in Ukrainian).
48. Morozov V.G., Tokarchuk M.V., Idzyk I.M., Kobryn A.E., Preprint of the Institute for Condensed Matter Physics, ICMP–96–15U, Lviv, 1996 (in Ukrainian).
49. Ignatyuk V.V., Condens. Matter Phys., 1999, **2**, No. 1(17), 37; doi:10.5488/CMP.2.1.37.

# 12

# Similarities and differences in the construction of dispersion laws for charge carriers in semiconductor crystals and adiabatic potentials in molecules

S.A. Bercha, V.M. Rizak

Uzhgorod National University, 54 Voloshyn St., 88000 Uzhgorod, Ukraine

Using the group theory and the method of invariants, it is shown how the vibronic potential can be written in a matrix form and the corresponding adiabatic potentials can be found. The molecule having $D_{3d}$ symmetry is considered herein as an example. The symmetries of normal vibrations active in Jahn-Teller's effect were defined. $E$–$E$ vibronic interaction was considered to obtain vibronic potential energy in a matrix form and thus the adiabatic potential. Significant differences are shown in the construction of a secular matrix $D(\vec{k})$ for defining a dispersion law for charge carriers in the crystals and the matrix of vibronic potential energy, which depends on the normal coordinates of normal vibrations active in Jahn-Teller's effect. Dispersion law of charge carriers in the vicinity of $\Gamma$ point of Brillouin zone of the crystal with $D^2_{3d}$ symmetry was considered as an example.

**Key words:** *Jahn-Teller's effect, method of invariants, dispersion law, adiabatic potential*

## 1. Introduction

The method of invariants was originally introduced in electronic theory of solids by Luttinger [1] while considering a Bloch's electron in the magnetic field $\vec{H}$. Luttinger presented a secular matrix $D(\vec{k},\vec{H})$, (where $\vec{k}$ is a small vector which starts at a high symmetry point of the Brillouin zone with wave vector $\vec{k}_0$) as a sum of invariants. These invariants are products of functions which depend on vector $\vec{k}$ components, intensity of magnetic field and operators of the momentum, which are presented in a matrix form.

G.E. Pikus had formalized the construction of secular matrices for defining the dispersion laws for charge carriers using group theory methods [2]. He had introduced a concept of basis matrices $A_{is}$, which can be built using the irreducible representations of $\vec{k}_0$ wave vector group.

Based on the group of $\vec{k}_0$ wave vector, in the vicinity of which one considers the dispersion law for charge carriers in [2], formulas were established to find irreducible representations $\tau_s$. Basic matrices, basic functions which depend on $k_x$, $k_y$, $k_z$, on the components of the strain tensor and on the components of a magnetic field all transform according to above mentioned representations $\tau_s$.

Basic matrixes which are used in constructing a secular matrix $D(\vec{k})$, form well known sets of matrices. For a double degenerated energy state in the $\vec{k}_0 = 0$ point, basic matrices are the Pauli's matrices and the unit matrix of the second rank. Triple degenerated states have got a set of basic matrices which correspond to momentum operators $\hat{P}^j$ for $j = 3/2$ written in a matrix form.

Thus, depending on the rank of a secular matrix $D(\vec{k})$, we have limited the number of the basic matrices. Moreover, they can be the same for different irreducible representations which describe crystals of different symmetry (space groups).

In this work we consider the possibility of applying the Pikus' method of invariants, formalized in the group theory terminology, in order to determine the vibronic potential energy and adiabatic potential in highly symmetric molecules.

We also analyze the similarities and differences in the construction of the vibronic potential energy of molecules and the matrix of the dispersion law for a solid obtained in the center of the Brillouin zone. For this reason, crystals with space group coinciding with the point group of a molecule are considered herein.

## 2. Jahn-Teller's effect

It is known that in highly symmetric molecules the Jahn-Teller's effect [3] is often observed. This effect causes the reduction of symmetry of a molecule due to electron-vibronic interaction. This interaction causes a split of a degenerated electronic term and a change of the configuration of a molecule. Energy of a molecule, as a function of the distance between the cores, should have a minimum for a stable configuration. Obviously, it means that the expansion of the energy of a molecule by small displacements of cores has no linear terms. Generally speaking, such terms appear when the adiabatic approximation is broken due to the so-called vibronic interaction.

Any complicated movement of cores of a molecule can be represented as a series of harmonic oscillations. Each of those is described by normal coordinates. The number of normal coordinates is equal to the number of degrees of freedom of a molecule.

Hamiltonian of an electronic subsystem now includes perturbational terms of vibronic potential energy. This potential energy is tailored by normal displacements as follows [4]:

$$U = \sum_{\alpha,i} V_{\alpha i}(q) Q_{\alpha i} + \sum_{\alpha,i,\beta,k} W_{\alpha i,\beta k}(q) Q_{\alpha i} Q_{\beta k} + \cdots . \tag{2.1}$$

As it was mentioned above, the linear part is the most important in Jahn-Teller's effect realization.

The first correction in perturbation theory is defined by a matrix element:

$$V_{\rho\sigma} = \sum_{\alpha i} Q_{\alpha i} \int \Psi_\rho^* \hat{V}_{\alpha i}(q) \Psi_\sigma \, \mathrm{d}q, \tag{2.2}$$

here, $\Psi_\rho$, $\Psi_\sigma$ are wave functions of a degenerate electron state and integration is performed over the electronic configuration space $\{q\}$.

From the invariance of the Hamiltonian which includes the linear $Q_{\alpha i}$ term it follows, that the coefficient $V_{\rho\sigma}$ transforms by the elements of symmetry of a molecule in the same way as the normal coordinates $Q_{\alpha i}$ do. In formulas (2.1) and (2.2), greek indices $\alpha$, $\beta$, ... mean the number of irreducible representation, and $i$ and $k$ — the number of base functions of this irreducible representation (taken in the form of normal coordinates).

It is known that the secular equation, built on Hamiltonian $D(\vec{k})$ in a matrix form, is used to find the dispersion law $E(\vec{k})$ for charge carriers in crystals.

The same procedure is used in case of a vibronic interaction in a molecule. Adiabatic potential can be found after solving the secular equation that is built on a matrix of the vibronic potential energy $D(Q_1, Q_2, \ldots)$. Here, $Q_1, Q_2, \ldots$ — are normal displacements of vibrations which are active in Jahn-Taller's effect.

In general, the adiabatic potential predicts that there can be several stable and metastable configurations of a molecule.

There are some fundamental differences in constructing the matrix $D(\vec{k})$ and vibronic interaction potential energy operator in a matrix form. The first one is the difference between coefficients at the components of $\vec{k}$ wave vector and coefficients at the components of normal displacement.

The construction of $D(\vec{k})$ matrix which is used to find $E(\vec{k})$, is based on $\vec{k} \cdot \hat{\vec{p}}$-approximation and on the method of perturbation theory [5]. It is obvious that coefficients of $D(\vec{k})$ matrix are integrals of $\int \Psi_i^* P_\alpha \Psi_j \, d\tau$ type. These expressions are of two kinds (let us denote them I and II, respectively) which corresponds to the first and second perturbation corrections. In terms of type I, $P_\alpha$ is a component of the operator of an impulse, $\Psi_i$ and $\Psi_j$ are functions that describe the *one degenerate electronic state*, for which the dispersion law $E(\vec{k})$ is investigated, while in terms of type II, these functions belong to different states of a crystal. This means that if we want to define whether the integral equals zero or not we should investigate the antisymmetrized product of an irreducible representation that is built only on functions of this degenerate electronic state. Antisymmetrization is connected with the imaginary nature of the momentum operator $\hat{\vec{p}}$ (the perturbing part of $\vec{k}\cdot\hat{\vec{p}}$-approximation includes the operator $\hat{\vec{p}}$) [5].

Potential energy of vibronic interaction (2.1) has two terms that are built on linear combinations of components of normal displacements and their quadratic terms. Constructing the matrix of potential energy of vibronic interaction [unlike the constructing of $D(\vec{k})$] requires only the first correction of perturbation theory. It means that the matrix element of $D(\vec{k})$ matrix is built only on the eigenfunctions of the chosen degenerate state. Potential energy is an operator of multiplication [$V_{\alpha i}$ and $W_{\alpha i}$ in equation (2.1)]. Matrix elements built on the functions of a degenerate term of this operator will be evaluated by constructing the character of a symmetrized product of irreducible representation that describes a degenerate electronic term. Irreducible representation (denoted by $\tau_s$) for transformation of the components of normal displacements and their quadratic combinations is determined by equation [6]:

$$\frac{1}{2n}\sum_{g\in G}\chi^{s}(g)\left\{[\chi(g)]^{2}+\chi\left(g^{2}\right)\right\}=1, \tag{2.3}$$

where $\chi(g)$ is taken from the table of irreducible representations of a group of symmetry $G$ of a molecule.

The above statements contain the main differences in constructing the matrix of vibronic potential energy and the matrix $D(\vec{k})$.

Furthermore, unlike the matrix of vibronic potential energy, in order to construct a secular matrix which consists of a sum of invariants (the product of basic matrixes and functions that depend on the components of a wave vector $k_x$, $k_y$, $k_z$), one needs to consider not only equation (2.3) but also the following formula [7, 8]:

$$\frac{1}{2n}\sum_{g\in G}\chi^{s}(g)\left\{[\chi(g)]^{2}-\chi\left(g^{2}\right)\right\}=1. \tag{2.4}$$

Equation (2.3) gives us $\tau_s$ for even combinations of components of a wave vector and equation (2.4) provides $\tau_s$ for odd combinations. Basic matrices that form the $D(\vec{k})$ matrix are defined from the obtained $\tau_s$.

## 3. Implementation of theory to ethane molecule

The symmetry of the ethane molecule is described by $D_{3d}$ point group which has two-dimensional irreducible representations. These representations correspond to double degenerated electronic states (see table 1). In a crystal belonging to a crystallographic class with the same point symmetry $D_{3d}$, we will consider the group of the wave vector $\vec{k}_0 = 0$. In table 1 both types of notations (i.e., molecular and for point Γ) for an irreducible representation are presented. Also in table 2 we present the matrices of two-dimensional irreducible representations $\Gamma_5$ and $\Gamma_6$ in the real form (unlike the complex one presented in the book by O.V. Kovalev [9]).

**Table 1.** Characters of irreducible representations of a point group $D_{3d}$ and a group of wave vector $\vec{k} = 0$ for the space group $D_{3d}^2$ (denotation of elements of symmetry is in correspondence with O.V. Kovalev, $h_{13}$ is the operation of inversion [9]).

| | $h_1$ | $h_3, h_5$ | $h_8, h_{10}, h_{12}$ | $h_{13}$ | $h_{15}, h_{17}$ | $h_{20}, h_{22}, h_{24}$ |
|---|---|---|---|---|---|---|
| $A_g, \Gamma_1$ | 1 | 1 | 1 | 1 | 1 | 1 |
| $A_u, \Gamma_2$ | 1 | 1 | 1 | -1 | -1 | -1 |
| $B_g, \Gamma_3$ | 1 | 1 | -1 | 1 | 1 | -1 |
| $B_u, \Gamma_4$ | 1 | 1 | -1 | -1 | -1 | 1 |
| $E_g, \Gamma_5$ | 2 | -1 | 0 | 2 | -1 | 0 |
| $E_u, \Gamma_6$ | 2 | -1 | 0 | -2 | 1 | 0 |

**Table 2.** Irreducible representations $\Gamma_5$ ($E_g$) and $\Gamma_6$ ($E_u$) written as real matrices.

| | $h_1$ | $h_3$ | $h_5$ | $h_8$ | $h_{10}$ | $h_{12}$ |
|---|---|---|---|---|---|---|
| $\Gamma_5$ | $\begin{pmatrix}1 & 0\\0 & 1\end{pmatrix}$ | $\begin{pmatrix}-\frac{1}{2} & -\frac{\sqrt{3}}{2}\\\frac{\sqrt{3}}{2} & -\frac{1}{2}\end{pmatrix}$ | $\begin{pmatrix}-\frac{1}{2} & \frac{\sqrt{3}}{2}\\-\frac{\sqrt{3}}{2} & -\frac{1}{2}\end{pmatrix}$ | $\begin{pmatrix}1 & 0\\0 & -1\end{pmatrix}$ | $\begin{pmatrix}-\frac{1}{2} & \frac{\sqrt{3}}{2}\\\frac{\sqrt{3}}{2} & \frac{1}{2}\end{pmatrix}$ | $\begin{pmatrix}-\frac{1}{2} & \frac{\sqrt{3}}{2}\\\frac{\sqrt{3}}{2} & \frac{1}{2}\end{pmatrix}$ |
| | $h_{13}$ | $h_{15}$ | $h_{17}$ | $h_{20}$ | $h_{22}$ | $h_{24}$ |
| $\Gamma_5$ | $\begin{pmatrix}1 & 0\\0 & 1\end{pmatrix}$ | $\begin{pmatrix}-\frac{1}{2} & -\frac{\sqrt{3}}{2}\\\frac{\sqrt{3}}{2} & -\frac{1}{2}\end{pmatrix}$ | $\begin{pmatrix}-\frac{1}{2} & \frac{\sqrt{3}}{2}\\-\frac{\sqrt{3}}{2} & -\frac{1}{2}\end{pmatrix}$ | $\begin{pmatrix}1 & 0\\0 & -1\end{pmatrix}$ | $\begin{pmatrix}-\frac{1}{2} & \frac{\sqrt{3}}{2}\\\frac{\sqrt{3}}{2} & \frac{1}{2}\end{pmatrix}$ | $\begin{pmatrix}-\frac{1}{2} & \frac{\sqrt{3}}{2}\\\frac{\sqrt{3}}{2} & \frac{1}{2}\end{pmatrix}$ |
| | $h_1$ | $h_3$ | $h_5$ | $h_8$ | $h_{10}$ | $h_{12}$ |
| $\Gamma_6$ | $\begin{pmatrix}1 & 0\\0 & 1\end{pmatrix}$ | $\begin{pmatrix}-\frac{1}{2} & -\frac{\sqrt{3}}{2}\\\frac{\sqrt{3}}{2} & -\frac{1}{2}\end{pmatrix}$ | $\begin{pmatrix}-\frac{1}{2} & \frac{\sqrt{3}}{2}\\-\frac{\sqrt{3}}{2} & -\frac{1}{2}\end{pmatrix}$ | $\begin{pmatrix}1 & 0\\0 & -1\end{pmatrix}$ | $\begin{pmatrix}-\frac{1}{2} & \frac{\sqrt{3}}{2}\\\frac{\sqrt{3}}{2} & \frac{1}{2}\end{pmatrix}$ | $\begin{pmatrix}-\frac{1}{2} & \frac{\sqrt{3}}{2}\\\frac{\sqrt{3}}{2} & \frac{1}{2}\end{pmatrix}$ |
| | $h_{13}$ | $h_{15}$ | $h_{17}$ | $h_{20}$ | $h_{22}$ | $h_{24}$ |
| $\Gamma_6$ | $\begin{pmatrix}-1 & 0\\0 & -1\end{pmatrix}$ | $\begin{pmatrix}\frac{1}{2} & \frac{\sqrt{3}}{2}\\-\frac{\sqrt{3}}{2} & \frac{1}{2}\end{pmatrix}$ | $\begin{pmatrix}\frac{1}{2} & -\frac{\sqrt{3}}{2}\\\frac{\sqrt{3}}{2} & \frac{1}{2}\end{pmatrix}$ | $\begin{pmatrix}-1 & 0\\0 & 1\end{pmatrix}$ | $\begin{pmatrix}\frac{1}{2} & -\frac{\sqrt{3}}{2}\\-\frac{\sqrt{3}}{2} & -\frac{1}{2}\end{pmatrix}$ | $\begin{pmatrix}\frac{1}{2} & -\frac{\sqrt{3}}{2}\\-\frac{\sqrt{3}}{2} & -\frac{1}{2}\end{pmatrix}$ |

Thus, we will consider the so-called $E-E$ vibronic bonding, because the vibrational states will obviously transform according to the same irreducible representations of $D_{3d}$ group.

To construct the vibronic potential energy matrix of the ethane molecule ($C_2H_6$) having a $D_{3d}$ point symmetry we will define normal vibrations active in Jahn-Teller's effect. These normal vibrations should be chosen from the following set: $3A_{1g}$, $1A_{1u}$, $2A_{2u}$, $3E_g$, $3E_u$ [4].

Calculations show that normal oscillations which are active in the Jahn-Teller's effect have $A_{1g}$ and $E_g$ symmetry. Normal oscillation $A_{1g}$ should be excluded whereas the configuration of the molecule does not change with such a normal displacement.

In case of the so-called $E_g-E_g$ vibronic bonding in electron-vibrational interaction, there participate a double degenerate electronic state with $E_g$ symmetry and a normal oscillation of the molecule with $E_g$ symmetry .

Hence, the matrix of potential energy of vibronic interaction depends on two variables $Q_1$ and $Q_2$. It can also include squared combinations of $Q_1$ and $Q_2$. The method of projective operator was used to find those squared combinations [10]. Calculations show that such combinations are functions $Q_1^2-Q_2^2$ and $2Q_1Q_2$.

The following matrices transform in accordance with representation $E_g$, i.e., basic matrixes $\sigma_x$ and $\sigma_z$, Pauli's matrices chosen from the set and the identity matrix of the second rank $\sigma_1$. Such a result is gained after applying matrix transformation rules under symmetry elements.

One can get a $D(Q_1,Q_2)$ matrix having built the invariants from basic matrices and functions

$$D(Q_1,Q_2)=\frac{1}{2}\omega^2\left(Q_1^2+Q_2^2\right)\sigma_1+VQ_1\sigma_x+WQ_1Q_2\sigma_x+VQ_2\sigma_z+W\left(Q_1^2-Q_2^2\right)\sigma_z, \tag{3.1}$$

here, $V$ and $W$ are coefficients of linear and quadratic parts of the operator of potential energy of vibronic interaction, $\frac{1}{2}\omega^2\left(Q_1^2+Q_2^2\right)$ is potential energy of normal oscillation of a molecule which is described by $E_g$ representation, without getting vibronic interaction to account.

We should note that the same matrix of potential energy of vibronic interaction was obtained by us in [11] for a molecule of methane ($CH_4$) whose symmetry is described by the point group $C_{3v}$.

Despite the identical matrices of vibronic potential energy, normal displacements $Q_1$ and $Q_2$ for symmetric molecule $C_2H_6$ and non-symmetric molecule $CH_4$ [11] differ significantly. The point is that normal displacements $Q_1$ and $Q_2$ for $C_2H_6$ molecule are even functions, while in case of $CH_4$ molecule they have undefined parity. Even functions $Q_1$ and $Q_2$ are the base for irreducible representation $E_g$ ($\Gamma_5$) of a point group $D_{3d}$. In case of $CH_4$ molecule (a point group $C_{3v}$) $Q_1$ and $Q_2$ are the base for representation $E$ ($\Gamma_3$).

## 4. Constructing the adiabatic potential

It is known that adiabatic potential is defined by solving the secular equation constructed on the matrix of vibronic potential energy. Adiabatic potential should reproduce the symmetry of a chosen molecule. To fulfill this condition, one needs to replace the basic functions $Q_1$ and $Q_2$ of $E_g$ representation with basic functions of an equivalent representation written in cartesian coordinates: $x^2-y^2, 2xy$. Due to the fact that $C_{3v}$ group does not include the operation of inversion, Cartesian coordinates $x$ and $y$ can additionally be the base for representation $E$ of this group (besides $x^2-y^2$ and $2xy$).

Thanks to the symmetrical matching of $Q_1$ and $Q_2$ with functions $x^2-y^2$, $2xy$, one can rewrite the matrix of vibronic potential energy for $D_{3d}$ group in the form of a dependence on Cartesian coordinates. Matrix elements of the above mentioned matrix are written as follows:

$$\begin{aligned} D_{11} &= \frac{1}{2}\omega^2\left(x^2+y^2\right)+2V_{xy}+W\left(x^4+y^4-6x^2y^2\right), \\ D_{22} &= \frac{1}{2}\omega^2\left(x^2+y^2\right)-2V_{xy}-W\left(x^4+y^4-6x^2y^2\right), \\ D_{12} &= D_{21} = V\left(x^2-y^2\right)+2W\left(x^2-y^2\right)2xy. \end{aligned} \tag{4.1}$$

Such a denotation of the matrix makes possible the transformation to polar coordinates: $x=\rho\cos\varphi$, $y=\rho\sin\varphi$.

Having solved the secular equation obtained from $D(\rho,\varphi)$ matrix one gets the adiabatic potential:

$$\varepsilon_{1,2}(\rho,\varphi) = \frac{\omega^2\rho^4}{2} \pm \left[V\rho^4+2VW\rho^6\sin 6\varphi+W^2\rho^8\right]^{\frac{1}{2}}. \tag{4.2}$$

The presence of $\sin 6\varphi$ in the expression for adiabatic potential indicates the six minima in its structure in contrast to the three minima in case of non-centrosymmetric molecule $CH_4$ [11]. The structure of adiabatic potential for the considered centrosymmetric molecule reflects its symmetry.

As a result of Jahn-Teller's effect, the lowering of symmetry can occur in two ways: the loss of centre of symmetry or the loss of elements of symmetry (rotations $C_3$ and $C_3^2$). Namely, the lowering of symmetry occurs from $D_{3d}$ to $C_{3v}$ group or from $D_{3d}$ to $C_{2h}$ group.

Let us consider constructing the secular matrix $D(\vec{k})$ in the vicinity of $\vec{k}_0=0$ for a crystal having a $D_{3d}^2$ symmetry. We will choose an irreducible representation $E_g$ ($\Gamma_5$), that describes a degenerate energy state. Such a choice is conditioned by the aim to analyze the similarities in a matrix of potential energy of vibronic interaction of molecules and a secular matrix of the energy spectrum of a crystal. According to Pikus' method of invariants, as it was mentioned, one can find $\tau_s$-representation, according to which the basic matrices as well as linear and square functions of the wave vector get transformed.

To find $\tau_s$, the following equation is used [7]:

$$n_s = \sum_{g\in G_{\vec{k}_0}} |\chi(g)|^2\chi_s(g). \tag{4.3}$$

Trial characters $\chi_s(g)$ are taken from the table of characters of a group of wave vector $\vec{k}_0=0$ (table 1). Calculation shows that $n_s \neq 0$ only when $\tau_s=\Gamma_1,\Gamma_3,\Gamma_5$.

Thus, we find that the basic matrices as well the $f(\vec{k})$ functions get converted by representations $\Gamma_1$, $\Gamma_3$, $\Gamma_5$.

There is another significant difference in constructing a $D(Q_1,Q_2)$ matrix and a secular matrix $D(\vec{k})$. From the selection rules [equations (2.3) and (2.4)] one gets different irreducible representations $\tau_s = \Gamma_1,\Gamma_3,\Gamma_5$ that describe the functions and basic matrices of invariants on which $D(\vec{k})$ matrices are built. In case of constructing the matrix of potential energy of vibronic interaction, only one normal displacement responsible for the vibronic interaction of electronic and vibronic states is chosen out of all possible normal displacements that were gained from selection rules of irreducible representations. In our case, it is the $E_g-E_g$ interaction.

We should note that irreducible representations $\tau_s$, gained from equation (4.3) should be redistributed to those that describe even and odd $f(\vec{k})$ functions. We use equations (2.3) and (2.4) for this purpose.

**Table 3.** Matching the representations $\Gamma_1$ and $\Gamma_5$ with $f(\vec{k})$ functions and basic matrices included in constructing the invariants.

| representation | $f(\vec{k})$ | | $A_{ls}$ |
|---|---|---|---|
| | $\gamma = 1$ | $\gamma = -1$ | |
| $\Gamma_1$ | $k_x^2 + k_y^2$, $k_z$ | - | $\begin{pmatrix} 1 & 0 \\ 0 & 1 \end{pmatrix}$ |
| $\Gamma_5$ | $k_x k_z$, $k_y k_z$ | - | $\begin{pmatrix} 1 & 0 \\ 0 & -1 \end{pmatrix}, \begin{pmatrix} 0 & 1 \\ 1 & 0 \end{pmatrix}$ |

Calculations show that a symmetrical squared character of representation $\Gamma_5$ [equation (4.3)] includes representations $\Gamma_1$ and $\Gamma_5$ while antisymmetrical one includes $\Gamma_3$ representation. Thus, antisymmetrical function of a wave vector should be transformed by $\Gamma_3$ representation, which is impossible (see table 1).

Using the method of projective operator, one gets combinations of components of a wave vector that correspond to representations $\Gamma_1$ and $\Gamma_5$.

It is clear that functions $k_x^2 + k_y^2$, $k_z^2$ and an identity matrix are transformed by $\Gamma_1$ representation and functions $k_x k_z$ and $k_y k_z$ are the base for representation $\Gamma_5$.

As it was shown before, in constructing the $D(Q_1, Q_2)$ matrix, the basic matrices that are transformed by representation $E_g$ are $\sigma_x$ and $\sigma_z$. As representation $E$ matches $\Gamma_5$, these matrices also correspond to representation $\Gamma_6$.

The calculated functions and matrices included in invariants are presented in table 3.

Based on data from table 3, we construct a secular matrix $D(\vec{k})$:

$$D(\vec{k}) = \begin{pmatrix} a\left(k_x^2 + k_y^2\right) + bk_z^2 + ck_x k_z & ck_y k_z \\ ck_y k_z & a\left(k_x^2 + k_y^2\right) + bk_z^2 - ck_x k_z \end{pmatrix}. \tag{4.4}$$

By solving a corresponding secular equation we obtain an expression for the dispersion law of charge carriers in point $\vec{k}_0 = 0$ for the state described by $\Gamma_5$ representation:

$$E(\vec{k}) = a\left(k_x^2 + k_y^2\right) + bk_z^2 \pm \sqrt{c^2 k_z^2 \left(k_x^2 + k_y^2\right)}. \tag{4.5}$$

From equations (4.2) and (4.5), we conclude that solutions of corresponding secular equations reproduce the point symmetry of the crystal and the molecule.

## 5. Conclusions

Thus, the secular matrix $D(k)$ as well as the matrix of vibronic potential energy are built from the sum of invariants. In both cases, each of these invariants is a product of the basis matrix (Pauli's matrices in our case) and the basis function which depends on corresponding variables.

In the case of a secular matrix, basis functions and basis matrices transform according to irreducible representations, which form symmetrized and antisymmetrized squares of the irreducible representation connected with an active normal vibration or with a corresponding degenerated electronic term for which the secular matrix is written down. Polynomials from which the basis functions are built are powers of wave vector's components. This small wave vector originates from point in the Brillouin zone in the vicinity of which one construct the dispersion law $E(k)$.

In the case of vibronic potential energy construction, the basis matrices and functions are built solely for irreducible representations which form a symmetrized square of the irreducible representation, describing the vibration which is active in Jahn-Teller's effect. Corresponding basis functions are also built on components of this vibration.

In conclusion, we should note that the construction of vibronic potential energy and the adiabatic potential can be achieved without using the method of invariants, solely by using the Clebsch-Gordan coefficients [12]. A correct solution of the adiabatic potential construction problem by means of group theory method and the method of invariants allows one to successfully apply adiabatic potentials for a qualitative explanation of a wide variety of phenomena connected with a vibronic interaction in molecules and crystals. Moreover, the method of adiabatic potential construction can be adapted to the investigation of peculiarities of phase transitions in crystals with Jahn-Teller centers (for example, the $CuInP_2S_6$ crystal [13]). The mentioned problem will be investigated in our next work.

## Acknowledgements

Author (B.S.A.) wishes to thank Dr. Glukhov K.E. for helpful discussions and critical reading of the manuscript.

## References

1. Luttinger J.M., Phys. Rev., 1956, **102**, No. 4, 1030; doi:10.1103/PhysRev.102.1030.
2. Pikus G.E., Zh. Eksp. Teor. Fiz., 1961, **41**, 1258 (in Russian).
3. Jahn H.A., Teller E., Proc. R. Soc. London, Ser. A, 1937, **161**, No. 905, 220; doi:10.1098/rspa.1937.0142.
4. Landau L.D., Lifshitz E., Course of Theoretical Physics Vol. 3. Quantum Mechanics: Non-Relativistic Theory, Pergamon Press, New York, 1965.
5. Shockley W., Phys. Rev., 1950, **78**, No. 2, 173; doi:10.1103/PhysRev.78.173.
6. Lyubarskii G.Y., The Application of Group Theory in Physics, Pergamon Press, New York, 1960.
7. Pikus G.E., Bir G.L., Symmetry and Strain-Induced Effects in Semiconductors, Wiley, New York, 1974.
8. Rashba E.I., Fiz. Tverd. Tela, 1959, **1**, 368 (in Russian).
9. Kovalev O.V., Representations of the Crystallographic Space Groups. Irreducible Representations, Induced Representations and Corepresentations, Brigham Young University, Provo, 1993.
10. Knox R.S., Gold A., Symmetry in the Solid State, W.A. Benjamin, New York, 1964.
11. Bercha S.A., Rizak V.M., Uzhhorod University Scientific Herald. Series Physics, 2013, No. 33, 15 (in Ukrainian).
12. Bersuker I.B., Polinger V.Z., Vibronic Interactions in Molecules and Crystals, Springer-Verlag, Berlin, 1989.
13. Bercha D.M., Bercha S.A., Glukhov K.E., Shnajder M., In: Book of Abstarcts of the VI Ukrainian Scientific Conference on Physics of Semiconductors (Chernivtsi, 2013), Ruta, Chernivtsi, 2013, 242–243 (in Ukrainian).

# 13

# Interlevel absorption of electromagnetic waves by nanocrystal with divalent impurity

V.I. Boichuk, R.Ya. Leshko*

Department of Theoretical Physics, Ivan Franko Drohobych State Pedagogical University,
3 Stryiska St., 82100 Drohobych, Ukraine

The energy spectrum of central divalent impurity is calculated using the effective mass approximation in a spherical quantum dot (QD). The dipole moment and oscillator strength of interlevel transition is defined. The dependence of linear absorption coefficient on the QD size and electromagnetic frequency is analyzed. The obtained results are compared with the results of univalent impurity.

**Key words:** *divalent impurity, linear absorption coefficient*

## 1. Introduction

The semiconductor quantum dots (QDs) are widely used in opto- and nanoelectronics due to their unique properties. Lasers, sources of light, LEDs are constructed based on nanosystems. Sources of terahertz radiation, which are constructed based on QDs, take a special place [1]. The feature of terahertz radiation lies in the fact that it practically does not ionize materials, contrary to the X-ray, and is capable of penetrating into materials. That is why this kind of radiation is widely used in medical tomography [2], in security systems, in producing high resolution images of microscopic objects [3]. The possibilities of developing high-speed THz communication systems are studied [4]. The detector of terahertz radiation was proposed based on QDs [5, 6]. Taking into consideration that the energy of interlevel transitions responds to the terahertz range, the study of interlevel transitions became the basis for theoretical description and prognostication of the properties of terahertz detectors and sources.

Single-electron states in the QD which definitely depend on the QD size, the presence of defects, especially impurities, are the basis of interlevel transition analysis. At present, the theory of shallow hydrogenic donor impurities is widely developed in the QD. An exact solution of Schrödinger equation for the central impurity was derived [7], the energy spectrum of the off-central impurity was obtained using different methods in spherical [8] and ellipsoidal [9] QDs. The cubic [10] QDs are analysed too. Since the QD can contain several impurities, the problem regarding the QD with two impurities was solved [11, 12]. Based on the obtained results, the linear and nonlinear optical properties of the QD with impurities [8–10, 12–14] were calculated using the density matrix and iteration method [15].

Experimental data show that QDs can be doped with impurities which are divalent [16]. In particular, in this work it was shown that the zinc impurities penetrate the CdS QDs. This leads to the changes of the optical properties which are connected with interband (high-energy) and interlevel intraband (low-energy) transitions.

The above mentioned as well as the lack of a consistent theory of central divalent impurities in spherical QDs, which could make possible the calculation of the ground and excited states, brings about the necessity to consider the divalent impurity in a spherical QD; to determine the energy spectrum of this impurity; to calculate interlevel transitions in the QD with divalent impurity; to compare the obtained results with the corresponding results of monovalent impurity.

* E-mail: leshkoroman@gmail.com

## 2. Eigenvalues and eigenfunctions

We consider a spherical nanosize heterosystem. It consists of a nanocrystal of radius $a$ having electron effective mass $m_1^*$, which is placed in a matrix having electron effective mass $m_2^*$. There is a divalent impurity in the center of the QD. Let the heterosystem be made of crystals that have the values close to dielectric permittivity. This makes it possible to introduce the average value of dielectric permittivity $\varepsilon$. The effective-mass Hamiltonian of this system can be written as follows:

$$\hat{H} = \hat{H}_1 + \hat{H}_2 + \frac{e^2}{4\pi\varepsilon_0\varepsilon r_{12}}, \tag{2.1}$$

where

$$\hat{H}_i = -\frac{\hbar^2}{2}\nabla_i \frac{1}{m^*(r_i)}\nabla_i + U(r_i) - \frac{Ze^2}{4\pi\varepsilon_0\varepsilon r_i} = \hat{H}_i^{(0)} - \frac{Ze^2}{4\pi\varepsilon_0\varepsilon r_i}, \tag{2.2}$$

$Z = 2$. The potential energy caused by the heterostructure band mismatch is given by:

$$U(r_i) = \begin{cases} 0, & r_i \leqslant a, \\ U_0, & r_i > a. \end{cases} \tag{2.3}$$

The Schrödinger equation with the Hamiltonian (2.1) cannot be solved exactly. Therefore, the Ritz variation method has been used herein. Since the electrons are fermi-particles, the wave function should be antisymmetric. The approach of [8, 17, 18] has been used for the chosen variation function. Nonetheless in [17, 18] there was calculated only the ground state energy of divalent impurity, and in [8] there was calculated the energy of the ground state and the first exited states of the monovalent impurity. In both cases, one variational parameter was used. To improve the accuracy, two variational parameters are introduced in the present paper in the coordinate wave functions of ground state and some exited states of divalent impurity:

$$\psi_1 = c_1 \left|1s,\vec{r}_1,\alpha_1\right\rangle \left|1s,\vec{r}_2,\beta_1\right\rangle, \tag{2.4}$$

$$\psi_2 = c_2 \left(\left|1s,\vec{r}_1,\alpha_2\right\rangle \left|1p,\vec{r}_2,\beta_2\right\rangle - \left|1p,\vec{r}_1,\alpha_2\right\rangle \left|1s,\vec{r}_2,\beta_2\right\rangle\right), \tag{2.5}$$

$$\psi_3 = c_3 \left(\left|1s,\vec{r}_1,\alpha_3\right\rangle \left|1p,\vec{r}_2,\beta_3\right\rangle + \left|1p,\vec{r}_1,\alpha_3\right\rangle \left|1s,\vec{r}_2,\beta_3\right\rangle\right), \tag{2.6}$$

$$\psi_4 = c_4 \left(\left|1s,\vec{r}_1,\alpha_4\right\rangle \left|1d,\vec{r}_2,\beta_4\right\rangle - \left|1d,\vec{r}_1,\alpha_4\right\rangle \left|1s,\vec{r}_2,\beta_4\right\rangle\right), \tag{2.7}$$

$$\psi_5 = c_5 \left(\left|1s,\vec{r}_1,\alpha_5\right\rangle \left|1d,\vec{r}_2,\beta_5\right\rangle + \left|1d,\vec{r}_1,\alpha_5\right\rangle \left|1s,\vec{r}_2,\beta_5\right\rangle\right), \tag{2.8}$$

where

$$\begin{aligned} \left|j,\vec{r}_i,\gamma_q\right\rangle &= R_j(r_i,\gamma_q)\,Y_{l_j}^{m_j}(\theta_i,\varphi_i) \\ &= A_j Y_{l_j}^{m_j}(\theta_i,\varphi_i)\begin{cases} \mathbf{j}_{l_j}\left(k_{n_j,l_j} r_i\right)\exp(-\gamma_q r_i), & r_i \leqslant a, \\ \mathbf{k}_{l_j}\left(\varkappa_{n_j,l_j} r_i\right)\exp\left\{-\gamma_q\left[\frac{m_2^*}{m_1^*}(a-r_i)-a\right]\right\}, & r_i > a, \end{cases} \end{aligned} \tag{2.9}$$

$j = 1s, 1p, 1d$; index $q = 1,2,3,4,5$ enumerates variational parameters for states (2.4)–(2.8); $\gamma = \alpha, \beta$ are variational parameters, index $i = 1,2$ enumerates electrons; $l_{1s} = 0$, $l_{1p} = 1$, $l_{1d} = 2$; $m_{1s} = 0$, $m_{1p} = -1,0,1$; $m_{1d} = -2,-1,0,1,2$. The spherical Bessel function of the first kind $j_b(z)$ and the modified spherical Bessel function of the second kind $k_b(z)$ are the solutions of a Schrödinger equation regarding the particle in the spherical potential well with the Hamiltonian $\hat{H}_i^{(0)}$,

$$k_{n_j,l_j} = \sqrt{\frac{2m_1^*}{\hbar^2}E_{n_j,l_j}^{(0)}}, \qquad \varkappa_{n_j,l_j} = \sqrt{\frac{2m_2^*}{\hbar^2}\left(U_0 - E_{n_j,l_j}^{(0)}\right)},$$

$n_{1s}$, $n_{1p}$, $n_{1d}$ enumerates the solutions of dispersion equation when $l$ is fixed. $A_j$ can be found from the normalization condition for the function (2.9). $\psi_1$, $\psi_3$, $\psi_5$ are functions of singlet states; $\psi_2$, $\psi_4$ are functions of triplet states. Orthogonality of total wave functions (the coordinate part and the spin part) are provided by the orthogonality of spin parts of wave functions and by the orthogonality of spherical harmonics. The single particle wave function ensures the implementation of a boundary condition.

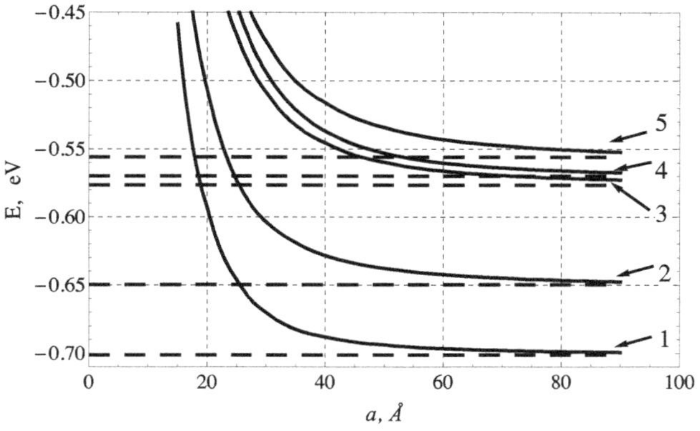


**Figure 1.** The energy of divalent impurity as a function of the QD radius. Numbers denote energies of respective states: 1 — $\psi_1$, 2 — $\psi_2$, 3 — $\psi_3$, 4 — $\psi_4$, 5 — $\psi_5$. Horizontal lines correspond to the energy of the divalent impurity in the bulk CdS.

After substitution (2.4)–(2.8) into the Schrödinger equation with Hamiltonian (2.1), the functional was found which depends on two variational parameters for excited states and depends on one variational parameter for the ground state. The performed procedure of numerical minimization makes it possible to get the corresponding energy states and find the values of variational parameters, and thus ultimately determine the wave functions.

Calculation of electron discrete energy was performed for heterostructure $CdS/SiO_2$ with the following parameters: $m_1^* = 0.2m_0$, $m_2^* = 0.42m_0$, $\varepsilon = (5.5 + 3.9)/2 = 4.7$, $U_0 = 2.7$ eV, where $m_0$ is free electron mass. The energy spectrum of a divalent impurity is presented in figure 1. Due to spherical symmetry, ground and excited states are degenerated by the magnetic quantum number. Figure 1 shows that an increase of the QD radius leads to a decrease of the energy of the ground state which quickly becomes saturated. For larger QD radius, the energy of excited states leads to the values corresponding to the values of the bulk crystal. Similar dependence was observed for a monovalent impurity [8]. This dependence is caused by a small effective Bohr radius $a_b^* = 12.44$ Å and a large confinement. Although the effective Bohr radius is small, the volume $a_b^{*3}$ approximately contains 10–12 elementary cells. This is the reason for using the Coulomb model potential interaction of electrons having an impurity.

An important characteristic of the QD having a divalent or monovalent impurity is the binding energy. In the case of a divalent impurity, $E_b$ is calculated by the similar formula [19]:

$$E_{b,II} = E_0 + E_{1s,Z=2} - E_1, \tag{2.10}$$

where $E_0$ is the electron energy of the QD without impurities, $E_{1s,Z=2}$ is the ground state energy of the QD having a singly ionized divalent impurity, $E_1$ is the energy of the state $\psi_1$ of the divalent impurity (2.4). In the case of an univalent impurity, the binding energy is defined by the formula:

$$E_{b,I} = E_0 - E_{1s,Z=1}, \tag{2.11}$$

where $E_{1s,Z=1}$ is the energy of the univalent impurity. If the QD radius reduces, the binding energy increases in both cases. For very small radii, $E_b$ decreases (figure 2). This is caused by an increase of the probability of location of the electrons outside the QD in both cases. However, if the QD has a divalent impurity, the binding energy is larger.

## 3. Optical properties

Energy spectrum and wave functions make it possible to calculate interlevel transitions. Selection rules by spin variables state that transitions are possible only between singlet-singlet and triplet-triplet states.

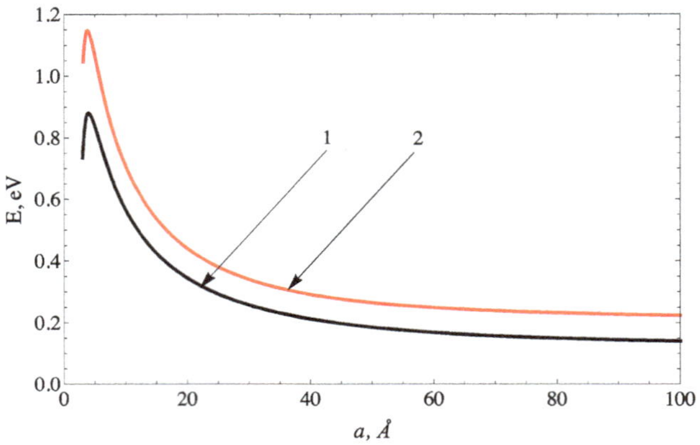


**Figure 2.** (Color online) The binding energy of the univalent impurity (curve 1) and of the divalent impurity (curve 2).

Let the QD be irradiated by the linearly polarized light along the $z$ direction. Then, in the dipole approximation, interlevel transitions are possible between the states $\psi_1$ and $\psi_3$; $\psi_2$ and $\psi_4$; $\psi_3$ and $\psi_5$. Dipole transition matrix elements between those states are given by:

$$d_{13} = \langle \psi_1 | ez | \psi_3 \rangle, \qquad d_{24} = \langle \psi_2 | ez | \psi_4 \rangle, \qquad d_{35} = \langle \psi_3 | ez | \psi_5 \rangle. \tag{3.1}$$

The dependence of the square of the dipole transition matrix element on the QD radius is presented in figure 3 with logarithmic scale. $|d_{1s-1p}/e|^2$ for the monovalent impurity is plotted too. Figure 3 shows that the corresponding values for a monovalent impurity are bigger than for the divalent one. This is due to the changes in the average distance of electrons in their respective states. Besides, it was established that all the curves for a large QD radii tend to the values that correspond to the values of the bulk crystal.

The oscillator strength of interlevel transitions is also defined

$$f_{mn} = \frac{2m_1^*}{\hbar^2 e^2}(E_n - E_m)|d_{mn}|^2. \tag{3.2}$$

The dependences are presented in figure 4 with logarithmic scale. The oscillator strength of interlevel transitions for a monovalent impurity in the center of the QD is plotted too. This is in agreement with the result of other works [8, 20]. Similarly to the dipole momentum, the oscillator strength of the divalent impurity is smaller than the oscillator strength of the monovalent impurity. This dependence is caused by the dependence of the dipole momentum and the transition energy $E_{tr}=E_n - E_m$ (figure 5).

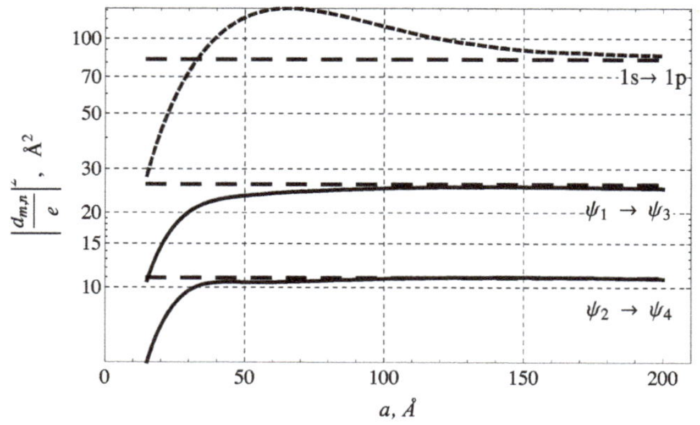


**Figure 3.** The square of the dipole momentum of interlevel transitions. Solid curves correspond to the divalent impurity. The dotted curve corresponds to the monovalent impurity. Horizontal dashed curve denotes the square of the dipole momentum of interlevel transitions of the monovalent and divalent impurity in the bulk crystal.

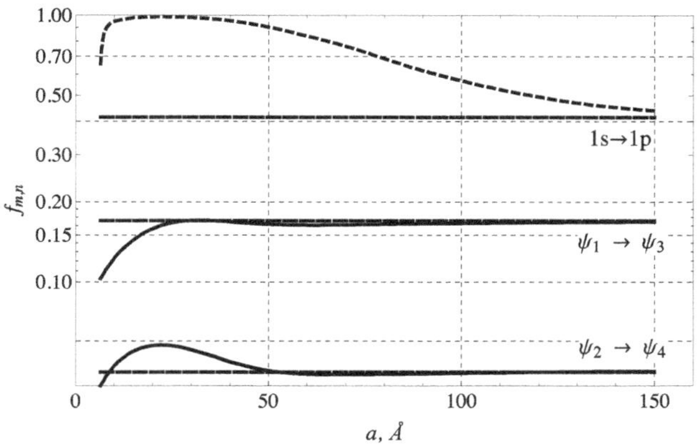


**Figure 4.** The oscillator strength of interlevel transitions. Solid curves correspond to the divalent impurity. The dotted curve corresponds to the monovalent impurity. Horizontal curves denote the oscillator strength of interlevel transitions of the monovalent and divalent impurity in the bulk crystal.

The above mentioned dependence of the dipole momentum and the transition oscillator strength effects the height of the absorption peaks. For a two-level system, the density matrix and iterative procedure were used to derive the absorption coefficient [13–15]. In this approach, the linear absorption coefficient can be expressed as follows:

$$\alpha_{m,n}(\omega) = \omega\sqrt{\frac{\mu_0}{\varepsilon_0\varepsilon}}\frac{N\left|d_{m,n}\right|^2\hbar\Gamma}{(E_n - E_m - \hbar\omega)^2 + (\hbar\Gamma)^2}, \tag{3.3}$$

where $\varepsilon_0$ is electric constant, $\mu_0$ is magnetic constant, $c$ is the speed of light, $N \approx 3\cdot 10^{16}$ cm$^{-3}$ is carrier concentration, $\hbar\Gamma$ is the scattering rate caused by the electron-phonon interaction and by some other factors of scattering. If $\hbar\Gamma$ limits to zero, one can obtain:

$$\alpha_{m,n}(\omega) = \lim_{\hbar\Gamma\to 0}\left(\omega\sqrt{\frac{\mu_0}{\varepsilon_0\varepsilon}}\frac{N\left|d_{m,n}\right|^2\hbar\Gamma}{(E_n - E_m - \hbar\omega)^2 + (\hbar\Gamma)^2}\right) = \omega\pi\sqrt{\frac{\mu_0}{\varepsilon_0\varepsilon}}N\left|d_{m,n}\right|^2\delta\left(E_n - E_m - \hbar\omega\right). \tag{3.4}$$

In practice, sets of QD are obtained which are located on both crystal and polymer matrix or in the solutions. Whatever method of cultivation is used, the set of QDs are always characterized by the size dispersion. Let the QD size distribution be approximated by the Gauss function:

$$g(s, \bar{a}, a) = \frac{1}{s\sqrt{2\pi}}\exp\left(-\frac{(a-\bar{a})^2}{2s^2}\right), \tag{3.5}$$

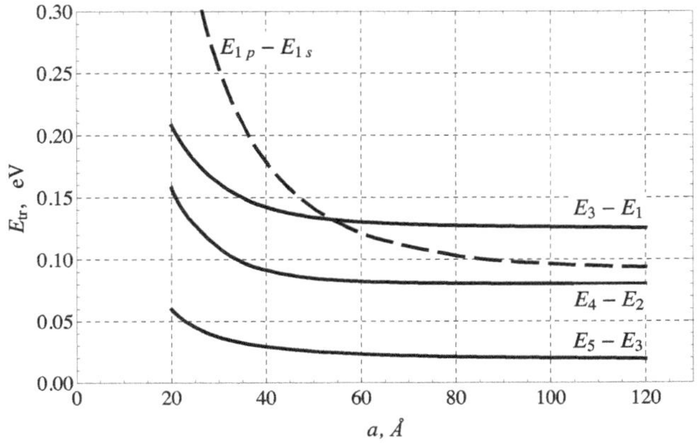


**Figure 5.** The transition energy. Solid curves correspond to the divalent impurity. Dashed curve corresponds to the monovalent impurity.

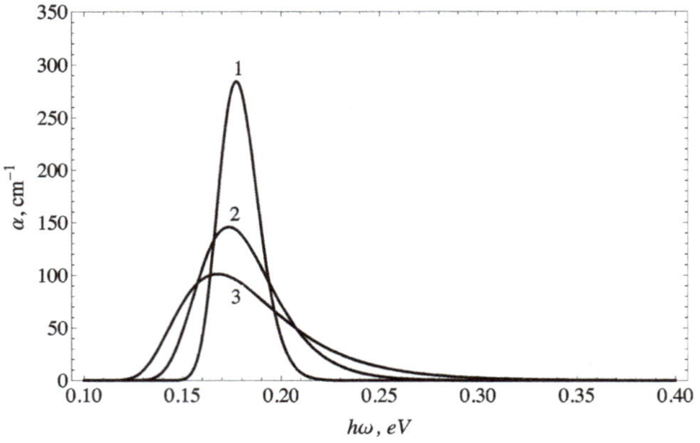


**Figure 6.** The absorption coefficient of the QD system with the average radius $\bar{a}$ = 40 Å. The curve 1 denotes the QD system with $\sigma = 5\%$, the curve 2 — $\sigma = 10\%$, the curve 3 — $\sigma = 15\%$.

where $a$ is the QD radius (variable), $s$ is a half-width of the distribution (3.5), which is expressed by the average radius $\bar{a}$ and the value $\sigma$ which is considered as the variance in the QD sizes expressed in percentage: $s = \bar{a}\sigma/100$. By regarding the QD dispersion (3.5), the absorption coefficient is obtained for the set of QDs:

$$\alpha_{m,n;\mathrm{system}}(\omega) = \omega\pi\sqrt{\frac{\mu_0}{\varepsilon_0\varepsilon}}N\int g(s,\bar{a},a)\,\left|d_{m,n}(a)\right|^2\delta\left(E_n(a) - E_m(a) - \hbar\omega\right)\mathrm{d}a.$$

Using delta-function properties we obtain:

$$\alpha_{m,n;\mathrm{system}}(\omega) = \omega\pi\sqrt{\frac{\mu_0}{\varepsilon_0\varepsilon}}N\int g(s,\bar{a},a)\,\left|d_{m,n}(a)\right|^2\sum_i\frac{\delta(a - a_{0i})}{\left|\frac{\mathrm{d}}{\mathrm{d}a}\left(E_n(a) - E_m(a) - \hbar\omega\right)\right|_{a=a_{0i}}}\mathrm{d}a, \tag{3.6}$$

where $a_{0i}$ are simple zeros of the function $F(a) = E_n(a) - E_m(a) - \hbar\omega$. Therefore,

$$\alpha_{m,n;\mathrm{system}}(\omega) = \omega\pi\sqrt{\frac{\mu_0}{\varepsilon_0\varepsilon}}N\sum_i\frac{g(s,\bar{a},a_{0i})\,\left|d_{m,n}(a_{0i})\right|^2}{\left|\frac{\mathrm{d}}{\mathrm{d}a}\left(E_n(a) - E_m(a) - \hbar\omega\right)\right|_{a=a_{0i}}}. \tag{3.7}$$

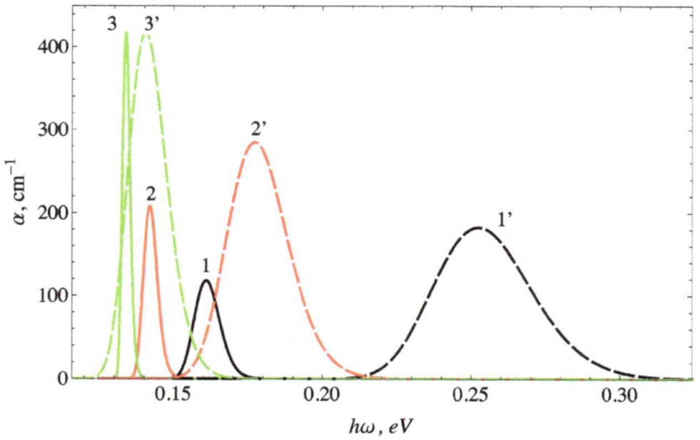


**Figure 7.** (Color online) The absorption coefficient of the QD system. Solid curves 1, 2, 3 denote the absorption coefficient of the QD with divalent impurity (transitions between singlet states $\psi_1$, $\psi_3$), dashed curves 1', 2', 3' denote the absorption coefficient of the QD with univalent impurity. 1, 1' — average radius is 30 Å; 2, 2' — average radius is 40 Å; 3, 3' — average radius is 50 Å.

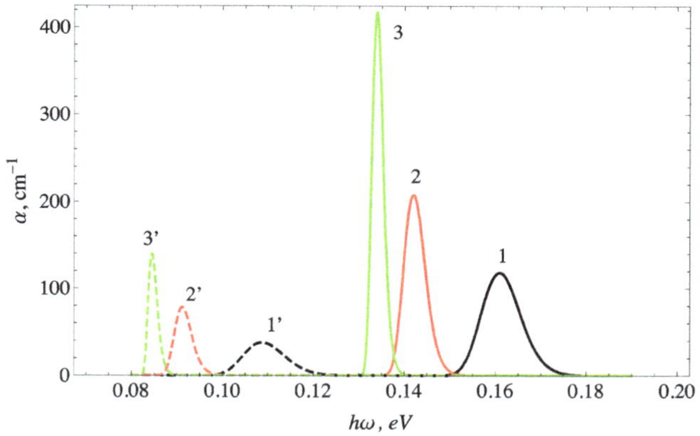


**Figure 8.** (Color online) Absorption coefficient of the system of QDs with divalent impurity. Solid curves 1, 2, 3 denote absorption between singlet states $\psi_1$–$\psi_3$ , dashed curves 1', 2', 3' denote absorption between triplet states $\psi_2$–$\psi_4$. Curves 1, 1' — average radius 30 Å; 2, 2' — 40 Å; 3, 3' — 50 Å.

The dependence of the absorption coefficient on the energy quant of light for different average radii and dispersion $\sigma$ was plotted using expression (3.7).

In figure 6 for an univalent impurity in the QD, the dependence of the QD absorption coefficient which is caused by the $1s-1p$ transition, was plotted for three different values of $\sigma$. The figure shows that for highly dispersed QDs, the height of the absorption peak decreases and the absorption band blurs. This leads to an overlap of absorption bands caused by transitions between other allowed states. For monodispersion systems or systems with low $\sigma$, those transitions are clearly seen. A similar situation exists for the bivalent impurity in the spherical QD. Thus, further we consider a system of QDs with $\sigma$=5%.

The absorption coefficient is plotted in figure 7. The figure shows that for the same average QD radius $\bar{a}$, $1s-1p$ transition in the QD having a monovalent impurity and the respective absorption coefficient are larger than corresponding values in the QD having a divalent impurity. This is caused by a larger oscillator strength and dipole momentum of the interlevel transition in the case of univalent impurity. Values of $|d_{m,n}|^2$ in the case of univalent impurity are larger, because $|\langle r_n\rangle - \langle r_m\rangle|$ is larger for the univalent impurity than for the divalent impurity. A similar explanation of the height of absorption bands is presented in our previous works [8, 12]. Both for the monovalent and divalent impurity, an increase of the average QD radius leads to the shift of absorption bands into the low-energy range. When the average QD radius is less than 55 Å, the absorption band caused by the transition $1s-1p$ with monovalent impurity is located in the high-energy range in comparison with the transition $\psi_1$–$\psi_3$ of the divalent impurity. For larger $\bar{a}$, this dependence reversed. Moreover, this can be seen in figure 5.

It should be noted that the absorption of electromagnetic waves by the system of QDs having a divalent impurity caused by the transitions between singlet states, is stronger than the corresponding absorption between triplet states (figure 8). In addition, the transition energy of triplet states $\psi_2$, $\psi_4$ is smaller than the transition energy of singlet states $\psi_1$, $\psi_3$. Therefore, the respective absorption bands are shifted into low-energy region. Both for the singlet-singlet and triplet-triplet transitions, the small $\sigma$ are provided without overlapping the absorption bands which are clearly identified.

## 4. Summary

The present paper studied optical properties of the QD heterosystem CdS/$SiO_2$ having a divalent impurity in the center of the QD, which made it possible:

- to determine the energy spectrum of the QD with a divalent impurity and to show that this energy is smaller than the energy of the monovalent impurity in the same QD;
- to calculate the dipole momentum and the oscillator strength of the interlevel transition and to find

out that the absorption between the singlet states is stronger than between the triplet states;

- to establish that the absorption bands of low dispersion systems with QD caused by a transition between the permitted states are clearly visible and do not overlap;
- to show that in the presence of a monovalent impurity, the absorption coefficient is larger than in the presence of a divalent impurity.

The results obtained are valid at very low temperatures. Their adjustments will be made by considering the temperature dependence. This will be implemented in our further works.

## References

1. Wu X.L., Xiong S.J., Liu Z., Chen J., Shen J.C., Li T.H., Wu P.H., Chu P.K., Nat. Nanotechnol., 2011, **6**, 103; doi:10.1038/nnano.2010.264.
2. Wang S., Zhang X.C., J. Phys. D: Appl. Phys., 2004, **37**, R1; doi:10.1088/0022-3727/37/4/R01.
3. Huber A.J., Keilmann F., Wittborn J., Aizpurua J., Hillenbrand R., Nano Lett., 2008, **8**, 3766; doi:10.1021/nl802086x.
4. Piesiewicz R., Jacob M., Koch M., Schoebel J., Kurner T., IEEE J. Quantum Electron., 2008, **14**, 421; doi:10.1109/JSTQE.2007.910984.
5. Wu W., Dey D., Mohseni H., J. Phys. D: Appl. Phys., 2010, **43**, 155101; doi:10.1088/0022-3727/43/15/155101.
6. Wu W., Dey D, Memis O.G., Mohseni H., Proc. SPIE, 2010, **7601**, 760109; doi:10.1117/12.841150.
7. Tkach M.V., Holovatsky V.A., Berezovsky Y.M., Phys. Chem. Solid State, 2003, **4**, 213 (in Ukrainian).
8. Boichuk V.I. Bilynskyi I.V., Leshko R.Ya., Turyanska L.M., Physica E, 2011, **44**, 476; doi:10.1016/j.physe.2011.09.025.
9. Sadeghi E., Avazpour A., Physica B, 2011, **406**, 241; doi:10.1016/j.physb.2010.10.051.
10. Rezaei G., Vahdani M.R.K., Barati M, J. Nanoelectron. Optoelectron., 2008, **3**, 159; doi:10.1166/jno.2008.208.
11. Holovatsky V., Frankiv I., J. Phys. Stud., 2012, **1/2**, 1706 (in Ukrainian).
12. Boichuk V.I. Bilynskyi I.V., Leshko R.Ya., Turyanska L.M., Physica E, 2013, **54**, 281; doi:10.1016/j.physe.2013.07.003.
13. Vahdani M.R.K., Rezaei G., Phys. Lett. A, 2009, **373**, No. 34, 3079; doi:10.1016/j.physleta.2009.06.042.
14. Rezaei G., Vahdani M.R.K., Vaseghi B., Curr. Appl. Phys., 2011, **11**, No. 2, 176; doi:10.1016/j.cap.2010.07.002.
15. Tang C.L., Fundamentals of Quantum Mechanics for Solid State Electronics and Optics, Cambridge University Press, Cambridge, 2005.
16. Korbutyak D.V., Tokarev S.V., Budzulyak S.I., Phys. Chem. Solid State, 2013. **1**, 222 (in Ukrainian).
17. Boichuk V.I., Bilynskyi I.V., Leshko R.Ya, Naukovy Visnyk Chernivetskogo Universitetu. Zbirnyk Naukovyh Prats. No. 420, Fizyka. Elektronika, Chernivtsi, 2008, p. 5 (in Ukrainian).
18. Boichuk V.I., Bilynskyi I.V., Leshko R.Ya., Condens. Matter Phys., 2008, **11**, 653; doi:10.5488/CMP.11.4.653.
19. Safwan S.A., Hekmat M.H, Asmaa A.S., Elmeshed N., Physica E, 2008, **41**, 150; doi:10.1016/j.physe.2008.06.016.
20. Holovatsky V.A., Makhanets O.M., Voitsekhivska O.M., Physica E, 2009, **41**, 1522; doi:10.1016/j.physe.2009.04.027.

# 14

# Iterated Prisoners Dilemma with limited attention

U. Çetin[1,2], H.O. Bingol[1]

[1] Department of Computer Engineering, Bogazici University, 34342 Bebek, Istanbul, Turkey

[2] Department of Computer Engineering, Istanbul Gelisim University, 34310 Avcilar, Istanbul, Turkey

How attention scarcity effects the outcomes of a game? We present our findings on a version of the Iterated Prisoners Dilemma (IPD) game in which players can accept or refuse to play with their partner. We study the memory size effect on determining the right partner to interact with. We investigate the conditions under which the cooperators are more likely to be advantageous than the defectors. This work demonstrates that, in order to beat defection, players do not need a full memorization of each action of all opponents. There exists a critical attention capacity threshold to beat defectors. This threshold depends not only on the ratio of the defectors in the population but also on the attention allocation strategy of the players.

**Key words:** *scarcity of attention, cooperation, memory size effect, iterated prisoners dilemma, social and economic models*

## 1. Introduction

Games and economic models are more interrelated than one can imagine [1]. This is also the case for social interactions. A simplistic virtual setting for simulating a trust in an e-commerce setting, would be the Iterated Prisoners Dilemma game which is, by its nature, very related to the evolution of a trust [2, 3]. Each transaction in an e-commerce setting can be viewed as a round in an iterated prisoner's dilemma game. Adherence to electronic contracts or providing services with good quality can be considered as cooperation while the temptation to act deceptively for immediate gain can be considered as deception.

Economy is the study of how to allocate scarce resources. According to Davenport, the scarcest resource of today is nothing but attention [4]. Attention scarcity is first stated by Herbert Simon. He says that, "What information consumes is rather obvious: it consumes the attention of its recipients" [5]. The new digital age has come with its vast amount of immediately available information that exceeds our information processing power. Thus, attention scarcity is a natural consequence of huge amount of information. Attention is very critical to any kind of interaction, especially in the era of digital technologies. Conventional Economy has been transforming itself to the Attention Economy [4, 6–8]. Games should do the same. Little work has been done on games with limited attention. How does attention scarcity effect a game? We will discuss attention games in a specific context of Iterated Prisoners Dilemma.

### 1.1. Iterated Prisoners Dilemma game

Prisoners Dilemma game is one of the commonly studied social experiments [2, 3, 9–12]. Two players should simultaneously select one of the two actions: cooperation or defection, and play accordingly with each other. Dependent on their choices, they receive different payoffs as seen in figure 1.

Payoff matrix can be described by the following simple rules. In the case of mutual cooperation, both players receive the *reward* payoff, $R$. If one cooperates, while the other defects, cooperator gets the *sucker's* payoff, $S$ while the defector gets *temptation* payoff, $T$. In the case of mutual defection, both get the *punishment* payoff $P$. Payoff matrix should satisfy the inequality $S < P < R < T$ and the additional constraint $T + S < 2R$ for repeated interactions. Rationality leads to defection, because $R < T$ and $S < P$

| | Cooperate | Defect |
|---|---|---|
| Cooperate | $R,\ R$ | $S,\ T$ |
| Defect | $T,\ S$ | $P,\ P$ |

**Figure 1.** Payoff matrix. We use $T = 5$, $R = 3$, $P = 1$, and $S = 0$.

makes defection better than cooperation. But, at the same time, $P < R$ implies that mutual cooperation is superior to mutual defection. So, rationality fails and this situation is referred to as a dilemma.

It is well known that the defection is the individually reasonable behavior that leads to a situation in which everyone is worse off [2]. On the other hand, cooperation results in the maximization of the joint outcomes [11].

If two players play prisoners dilemma more than once and they remember previous actions of their opponent and change their strategy accordingly, the game is called *Iterated Prisoners Dilemma* (*IPD*) [12]. Despite its level of abstraction, a large variety of situations starting from daily life (i.e., stop or go on when the red light is on?) to socio-economic relations (i.e., fulfill or renege on trade obligations?) may be represented as an IPD game. It is shown that repeated encounters between the same individuals foster cooperation. This is often referred to as the *shadow of the future*. If individuals are likely to interact again in the future, this allows for the return of an altruistic act [2, 10].

### 1.2. Attention in games

In general, a player is not capable of knowing all the players in an interacting environment and usually acts based on a limited information. One reason could be the huge number of players, or another could be that the players may have a very limited memory size to be informed of all the others [13, 14]. For example, in real life, a market has a few market leaders and many small brands whose number, in general, is simply too large for consumer to remember all of them. Therefore, a consumer can only have access to a limited number of service providers. The essence of any game is to interact with other players and get a chance to improve the payoff one gets. To interact with others, one should first capture their attention in a positive manner. When we give our attention to something, we always take it away from something else. We can think of having attention as owning a kind of property. This property is located in the memory of a player.

### 1.3. IPD game under limited attention

In many studies related to IPD game, it is assumed that there exists enough memory to remember all the previously encountered players and their actions. Memory is an important aspect, because knowing the identity and history of an opponent allows one to respond in an appropriate manner. We use the term *limited attention* to indicate the existence of an upper bound on how many distinct encounters are remembered by a player. We ask the following reasonable question, as in reference [14], what if the memory size is limited? The same question can be reformulated as follows: what if attention capacity is limited? In this study, we introduce attention capacity as an important parameter to investigate the dynamics of the mentioned game.

## 2. The model

Researcher Tesfatsion introduced the notions of choice and refusal into IPD games [3]. In order to choose or refuse an opponent, players should be able to remember the identity of each player and their past behaviors. It is known that the choice helps players to find cooperation while refusal lets them escape from defection [3]. In our very simplistic model, we consider that there exist two type of players: *cooperators*, who always cooperate, and *defectors*, who always defect. We combine these pure strategies with a simple choice-and-refusal rule: If a player knows that the opponent is a defector, then he or she refuses to play. Otherwise he or she plays.

Each round of the IPD game consumes a limited attention of its players. We assume that every player has the same *attention capacity* $M$. When a player encounters an opponent, he stores the necessary information related to the opponent's action in his memory. After playing with $M$ different opponent, the attention capacity fills up. As the player encounters more opponents, he will have the problem of attention scarcity. He has to forget the previously encountered ones. To use ones memory efficiently, one needs to decide whom to forget? In this respect, in section 4.1 we will discuss 5 different attention allocation strategies. Like the rest of the literature, we focus on the conditions under which "cooperative move" becomes more favorable. However, our research considers that the game takes place in a world with a limited attention.

The personality of a player (cooperator or defector) is randomly set. Remember that once the personality is set, it never changes. In each iteration, two individuals are randomly chosen to play the game. In this respect, there is no spatial pattern. One considers that the underlying interaction graph is a complete graph.

Let $C$ and $D$ denote the sets of cooperator and defector players, respectively. Let $\mathcal{N}$ denote the set of all players, that is, $\mathcal{N} = C \cup D$. The number of defectors is denoted by $|D|$. Thus, the remaining $|C| = N - |D|$ players are the cooperators, where $N = |\mathcal{N}|$. We define our model parameters *attention capacity ratio* and *defector ratio* as $\mu = M/N$ and $\delta = |D|/N$, respectively. Hence, we have $0 \leqslant \mu \leqslant 1$ and $0 \leqslant \delta \leqslant 1$.

We use the de facto payoff values of $T = 5$, $R = 3$, $P = 1$, and $S = 0$ throughout this study.

## 3. Evaluation metrics

Social welfare can be measured by the average payoff of players. The payoffs of all the encounters are added up to have the final outcome of each player. To make a comparison between the defectors and the cooperators, we take the average outcome of each. Let $c_i$ and $d_i$ be the numbers of games, where the player $i$ plays with cooperators and defectors, respectively. We use the payoff matrix given in figure 1 to calculate the total payoff of the player $i$ as follows:

$$\text{payoff}(i) = \begin{cases} Rc_i + Sd_i, & i \in C, \\ Tc_i + Pd_i, & \text{otherwise.} \end{cases}$$

We evaluated our results by a comparison between the average performances of the cooperators and the average performances of the defectors. Our performance metrics are as follows:

$$\bar{P}_C = \frac{1}{|C|} \sum_{i \in C} \text{payoff}(i) \quad \text{and} \quad \bar{P}_D = \frac{1}{|D|} \sum_{i \in D} \text{payoff}(i).$$

Although further investigations call for simulations, some analytical investigation of average performances is possible.

### 3.1. Cooperator's average performance

Cooperator's average performance of $\bar{P}_C$ can be analytically found. For a cooperator, to play with a defector means no gain, since sucker's payoff is equal to zero, that is, $S = 0$. $\bar{P}_C$ can only increase if two cooperators play a round with each other. When two cooperators are selected to play with each other, each cooperator gets $R = 3$ points. The probability of matching two cooperators is equal to $(1-\delta)^2$. Among $\mathcal{T} = \tau N^2/2$ rounds, only $(1-\delta)^2\mathcal{T}$ of them is expected to pass between two cooperators. As a result, $|C| = (1-\delta)N$ cooperators share $(R+R)(1-\delta)^2\tau N^2/2$ payoffs. In other words,

$$\bar{P}_C = \frac{2R(1-\delta)^2\tau\frac{N^2}{2}}{(1-\delta)N} = R(1-\delta)\tau N.$$

Without any further investigation, we can conclude that increasing $\tau$, $N$ and $R$ is favorable for $\bar{P}_C$ while increasing $\delta$ is not. Note that neither attention capacity $M$ nor any attention allocation strategy has effect in this setting. If the population is composed of only cooperators, that is $|C| = N$ and $\delta = 0$, $\bar{P}_C$ will be $R\tau N$.

## 3.2. Defector's average performance

Due to the choice and refusal rule, if an opponent is known to be a defector, no player plays with him. Therefore, in order to obtain the defector's average performance of $\bar{P}_D$, we need the probability of a defector $j \in D$ to be unknown by player $i \in \mathcal{N}$. This probability cannot be analytically found except for the special cases of players without memory and players with unlimited memory.

### 3.2.1. Players without memory

When players have no memory, i.e., attention capacity is zero, they are totally forgetful and remember nothing. Note that this case actually corresponds to a player playing prisoners dilemma without realizing that they are playing repeatedly. As a result, players continue to play with defectors in spite of the choice and refusal rule. The probability of matching a defector with a cooperator is equal to $2\delta(1-\delta)$ while matching the two defectors is equal to $\delta^2$. Therefore, for a special case of $\mu = 0$, we have

$$\bar{P}_D = \frac{\left[T\,2\delta(1-\delta) + 2P\,\delta^2\right]\tau\frac{N^2}{2}}{\delta N} = [T\,(1-\delta) + P\,\delta]\,\tau N = (5-4\delta)\tau N$$

for $T = 5$ and $P = 1$. We observe that increasing the number of defectors is not favorable even for defectors. Nevertheless, it is easy to verify that for $\mu = 0$, $\bar{P}_D$ is always greater than $\bar{P}_C$ which can be stated as *defection is a favorable action against the players with no memory*.

### 3.2.2. Players with unlimited memory

For a special case of $M \geqslant N$, the players are no longer forgetful and they are able to remember each opponent's last action. Due to the choice and refusal system, any defector can play at most $|C|$ rounds with cooperators and $|D|-1$ rounds with defectors. Therefore, for a sufficiently large $\tau$, we have

$$\bar{P}_D = T|C| + P(|D|-1) = (P-T)|D| + TN - P.$$

We can conclude that as we increase the number of defectors in this setting, the average payoff of the defectors again decreases.

# 4. Simulations

The dynamics of a system is further investigated by simulation while the attention capacity ratio $\mu$ and the defectors ratio $\delta$ vary. The model is simulated for every possible attention capacity values of $M$ (from 0 to $N$) and for every possible number of defectors (from 0 to $N$). We study a population of $N = 100$.

The number of iterations, $\mathcal{T}$, is another critical issue. It is set to $\mathcal{T} = \tau \times N^2/2$ since there are $\binom{N}{2}$ pairs, where $\tau$, being the third model parameter, is the number of plays for a pair of players. Note that, when $\tau = 1$, no two players are expected to meet again during the simulation. This situation corresponds to a non-iterated version of the game. In order to see the effect of time, $\tau$ is set to 2 and 5. The results were averaged over 20 independent realizations for every combination of parameter values.

## 4.1. Attention allocation strategies

Some people are positive and remember only good memories. On the contrary, some remember bad events and live to get their revenge. Motivated by these, we make a comparison of 5 simple attention allocation strategies based on forget mechanisms: (i) Players that prefer to forget only cooperators, denoted by FOC. (ii) Players that prefer to forget only defectors, denoted by FOD. (iii) When players have no preference, they can select someone, uniformly at random, to forget. We call this strategy as FAR. (iv) Players may also prefer to use coin flips to decide which type, namely, cooperators or defectors, of a player to forget. Once the type is decided, someone among this type is randomly selected and forgotten. Let FEQ denote this "equal probability" to types approach. (iv) If the knowledge of which type has the majority is available, this extra information can be used in devising a strategy. One possible effective strategy could

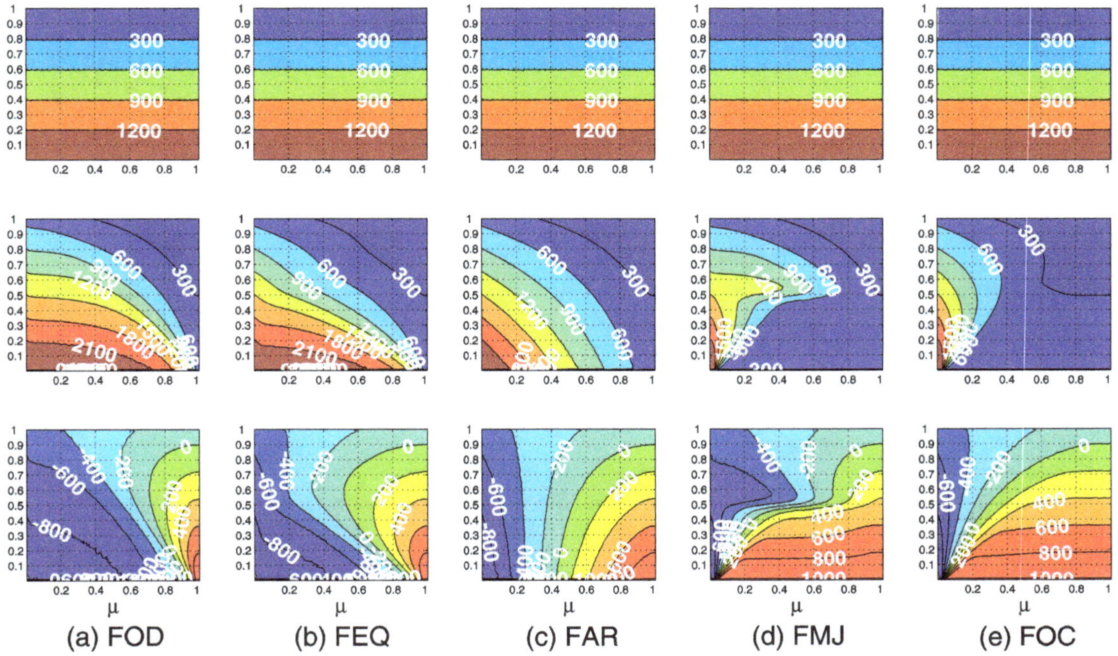


**Figure 2.** (Color online) Average performances as a function of attention capacity ratio $\mu$ and defector ratio $\delta$. The columns represent five strategies. The rows represent $\bar{P}_C$, $\bar{P}_D$ and $\bar{P}_C - \bar{P}_D$ values, respectively.

be to assume that the opponent is of the type of majority, hence, pay attention to the minorities only. That is, one prefers to forget the majority which we call FMJ strategy.

We investigate the average performances of cooperators and defectors when they use the same strategy.

## 5. Observations

In this section, for a more general view, we present our observations based on our simulation data. With our essential parameters of $\mu$, $\delta$, and $\tau$ along with the different attention allocation strategies, we can determine the conditions under which cooperation is more favorable than defection.

Simulation results for various values of attention capacity ratio $\mu$ and defector ratio $\delta$ are given in figure 2. Columns of figure 2 correspond to five strategies. Within a column, the top plot provides the average performance of cooperators, $\bar{P}_C$, as a function of $\mu$ and $\delta$. Similarly, the middle plot gives the average performance of defectors. The bottom plot is the difference of the averages. Note that being a cooperator is better when $\bar{P}_C - \bar{P}_D > 0$. For the sake of comparison, $\bar{P}_C - \bar{P}_D = 0$ curves for different attention allocation strategies are superposed in figure 3(b).

### 5.1. Average performance of cooperators

Findings from the first row of figure 2 are as follows: (i) Interestingly, cooperator's average payoff does not significantly change neither by attention capacity ratio nor by attention allocation strategy. (ii) However, the defector ratio has a negative effect on the average performances of cooperators. Our analytical explanation given in section 3.1 is in agreement with these findings. For any $\delta$ values, $\bar{P}_C = R(1-\delta)\tau N$ gives exactly the same results seen in the first row of figure 2.

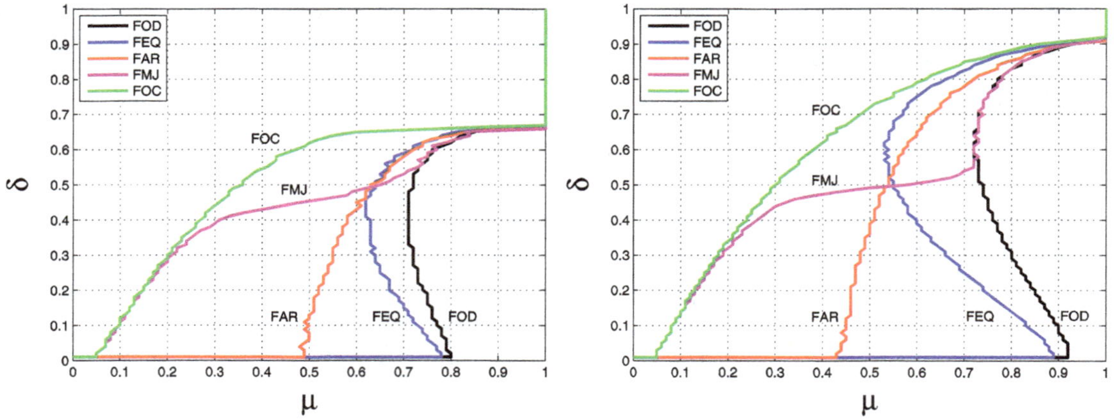


(a) $\tau = 2$

(b) $\tau = 5$

**Figure 3.** (Color online) Attention boundaries of different allocation strategies are visualized in the same figure for the sake of comparison.

## 5.2. Average performance of defectors

The second row of figure 2 can be interpreted as follows: (i) Greater attention capacity, i.e., an increase in $\mu$, helps players to remember the defectors. As a result, defectors experience social isolation and their average payoff severely diminishes. (ii) An increase in the number of defectors, i.e., an increase in $\delta$, leads a competition among them. Thus, defectors' average payoff again diminishes. (iii) Note that all five plots are in agreement with our discussion in section 3.2.1 and section 3.2.2 for special cases of $\mu = 0$ and $\mu = 1$.

## 5.3. Attention boundaries

We refer to the $\bar{P}_C - \bar{P}_D = 0$ contour lines, seen in the third row of figure 2, as the *attention boundaries.* An attention boundary determines a favorable action. If a pair of $(\mu, \delta)$ remains inside the attention boundary, it means $\bar{P}_C - \bar{P}_D > 0$ and cooperation is a favorable action, otherwise defection is a favorable action. Attention boundaries for five different attention allocation strategies seen in figure 2, are visually superposed in figure 3(b) for the sake of comparison.

For a given defector ratio, we observe that there is a critical threshold for attention capacity, below which defection is advantageous, and above which cooperation becomes a favorable action. With lesser attention capacity, defectors can be easily overlooked. Greater attention capacity along with the choice-and-refusal rule do not let defectors improve their payoffs. Due to the degrading of defector's performance, the average payoff of cooperators manages to exceed that of defectors when players have a greater attention capacity.

## 5.4. Attention allocation strategies

We consider a strategy *better* if it has a larger area, where cooperators are doing better than defectors, in the $\mu, \delta$ plain. That is, a better strategy has more $(\mu, \delta)$ pairs below its attention boundary. From this perspective, the best strategy is FOC, and the worst one is FOD. All the remaining strategies are located in between these two strategies.

The forget majority, FMJ, is a mixed strategy. When $0.5 < \delta$, defectors are the majority and FMJ acts as if they forget only the defectors. When $\delta < 0.5$, cooperators are the majority. Thus FMJ switches to forget only cooperator. Therefore, its plot is similar to that of FOD for $0 < \delta < 0.5$ and that of FOC for $0.5 < \delta < 1$. FMJ strategy can be put differently as allocation of the minority. One can think that this strategy is better than the rest, since scarcity, in general, triggers the perception of greater importance.

Nevertheless, figure 3(b) is against this intuition. The optimal strategy is to forget only the cooperators. By doing so, players manage to allocate their memories only for defectors. In other words, they keep their enemies closer. Thus, they become more prudent to the defectors. On the other hand, forgetting defectors seems to be the most wasteful and carefree attention consuming habit. We observe that the necessary information for refusing the defectors is dismissed while applying the FOD strategy.

The critical value of $\delta = 0.5$ determines which strategy is superior, except for the two extreme strategies of FOD and FOC. FEQ does better than FAR when $0.5 < \delta$ and FAR does better than FEQ when $\delta < 0.5$. Even if FAR strategy seems identical to FEQ strategy, there exists a slight difference between them. Notice that, forgetting at random depends on the content of the memory, while forgetting with equal probability does not. Higher defector's ratio, that is $0.5 < \delta$, causes one to encounter more defectors. In that case, memories of the players would be plentiful with defective experiences. Thus, forgetting at random would be more biased towards FOD. Similarly, forgetting at random would be more biased towards FOC when $\delta < 0.5$.

### 5.5. Effect of time

Literature on IPD game suggests that as the number of iterations increases, the cooperative behavior also increases among the players [2, 10]. This is also verified by our simulations. The shadow of the future can be quantified by the parameter of $\tau$. A short shadow of the future (lesser $\tau$), hinders the detection of the defectors. When the future of the shadow is longer, lesser attention capacity would be sufficient for cooperators to beat the defectors. As $\tau$ increases, defector's performance gets worse in comparison with cooperators. Attention boundaries obtained by setting $\tau = 2$ and $\tau = 5$ are given in figure 3(a) and figure 3(b), respectively. The area inside the attention boundaries is much larger in figure 3(b) than in figure 3(a). This finding suggests that the shadow of the future fosters cooperation.

## 6. Conclusions

We observe that as the proportion of the defectors increases, the average payoff for any player decreases. On the other hand, an increase in the attention capacity has different outcomes for cooperators and defectors. As attention capacity increases, the change in the cooperators overall performance is almost negligible, but the defectors' performance significantly diminishes. The rule of choice-and-refusal plays an important role in this situation. Nevertheless, it is worth pointing out that even the choice-and-refusal alone cannot fulfill the desired goal without passing some threshold value of attention capacity. As the attention capacity increases, or the shadow of the future gets longer, the detection of the defectors gets feasible, consequently the defectors face a social isolation due to the rule of choice-and-refusal. As a result, the cooperators' performance exceeds the defectors' performance. Thus, cooperation becomes a favorable action. This work demonstrates that in order to beat a defection, players do not need a full memorization of each action of all opponents. This finding is really important especially in the world of a limited attention. We also investigate five different attention allocation strategies and we find out that the best strategy is "forgetting only the cooperators". By applying this strategy, one becomes more prudent to the deceptive actions. In conclusion, attention should be selective, and it should be directed towards the defectors and towards their defective moves.

In the present work, players are pure cooperators or pure defectors. They never change their character. Various forgetting strategies are investigated but both cooperators and defectors use the same strategy in a game. The situation of cooperators using one strategy and defectors another is left for the future work. It would be also interesting to study the effect of a biased payoff matrix. As a future work, we plan to investigate other means for fostering cooperation, even in the conditions of attention scarcity. To achieve this goal, we can make use of other experiences by taking recommendations to determine with whom to play. But from whom to take advice is very critical and must be well studied to clarify which collaboration strategy is better. We will also extend our work to the mixed strategies for interaction, such as "mostly defect" and "mostly cooperate".

## References

1. Kreps D., Game Theory and Economic Modelling, Clarendon Press, Oxford, 1990.
2. Axelrod R., The Evolution of Cooperation, Basic Books, New York, 2006.
3. Tesfatsion L., In: Computational Approaches to Economic Problems, Vol. 6, Amman H., Rustem B., Whinston A. (Eds.), Kluwer Academic Publishers, Dordrecht, 1997, 249–269; doi:10.1007/978-1-4757-2644-2_17.
4. Davenport T.H., Beck J.C., The Attention Economy: Understanding the New Currency of Business, Harvard Business School Press, Boston, 2001; doi:10.1103/PhysRevE.77.036118.
5. Simon H.A., Designing Organizations for an Information-Rich World, John Hopkins University Press, Baltimore, 1971.
6. Falkinger J., J. Econ. Theory, 2007, **133**, 266; doi:10.1016/j.jet.2005.12.001.
7. Falkinger J., Econ. J., 2008, **118**, 1596; doi:10.1111/j.1468-0297.2008.02182.x.
8. Goldhaber M.H., First Monday, **2**, No. 4, 1997; doi:10.5210/fm.v2i4.519.
9. Axelrod R., Hamilton W., Science, 1981, **211**, 1390; doi:10.1126/science.7466396.
10. Axelrod R., The Complexity of Cooperation: Agent-Based Models of Competition and Collaboration, Princeton University Press, Princeton, 1997.
11. Kollock P., Annu. Rev. Sociol., 1998, **24**, 183; doi:10.1146/annurev.soc.24.1.183.
12. Rapoport A., Prisoner's Dilemma: A Study in Conflict and Cooperation, University of Michigan Press, Ann Arbor, 1965.
13. Bingol H., In: Computer and Information Sciences — ISCIS 2005, Lecture Notes in Computer Science Series, Vol. 3733, Yolum P., Güngör E., Gürgen F., Özturan C. (Eds.), Springer, Berlin, 2005, 294–303; doi:10.1007/11569596_32.
14. Bingol H., Phys. Rev. E, 2008; 77, 036118; doi:10.1103/PhysRevE.77.036118.

# 15

# Lamplighter model of a random copolymer adsorption on a line

L.I. Nazarov[1], S.K. Nechaev[2,3,4], M.V. Tamm[1,4]

[1] Physics Department, M.V. Lomonosov Moscow State University, 119991 Moscow, Russia

[2] Université Paris-Sud/CNRS, LPTMS, UMR8626, Bât. 100, 91405 Orsay, France

[3] P.N. Lebedev Physical Institute, RAS, 119991 Moscow, Russia

[4] Department of Applied Mathematics, National Research University "Higher School of Economics", 101000 Moscow, Russia

We present a model of an AB-diblock random copolymer sequential self-packaging with local quenched interactions on a one-dimensional infinite sticky substrate. It is assumed that the A-A and B-B contacts are favorable, while A-B are not. The position of a newly added monomer is selected in view of the local contact energy minimization. The model demonstrates a self-organization behavior with the nontrivial dependence of the total energy, $E$ (the number of unfavorable contacts), on the number of chain monomers, $N$: $E \sim N^{3/4}$ for quenched random equally probable distribution of A- and B-monomers along the chain. The model is treated by mapping it onto the "lamplighter" random walk and the diffusion-controlled chemical reaction of $X + X \to 0$ type with the subdiffusive motion of reagents.

**Key words:** *local optimization, heteropolymer folding, lamplighter random walk, subdiffusive chemical reaction*

## 1. Introduction

In this letter we propose a simple one-dimensional model of stochastic dynamic system possessing a local optimization in quenched environment. The question of choosing an optimal strategy for a dynamic system if only *local* optimization is accessible, is a generic problem far beyond the scope of natural science. Let us imagine that someone knows nothing about the future and adapts his or her own behavior in each current time moment only on the basis of the knowledge about the best solution at a given narrow time slice. Such a behavior leads to an optimal local strategy when only a partial (current) knowledge is accessible, though it might be far from generic optimal one if the knowledge about the future is available.

As an example of physical problems sharing the properties of local and global optimization, we can mention the problems of protein secondary structure formation and DNA packaging in a viral capsid. If in the course of protein folding, the chain forms a "frozen" network of contacts, then the optimal conformation is determined only by a sequential step-by-step optimization of the added heteropolymer fragments. However, if the whole protein is capable of "adjusting" the structure of a contact network by breaking some bonds and creating other bonds, then the global optimization might essentially change the folding picture.

The system considered below should not be regarded as a model of any specific physical system (though it has some features of a protein folding), but it rather highlights the principles of local optimization of a specific one-dimensional stochastic system in quenched random environment, which could lead to a nontrivial self-organization.

Our model resembles the "lamplighter random walk" proposed in 1973 by A.M. Vershik in the afterword to the Russian edition [1] of the book "Invariant means on topological groups and their applications" by F.P. Greenleaf as an example of nontrivial estimate for the growth of various numerical characteristics of groups. The object known as a "lamplighter group" (the name was also given by A.Vershik) became very popular among probabilists after the pioneering work by A. Vershik and V. Kaimanovich [2].

## 2. The model

The toy model presented below imitates the packaging of a random diblock copolymer with quenched sequence of monomers on a one-dimensional substrate. Schematically it can be viewed as a tissue folding getting out of the roll, as depicted in figure 1 (a), where the size of folds depends on local heterogenous interactions between different parts of the tissue. Such a schematic view might be compared with the specific secondary structure formation of a linear polypeptide chain in figure 1 (b), where two main types of secondary structure, $\alpha$-helices and $\beta$-sheets, are shown and the secondary structure is stabilized by the hydrogen bonds.

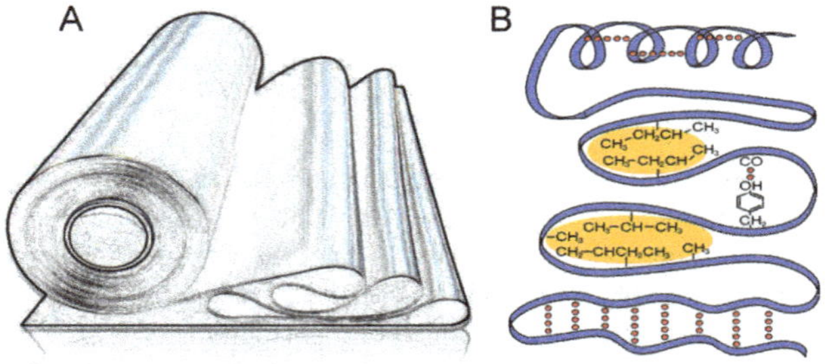


**Figure 1.** (Color online) (a) Folds of tissue getting out of the roll; (b) Secondary structure formation in a polypeptide heteropolymer chain.

To pass from these intuitive pictures to a more specific description, consider a $N$-monomer chain with quenched random primary sequence of two types of links, A and B, placed on an infinite line, which is sticky for links A and repulsive for links B. We suppose that the energy of the A-A or B-B contacts is 0, while the energy of the A-B contact is $+1$ (in dimensionless units), i.e., the monomers of the same types are indifferent to each other, while the monomers of different types repulse.

The heteropolymer folding (packaging) on an infinite line is an irreversible sequential process of adding monomers to the already existing frozen environment. Let us recursively describe the folding process. For definiteness, suppose that the line is initially uniformly covered with monomers of type A and let the first chain monomer be always of type A. Put the 1st monomer at a position $x_0 = 0$ on the line. The 2nd monomer could be placed either next to the 1st monomer on a line in the *trans* state, or it can be put on top of the 1st monomer in the *gauche* state, making a "hook" as it is shown in figure 2. The newly added monomer interacts with the one located below it in the projection to the line.

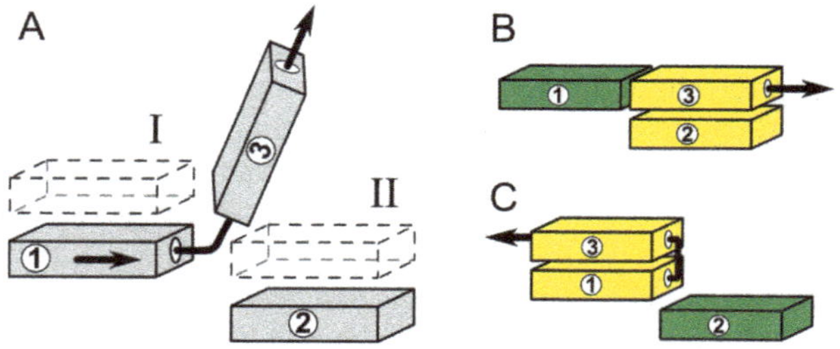


**Figure 2.** (Color online) Local rules of the monomers freezing to the structure. (a) The monomer 3 has two possible locations I and II. The selection of the position depends on the types of all three monomers. (b) Monomers 2 and 3 are identical, while the monomer 1 differs from 2 and 3. (c) The inverse situation with respect to (b).

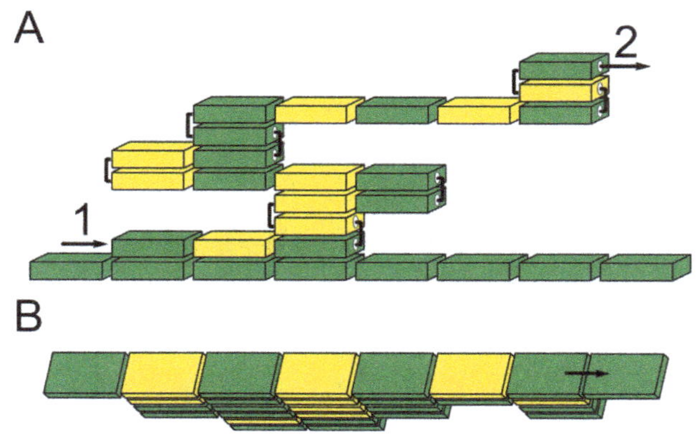


**Figure 3.** (Color online) The 20-step of block copolymer package on the plate with the uniform monomer type distribution [(a) The side view; (b) The top view].

The selection between *trans-* and *gauche-* states is made according to the following rules: i) if one of two possible new (*trans-* or *gauche-*) states is favorable and the other is not, then the favorable state is always selected; ii) if new *trans-* and *gauche-* states are both favorable or both unfavorable, then a new conformation (*trans-* or *gauche-*) is chosen randomly with the probability $1/2$. For any "defect" (i.e., unfavorable pairing) the energy penalty $+1$ is added. The sample structure of the first 20 steps is depicted in figure 3.

The position of a newly added monomer is selected in view of the irreversible local contact energy minimization, while for equal energies, the choice of *trans-* or *gauche-* states is random with equal probabilities. Let us emphasize that only the uppermost monomers or the "roof" of the structure [i.e., the ones visible from the top of the picture in figure 3 (b)] take part in the play: the monomers that are underneath do not participate in a further optimization process. Therefore, the folding is an effective Markovian process, whose states are specified by i) the set of monomers visible currently from the top of the structure (i.e., by the "roof"), and ii) the position and direction of the last step, which defines the positions where the next monomer can be added.

## 3. Results

In this section we consider the properties of the described locally optimal folding averaged over a set of completely random initial primary sequences of a polymer, i.e., we assume A and B monomers to appear in the chain with equal probabilities without any correlations between sequential letters. The most natural question to ask about this folding is how the average total energy penalty, $\langle E\rangle$, depends on the chain length, $N$, i.e., how many unfavorable A-B contacts one has on average in the $N$-monomer chain. Unexpectedly, it turns out that $\langle E\rangle$ grows sub-linearly with $N$, namely as $\langle E\rangle \sim N^{3/4}$ [see figure 5 (center)].

To understand this result, one should notice that in terms of the evolution of the roof of the structure defined above, the system has an adsorbing state. Indeed, if the roof is represented by an alternating sequence, i.e., ...-A-B-A-B-..., every newly added monomer can be positioned on the top of this periodic sequence *without any energetic penalty*, and the resulting structure of the roof will stay unchanged. Moreover, if some part of the roof is alternating [as in figure 3 (b)], this part will remain alternating and any sequence can be adsorbed on it without penalty. Thus, the heteropolymer folding in our model can be considered as a dynamic process of approaching the adsorbing alternating state. This dynamics consists in sequential cancellation of "defects", represented by non-alternating pairs of A-A or B-B of consecutive monomers on the substrate as it is shown in figure 4 (a).

To formalize this idea, we introduce an Ising variable, $s(x)$, which labels the state of the roof in position $x$:

$$s(x)=\begin{cases} +1 & \text{for monomer A in position } x, \\ -1 & \text{for monomer B in position } x. \end{cases} \tag{3.1}$$

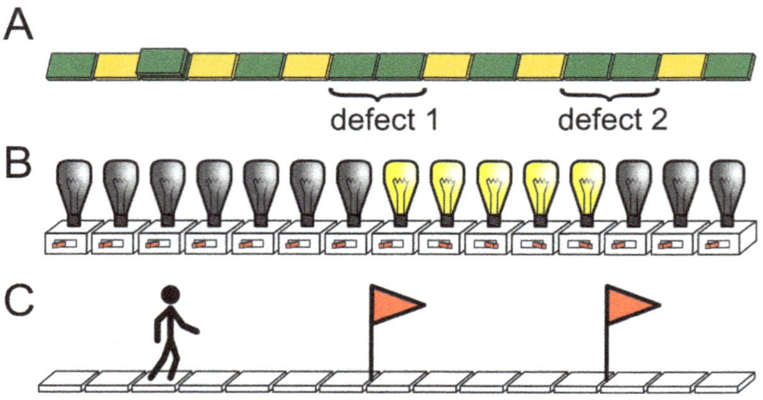


**Figure 4.** (Color online) The representation of two arbitrary packages in a context of 1D walks on an adjustable substrate. Periodic regions, shown in (a) are mapped (depending on the parity) to the domains of lamps in "off" or "on" positions, shown in (b). The white segments, shown in (c), correspond to the domains of lamps, while red flags mark the domain walls. The human figure designates the tip of the polymer chain, which performs a random walk and can push flags to the left or to the right when passing by them.

Now, we define another Ising variable, $\sigma(x)$:

$$\sigma(x) = (-1)^x s(x). \tag{3.2}$$

The change of variables form $s(x)$ to $\sigma(x)$ essentially simplifies the picture, because in terms of the variable, $\sigma$, the adsorbing state is homogenous. The state of the chain in terms of variables $\sigma$ is depicted in figure 4 (b) by lamps switched on (if $\sigma = 1$) and off (if $\sigma = -1$). Each defect (i.e., the pair of segments A-A or B-B) is a "domain wall" separating two states: to the left of it, the lamps are off, while to the right of it, the lamps are on. A configuration with two defects is depicted in figure 4 (b). The defects (domain walls) are shown in figure 4 (c) by flags. In terms of variables $\sigma$, the uniform adsorbing state corresponds to the periodic adsorbing state of our initial Ising model, and the dynamics of folding in terms of $\sigma$ means coarsening of the Ising domains.

We can easily formulate the dynamic rules describing the evolution of the system. The tip of the original polymer chain, shown as a human figure ("lamplighter") in figure 4 (c) performs a walk on a one-dimensional lattice. In the absence of any energy gain, the *trans*– and *gauche*–states are equiprobable. Since the concentrations of monomers of types A and B in the chain are the same, the lamplighter performs a symmetric random walk on a one-dimensional discrete line as it is shown in figure 4 (c).

The initial state for the spins $\sigma$ is an infinite one-dimensional lattice with all domains of length 1. Using the notations of figure 4 (c), the flags are located at each boundary between adjacent monomers, while a single random walker (lamplighter) is initially positioned at $x = 0$. As the time (measured in the number of heteropolymer monomers, $N$) runs, the lamplighter performs a discrete symmetric random walk with the following properties: i) if the lamplighter stays within a domain of lamps which all are "on" or "off", he does nothing with the lamps, ii) if the lamplighter crosses the domain wall, marked by a flag, he may (with the probability 1/2) randomly shift the flag on one lattice site to the left or to the right. If two flags meet each other at one site, they annihilate. The energy of the folding configuration is equal to the number of flag shifts. As the flags annihilate and their concentration decreases, the energy of the folding grows slower.

To consider the dynamics more quantitatively, we have first checked our conjectures computing numerically the span, $\langle R^2(N)\rangle$, of the lamplighter's $N$-step walk. As shown in figure 5 (left), we have got

$$\langle R^2(N)\rangle \sim N \tag{3.3}$$

as expected for the standard Brownian random walk. In terms of the original problem, this means that the $N$-step random block-copolymer has a typical span of size of order $\sqrt{N}$ on the 1D substrate.

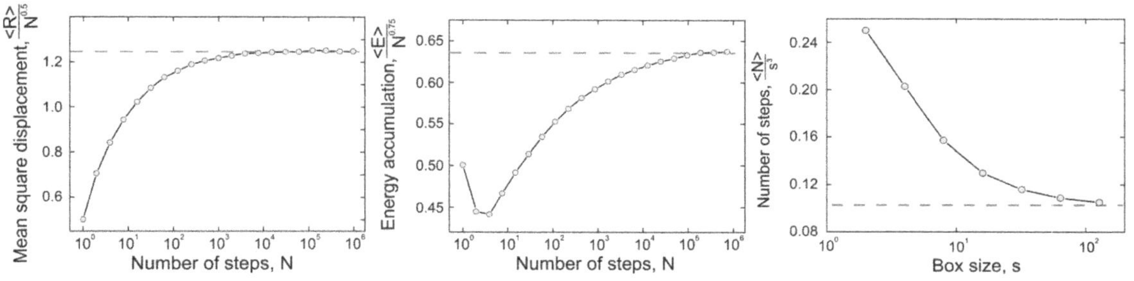


**Figure 5.** (Color online) (left): The average normalized span, $\langle R^2(N)\rangle$, of the folded trajectory on an infinite substrate, as a function of the number of steps, $N$; (center): The normalized cumulative folding energy, $E(N)/N^{3/4}$, as a function of the number of steps, $N$; (right): The typical normalized demixing time, $\tau/s^3$, of the folded trajectory as a function of the bounding box size, $s$.

Now, let us pay attention to the dynamics of a single flag. We have already mentioned that the lamplighter passing across the flag, moves it on one lattice site to the left or to the right with equal probabilities. This means that the flag itself performs a random walk with the mean-square displacement $\langle \Delta x^2(s)\rangle \sim m$, where $m$ is the number of jumps the flag has made (i.e., the number of times the lamplighter returned to the flag). Since the number of returns to a given point (the position of a flag), $m$, scales as $m \sim t^{1/2}$ for a one-dimensional random walk at large $t$, we immediately conclude that the mean-square displacement of the flag obeys the following scaling dependence

$$\langle \Delta x^2(t)\rangle \sim m \sim t^{1/2}. \tag{3.4}$$

The waiting times between sequential visits of a given flag by the lamplighter are distributed as the first-return times of a one-dimensional random walk, with a tail $t^{-3/2}$ at long times. This makes the movement of a single flag exactly the continuous-time random walk (CTRW) with the exponent 1/2 [3]. Note that since the movement of different flags is governed by a *single* lamplighter, they are correlated. However, in the first approximation one expects that it is possible to neglect these correlations and assume that the mean square distance, $d = \sqrt{\langle (x_{i+1} - x_i)^2\rangle}$ between two neighboring flags $i$ and $i+1$ located at the points $x_i$ and $x_{i+1}$, behaves as in (3.4), i.e. $d^2(t) \sim t^{1/2}$.

The process of annihilation of flags is described by the equation of a one-dimensional chemical kinetics $X + X \to 0$, where the reagent $X$ designates a flag, and $X$ experiences a sub-diffusive random walk at time $t$ with the mean-square displacement $\langle \Delta x^2(t)\rangle \sim t^{1/2}$. Such a process has been widely studied in chemical kinetics — see, for example, [4].

Let $c(t)$ be the typical concentration of $X$ at time $t$ within a region visited by the lamplighter. We are interested in the survival probability of $X$ in time, i.e., in the scaling behavior $c(t) \sim t^{-\alpha}$ at $t \to \infty$. Each time the two consecutive flags touch each other, they disappear. Since the flags cannot pass through each other, the typical distance between the reagents $X$ (flags) at time $t$ is of the order of the explored territory up to the time $t$. Therefore, $c(t) \sim d^{-1}(t)$ or

$$c(t) \sim t^{-1/4} \tag{3.5}$$

giving $\alpha = 1/4$.

The energy of the folded chain increases by one with the probability 1/2 every time the lamplighter visits a flagged site. There are two typical situations when it happens: a) when the lamplighter reaches the boundary of the domain visited (recall that outside the domain visited all sites are flagged), and b) when it encounters a flag in the bulk of the already visited region. It is easy to estimate the frequencies of both these events. Indeed, according to (3.3), the size of the domain visited grows as $t^{1/2}$, and the resulting energy increment scales as

$$\left.\frac{\mathrm{d}E(t)}{\mathrm{d}t}\right|_{\text{boundary}} \sim \frac{\mathrm{d}t^{1/2}}{\mathrm{d}t} \sim t^{-1/2}. \tag{3.6}$$

On the other hand, the probability for a random walk to meet a flag at time $t$ in the domain's bulk, scales as $c(t)$ and, therefore,

$$\left.\frac{\mathrm{d}E(t)}{\mathrm{d}t}\right|_{\text{bulk}} \sim d^{-1}(t) \sim t^{-1/4}. \tag{3.7}$$

We see that, at least on large timescales (i.e., for long chains), the second mechanism dominates, and the energy is mostly accumulated via interaction of the lamplighter with the flags in the bulk. Integrating (3.7) up to the total chain length, $N$, we end up with the scaling behavior of the average energy accumulation, $E(N)$ up to the length $N$ ($N \gg 1$):

$$\langle E(N)\rangle \sim \int_0^N t^{-1/4}\,\mathrm{d}t \sim N^{3/4}. \tag{3.8}$$

This scaling behavior is clearly seen in the numeric simulations shown in figure 5 (center), where we have plotted the normalized cumulative energy, $E(N)/N^{3/4}$, as a function of $N$. The results are obtained by averaging over $10^6$ random block-copolymer packaging attempts on the infinite line.

Apart from the chain adsorption on an *infinite* one-dimensional substrate, it seems instructive to consider the case when the total size of a substrate is *finite*, e.g., the chain is trapped in a cage with periodic boundary conditions. In this case (for even total number of sites, $L$, in the cage), a long enough chain, once again, finally forms a state with completely regular ... A-B-A-B ... roof, which means, in terms of the Ising domain representation, that a single Ising domain is formed. The natural questions to ask are as follows: i) how long (on average) does it take for a perfect domain structure to be formed in the cage of $L$ sites (i.e., which is the typical "demixing time", $\tau$, as a function of $L$), and ii) what is the typical total energy, $\langle E(L)\rangle$ (the number of defects in the structure), as a function of the cage size, $L$. It turns out that the question ii) is easier to answer. Indeed, the energy $E$ is the total number of the flag moves, regardless of the waiting times between them. Since the moves of the flags in both directions are equiprobable, the typical number of the moves needed for two adjacent flags to annihilate, scales as $d^2$, where $d$ is the initial distance between them. Therefore, one expects the total accumulated energy to scale as

$$\langle E(L)\rangle \sim \sum_{n=0}^{N} d^2(n), \tag{3.9}$$

where $n$ ($1 \leqslant n \leqslant L$) is the number of surviving flags at a given time (all numeric constants are omitted). Since the typical distance between the flags is $d(n) \sim L/n$, we can evaluate the sum in (3.9), getting finally

$$\langle E(L)\rangle \sim L^2 \tag{3.10}$$

for large $L$, which is perfectly confirmed by numeric simulations.

For the demixing time, we have not yet obtained the analytic scaling estimate, but according to the numeric simulations shown in figure 5 (right), the demixing time $\tau(L)$ scales as

$$\tau(L) \sim L^3. \tag{3.11}$$

Combining (3.10) and (3.11), we get a nontrivial prediction $\langle E(\tau)\rangle \sim \tau^{2/3}$, different from the result of the mean-field theory (3.8) in the infinite domain ($\tau$ plays the role of the chain length). Therefore, it turns out that in a bounded domain, the shifts of sequential flags are highly correlated since they are governed by a single lamplighter which cannot escape the cage. Indeed, when only several flags are left in a cage, the correlations between their shifts become very important. A more detailed consideration of this phenomenon will be provided elsewhere.

## 4. Discussion

We have already mentioned that our model is reminiscent to the so-called "lamplighter random walk", widely considered in the mathematical literature (see, for example, [1, 2, 5–11]). Here is the definition of the *lamplighter graph*, taken from [11]: "...Think of a (typically infinite) connected graph $X$, where in each vertex there is a lamp that may be switched off (state 0), or switched on with $q-1$ different intensities (states $1,\ldots,q-1$). Initially, all lamps are turned off, and a lamplighter starts at some vertex of $X$ and walks around. When he visits a vertex, he may switch the lamp sitting there into one of its $q$ different states (including "off"). Our information consists of the position $x \in X$ of the lamplighter and of the finitely supported configuration $\eta : X \to \mathbb{Z}_q = \{0, \ldots, q-1\}$ of the lamps that are switched on, including their respective intensities. The set $\mathbb{Z}_q \wr X$ of all such pairs $(\eta, x)$ can be in several ways equipped with a naturally connected graph structure, giving rise to a *lamplighter graph...*"

The key difference between the lamplighter model considered in the mathematical literature (math-LL) and the model analyzed in our work, consists in the following: in math-LL all lamps (governed by a single random walk) are immobile, are located at every lattice site, and do not interact with each other, while in our model, the flags (also governed by a single random walk) can perform random local shifts and interact (i.e., annihilate while in a contact). Despite the mentioned difference, we believe that some important questions, such as the computation of the "mixing time" [10] on lamplighter graphs, could be addressed in our model as well.

It seems interesting to compare the results of the local step-by-step optimization discussed at length in this work with the global optimization of the chain configuration. Namely, assume that after the described packaging, the frozen configuration of folds is annealed, i.e., the existing folds are allowed to retract and the new folds can be formed (for unchanged sequence of monomers). Clearly, the resulting ground state energy will be lower than the one obtained in our local optimization process. The partition function of such a system having an annealed secondary structure, apparently admits the recursive representation similar to the dynamic programming algorithm for RNA secondary structure [12]. In the forthcoming works we plan to provide a detailed analysis of the global optimization and its comparison with the results of local optimization for the lamplighter random walk.

## Aknowledgements

The authors are grateful to V. Avetisov, P. Krapivsky and R. Metzler for encouraging discussions. This work was partially supported by the grants ANR-2011-BS04-013-01 WALKMAT and the IRSES project FP7-PEOPLE-2010-IRSES 269139 DCP-PhysBio. M.T. and S.N. acknowledge the financial support of the Higher School of Economics program for Basic Research.

## References

1. Vershik A.M., In: Invariant Means on Topological Groups and Their Applications Greenleaf F.P., afterword to Russian edn., Mir, Moscow, 1973 (in Russian).
2. Kaimanovich V.A., Vershik A.M., Ann. Probab., 1983, **11**, 457–490; doi:10.1214/aop/1176993497.
3. Metzler R., Klafter J., Phys. Rep., 2000, **339**, 1–77; doi:10.1016/S0370-1573(00)00070-3.
4. Yuste S.B., Lindenberg K., Chem. Phys., 2002, **284**, 169–180; doi:10.1016/S0301-0104(02)00546-3.
5. Lyons R., Pemantle R., Peres Y., Ann. Probab., 1996, **24**, 1993–2006; doi:10.1214/aop/1041903214.
6. Grigorchuk R.I., Zuk A., Geometriae Dedicata, 2001, **87**, 209–244; doi:10.1023/A:1012061801279.
7. Kaimanovich V.A., Woess W., Ann. Probab., 2002, **30**, 323–363; doi:10.1214/aop/1020107770.
8. Revelle D., Ann. Probab., 2003, **31**, 1917–2300; doi:10.1214/aop/1068646371.
9. Revelle D., Electr. Commun. Probab., 2003, **8**, 142–154; doi:10.1214/ECP.v8-1092.
10. Peres Y., Revelle D., Electron. J. Probab., 2004, **9**, 825–845; doi:10.1214/EJP.v9-198.
11. Woess W., Comb. Probab. Comput., 2005, **14**, 415–433; doi:10.1017/S0963548304006443.
12. Nechaev S.K., Tamm M.V., Valba O.V., J. Phys. A: Math. Theor., 2011, **44**, 195001; doi:10.1088/1751-8113/44/19/195001.

# 16

# $N$-point free energy distribution function in one dimensional random directed polymers

V. Dotsenko[1,2]

[1] LPTMC, Université Paris VI, 75252 Paris, France

[2] L.D. Landau Institute for Theoretical Physics, 119334 Moscow, Russia

Explicit expression for the $N$-point free energy distribution function in one dimensional directed polymers in a random potential is derived in terms of the Bethe ansatz replica technique. The obtained result is equivalent to the one derived earlier by Prolhac and Spohn [J. Stat. Mech., 2011, P03020].

**Key words:** *directed polymers, random potential, replicas, fluctuations, distribution function*

## 1. Introduction

In this paper we consider the model of one-dimensional directed polymers in a quenched random potential. This model is defined in terms of an elastic string $\phi(\tau)$ directed along the $\tau$-axis within an interval $[0,t]$ which passes through a random medium described by a random potential $V(\phi,\tau)$. The energy of a given polymer's trajectory $\phi(\tau)$ is

$$H[\phi(\tau),V] = \int_0^t d\tau \left\{ \frac{1}{2}\left[\partial_\tau \phi(\tau)\right]^2 + V[\phi(\tau),\tau] \right\}, \tag{1}$$

where the disorder potential $V[\phi,\tau]$ is described by the Gaussian distribution with a zero mean $\overline{V(\phi,\tau)} = 0$ and the $\delta$-correlations: $\overline{V(\phi,\tau)V(\phi',\tau')} = u\delta(\tau-\tau')\delta(\phi-\phi')$ The parameter $u$ describes the strength of the disorder.

The system of this type as well as the equivalent problem of the KPZ-equation [1] describing the growth of an interface with time in the presence of noise have been the subject of intense investigations for about the last three decades (see e.g. [2–13]). Such a system exhibits numerous non-trivial features due to the interplay between elasticity and disorder. In particular, in the limit $t \to \infty$, the polymer mean squared displacement exhibits a universal scaling form $\overline{\langle \phi^2 \rangle} \propto t^{4/3}$ (where $\langle \dots \rangle$ and $\overline{(\dots)}$ denote the thermal and the disorder averages) while the typical value of the free energy fluctuations scales as $t^{1/3}$. Note that in the corresponding pure system (with $V(\phi,\tau) \equiv 0$) $\langle \phi^2 \rangle \propto t$ while the free energy is proportional to $\ln(t)$.

A few years ago, an exact solution for the free energy probability distribution function (PDF) has been found [14–27]. It was shown that depending on the boundary conditions, this PDF is given by the Tracy-Widom (TW) distribution [28] either of the Gaussian Unitary Ensemble (GUE) or of the Gaussian Orthogonal Ensemble (GOE) or of the Gaussian Simplectic Ensemble (GSE). Besides, recently the two-point free energy distribution function which describes the joint statistics of the free energies of the directed polymers coming to two different endpoints has been derived in [29–31].

For fixed boundary conditions, $\phi(0) = 0$; $\phi(t) = x$, the partition function of the model (1) is

$$Z_t(x) = \int_{\phi(0)=0}^{\phi(t)=x} \mathcal{D}\phi(\tau)\, e^{-\beta H[\phi]} = \exp\left[-\beta F_t(x)\right], \tag{2}$$

where $\beta$ is the inverse temperature and $F_t(x)$ is the free energy. In the limit $t \to \infty$, the free energy scales as

$$\beta F_t(x) = \beta f_0 t + \beta x^2/2t + \lambda_t f(x), \tag{3}$$

where $f_0$ is the selfaveraging free energy density and

$$\lambda_t = \frac{1}{2}\left(\beta^5 u^2 t\right)^{1/3} \propto t^{1/3}. \tag{4}$$

It is the statistics of rescaled free energy fluctuations $f(x)$ which in the limit $t \to \infty$ is expected to be described by a non-trivial universal distribution $W(f)$. In fact, the first two trivial terms of this free energy can be easily eliminated by simple redefinition of the partition function:

$$Z_t(x) \to \exp\left\{-\beta f_0 t - \beta x^2/2t\right\} \tilde{Z}_t(x) \tag{5}$$

so that

$$\tilde{Z}_t(x) = \exp\left\{-\lambda_t f(x)\right\}. \tag{6}$$

The aim of the present work is to study the $N$-point free energy probability distribution function

$$W(f_1,\dots,f_N;x_1,\dots,x_N) \equiv W(\mathbf{f};\mathbf{x}) = \lim_{t\to\infty} \text{Prob}\left[f(x_1) > f_1,\dots,f(x_N) > f_N\right], \tag{7}$$

which describes the joint statistics of the free energies of $N$ directed polymers coming to $N$ different endpoints. Some time ago the result for this function has been derived in terms of the Bethe ansatz replica technique under a particular decoupling assumption [32]. Here, I am going to recompute this function using somewhat different computational tricks which do not require any supplementary assumptions and which permit to represent the final result in somewhat more explicit form.

## 2. $N$-point distribution function

The probability distribution function, equation (7) can be defined as follows:

$$W(\mathbf{f};\mathbf{x}) = \lim_{\lambda\to\infty} \sum_{L_1,\dots,L_N=0}^{\infty} \prod_{k=1}^{N} \left[\frac{(-1)^{L_k}}{L_k!} \exp\left(\lambda L_k f_k\right)\right] \overline{\left(\prod_{k=1}^{N} \tilde{Z}_t(x_k)\right)}, \tag{8}$$

where $\overline{(\dots)}$ denotes the average over random potentials. Indeed, substituting here equation (6) we get

$$W(\mathbf{f};\mathbf{x}) = \lim_{\lambda\to\infty} \overline{\left(\prod_{k=1}^{N} \exp\left\{-\exp\left[\lambda_t\left(f_k - f(x_k)\right)\right]\right\}\right)} = \overline{\left[\prod_{k=1}^{N} \theta\left(f(x_k) - f_k\right)\right]} \tag{9}$$

which coincides with the definition (7).

Performing the standard averaging over random potentials in equation (8) one obtains (for details see e.g. [20])

$$W(\mathbf{f};\mathbf{x}) = \lim_{\lambda\to\infty} \sum_{L_1,\dots,L_N=0}^{\infty} \prod_{k=1}^{N} \left[\frac{(-1)^{L_k}}{L_k!} \exp\left(\lambda L_k f_k\right)\right] \Psi\big(\underbrace{x_1,\dots,x_1}_{L_1}, \underbrace{x_2,\dots,x_2}_{L_2}, \dots, \underbrace{x_N,\dots,x_N}_{L_N}; t\big), \tag{10}$$

where the time dependent $n$-point wave function $\Psi(x_1,\dots,x_n;t)$ $(n = \sum_{k=1}^{N} L_k)$ is the solution of the imaginary time Schrödinger equation

$$\beta\partial_t \Psi(\mathbf{x};t) = \left[\frac{1}{2}\sum_{a=1}^{n} \partial_{x_a}^2 + \frac{1}{2}\kappa \sum_{a\neq b}^{n} \delta(x_a - x_b)\right] \Psi(\mathbf{x};t) \tag{11}$$

with $\kappa = \beta^3 u$ and the initial condition

$$\Psi(\mathbf{x}; t=0) = \prod_{a=1}^{n} \delta(x_a). \tag{12}$$

A generic eigenstate of such a system is characterized by $n$ momenta $\{Q_a\}$ $(a = 1,\dots,n)$ which split into $M$ $(1 \leqslant M \leqslant n)$ clusters described by continuous real momenta $q_\alpha$ $(\alpha = 1,\dots,M)$ and having $n_\alpha$ discrete imaginary parts

$$Q_a \equiv q_r^\alpha = q_\alpha - \frac{\mathrm{i}\kappa}{2}(n_\alpha + 1 - 2r), \qquad (r = 1,\dots,n_\alpha), \tag{13}$$

with the global constraint

$$\sum_{\alpha=1}^{M} n_\alpha = n. \tag{14}$$

The time dependent solution $\Psi(\mathbf{x}, t)$ of the Schrödinger equation (11) with the initial conditions, equation (12), can be represented in the form of a linear combination of eigenfunctions $\Psi_{\mathbf{Q}}^{(M)}(\mathbf{x})$:

$$\Psi(\mathbf{x}; t) = \sum_{M=1}^{N} \frac{1}{M!} \prod_{\alpha=1}^{M} \left[ \int_{-\infty}^{+\infty} \frac{\mathrm{d}q_\alpha}{2\pi} \sum_{n_\alpha=1}^{\infty} \right] \boldsymbol{\delta}\left( \sum_{\alpha=1}^{M} n_\alpha, n \right) \frac{\kappa^N |C_M(\mathbf{q}, \mathbf{n})|^2}{N! \prod_{\alpha=1}^{M} (\kappa n_\alpha)} \Psi_{\mathbf{Q}}^{(M)}(\mathbf{x}) \Psi_{\mathbf{Q}}^{(M)*}(\mathbf{0}) \exp\{-E_M(\mathbf{q}, \mathbf{n}) t\}. \tag{15}$$

Here, $\boldsymbol{\delta}(k, m)$ is the Kronecker symbol, the normalization factor

$$|C_M(\mathbf{q}, \mathbf{n})|^2 = \prod_{\alpha<\beta}^{M} \frac{\left| q_\alpha - q_\beta - \frac{\mathrm{i}\kappa}{2}(n_\alpha - n_\beta) \right|^2}{\left| q_\alpha - q_\beta - \frac{\mathrm{i}\kappa}{2}(n_\alpha + n_\beta) \right|^2} \tag{16}$$

and the eigenvalues:

$$E_M(\mathbf{q}, \mathbf{n}) = \sum_{\alpha=1}^{M} \left( \frac{1}{2\beta} n_\alpha q_\alpha^2 - \frac{\kappa^2}{24\beta} n_\alpha^3 \right). \tag{17}$$

For a given set of integers $\{M; n_1, \dots, n_M\}$, the eigenfunctions $\Psi_{\mathbf{Q}}^{(M)}(\mathbf{x})$ can be represented as follows (for details see [33–37]):

$$\Psi_{\mathbf{q}}^{(M)}(\mathbf{x}) = \sum_{\mathscr{P}} \prod_{a<b}^{n} \left[ 1 + \mathrm{i}\kappa \frac{\mathrm{sgn}(x_a - x_b)}{Q_{\mathscr{P}_a} - Q_{\mathscr{P}_b}} \right] \exp\left( \mathrm{i} \sum_{a=1}^{n} Q_{\mathscr{P}_a} x_a \right), \tag{18}$$

where the summation goes over $n!$ permutations $\mathscr{P}$ of $n$ momenta $Q_a$, equation (13), over $n$ particles $x_a$.

Substituting equations (15)–(18) into equation (10) we get

$$W(\mathbf{f}; \mathbf{x}) = 1 + \lim_{\lambda \to \infty} \left\{ \sum_{L_1 + \dots + L_N \geqslant 1}^{\infty} \prod_{k=1}^{N} \left[ \frac{(-1)^{L_k}}{L_k!} \exp(\lambda L_k f_k) \right] \right.$$
$$\times \sum_{M=1}^{L_1+\dots+L_N} \frac{1}{M!} \prod_{\alpha=1}^{M} \left[ \sum_{n_\alpha=1}^{\infty} \int_{-\infty}^{+\infty} \mathrm{d}q_\alpha \frac{\kappa n_\alpha}{2\pi \kappa n_\alpha} \exp\left( -\frac{t}{2\beta} n_\alpha q_\alpha^2 + \frac{\kappa^2 t}{24\beta} n_\alpha^3 \right) \right] \boldsymbol{\delta}\left( \sum_{\alpha=1}^{M} n_\alpha, \sum_{k=1}^{N} L_k \right) |C_M(\mathbf{q}, \mathbf{n})|^2$$
$$\left. \times \sum_{\mathscr{P}^{(L_1,\dots,L_N)}} \prod_{k=1}^{N} \left[ \sum_{\mathscr{P}^{(L_k)}} \right] \prod_{k<l}^{N} \prod_{a_k=1}^{L_k} \prod_{a_l=1}^{L_l} \left( \frac{Q_{\mathscr{P}_{a_k}^{(L_k)}} - Q_{\mathscr{P}_{a_l}^{(L_l)}} - \mathrm{i}\kappa}{Q_{\mathscr{P}_{a_k}^{(L_k)}} - Q_{\mathscr{P}_{a_l}^{(L_l)}}} \right) \exp\left( \mathrm{i} \sum_{k=1}^{N} x_k \sum_{a_k=1}^{L_k} Q_{\mathscr{P}_{a_k}^{(L_k)}} \right) \right\}. \tag{19}$$

In the above expression, the summation over permutations of $n = L_1 + \dots + L_N$ momenta $Q_a$ split into the internal permutations $\mathscr{P}^{(L_k)}$ of $L_k$ momenta [taken at random out of the total list $\{Q_a\}$ $(a = 1,\dots,n)$] and the permutations $\mathscr{P}^{(L_1,\dots,L_N)}$ of the momenta among the groups $L_k$. It is evident that due to the symmetry of the expression in equation (19), the summations over $\mathscr{P}^{(L_k)}$ give just the factor $L_1! \dots L_N!$. On the other hand, the structure of the Bethe ansatz wave functions, equation (18), is such that for the positions of ordered particles in the summation over permutations, the momenta $Q_a$ belonging to the same cluster also remain ordered (for details see e.g. [37]). Thus, in order to perform the summation over the permutations $\mathscr{P}^{(L_1,\dots,L_N)}$ in equation (19) it is sufficient to split the momenta of each cluster into $N$ parts:

$$\{\underbrace{q_1^\alpha,\dots,q_{m_\alpha^1}^\alpha}_{m_\alpha^1};\ \underbrace{q_{m_\alpha^1+1}^\alpha,\dots,q_{m_\alpha^1+m_\alpha^2}^\alpha}_{m_\alpha^2};\ \dots;\ \underbrace{q_{\sum_{k=1}^{N-1} m_\alpha^k+1}^\alpha,\dots,q_{\sum_{k=1}^{N} m_\alpha^k}^\alpha}_{m_\alpha^N}\}, \tag{20}$$

where the integers $m_\alpha^k = 0,1,\dots,n_\alpha$ are constrained by the conditions

$$\sum_{k=1}^{N} m_\alpha^k = n_\alpha, \tag{21}$$

$$\sum_{\alpha=1}^{M} m_\alpha^k = L_k, \tag{22}$$

and the momenta of every group $\left\{q^\alpha_{\sum_{l=1}^{k-1} m_\alpha^l+1},\ \dots,\ q^\alpha_{\sum_{l=1}^{k} m_\alpha^l}\right\}$ all belong to the particles whose coordinates are all equal to $x_k$. Let us redefine:

$$q^\alpha_{\sum_{l=1}^{k-1} m_\alpha^l+r} \equiv q^\alpha_{k,r} = q_\alpha + \frac{\mathrm{i}\kappa}{2}\left(n_\alpha + 1 - 2\sum_{l=1}^{k-1} m_\alpha^l - 2r\right). \tag{23}$$

In this way, the summation over $\mathscr{P}^{(L_1,\dots,L_N)}$ is changed by the summation over the integers $\{m_\alpha^k\}$. Substituting equations (20)–(23) into equation (19) after simple algebra, we find

$$W(\mathbf{f};\mathbf{x}) = 1 + \lim_{\lambda\to\infty}\left(\sum_{M=1}^{\infty}\frac{(-1)^M}{M!}\prod_{\alpha=1}^{M}\left\{\sum_{\sum_k^N m_\alpha^k\geqslant 1}(-1)^{\sum_k^N m_\alpha^k-1}\int_{-\infty}^{+\infty}\frac{dq_\alpha}{2\pi\kappa\left(\sum_k^N m_\alpha^k\right)}\right.\right.$$
$$\times\exp\left[\lambda\sum_{k=1}^{N} m_\alpha^k f_k + \mathrm{i}\sum_{k=1}^{N} m_\alpha^k x_k q_\alpha - \frac{1}{4}\kappa\sum_{k,l=1}^{N} m_\alpha^k m_\alpha^l |x_k - x_l| - \frac{t}{2\beta}q_\alpha^2\sum_{k=1}^{N} m_\alpha^k + \frac{\kappa^2 t}{24\beta}\left(\sum_{k=1}^{N} m_\alpha^k\right)^3\right]\Bigg\}$$
$$\left.\times\left|C_M\left(\mathbf{q};\{m_\alpha^k\}\right)\right|^2 G_M\left(\mathbf{q};\{m_\alpha^k\}\right)\right), \tag{24}$$

where the normalization constant $\left|C_M\left(\mathbf{q};\{m_\alpha^k\}\right)\right|^2$ is given in equation (16) (with $n_\alpha = \sum_{k=1}^N m_\alpha^k$) and

$$G_M\left(\mathbf{q};\{m_\alpha^k\}\right) = \prod_{\alpha=1}^{M}\prod_{k<l}^{N}\prod_{r=1}^{m_\alpha^k}\prod_{r'=1}^{m_\alpha^l}\left(\frac{q^\alpha_{k,r} - q^\alpha_{l,r'} - \mathrm{i}\kappa}{q^\alpha_{k,r} - q^\alpha_{l,r'}}\right)\prod_{\alpha<\beta}^{M}\prod_{k=1}^{N}\prod_{l=1}^{N}\prod_{r=1}^{m_\alpha^k}\prod_{r'=1}^{m_\alpha^l}\left(\frac{q^\alpha_{k,r} - q^\alpha_{l,r'} - \mathrm{i}\kappa}{q^\alpha_{k,r} - q^\alpha_{l,r'}}\right). \tag{25}$$

Substituting the expressions for $q^\alpha_{k,r}$, equation (23), one can find an explicit formula for the above factor $G_M$ which is rather cumbersome: it contains the products of all kinds of the Gamma functions of the type $\Gamma\left[1+\frac{1}{2}\left(\sum_k^N(\pm)m_\alpha^k + \sum_l^N(\pm)m_\beta^l\right) \pm \frac{i}{\kappa}(q_\alpha - q_\beta)\right]$ [the example of this kind of the product is given in [38], equation (A17)]. We do not reproduce it here as it turns out to be irrelevant in the limit $t\to\infty$ (see below).

After rescaling

$$q_\alpha \to \frac{\kappa}{2\lambda} q_\alpha, \tag{26}$$

$$x_k \to \frac{2\lambda^2}{\kappa} x_k, \tag{27}$$

with

$$\lambda = \frac{1}{2}\left(\frac{\kappa^2 t}{\beta}\right)^{1/3} = \frac{1}{2}\left(\beta^5 u^2 t\right)^{1/3} \tag{28}$$

the normalization factor $|C_M(\mathbf{q};\{m_\alpha^k\})|^2$, equation (16) (with $n_\alpha = \sum_k^N m_\alpha^k$), can be represented as follows:

$$
\begin{aligned}
|C_M(\mathbf{q};\{m_\alpha^k\})|^2 &= \prod_{\alpha<\beta}^{M} \frac{|\lambda\sum_k^N m_\alpha^k - \lambda\sum_k^N m_\beta^k - \mathrm{i}q_\alpha + \mathrm{i}q_\beta|^2}{|\lambda\sum_k^N m_\alpha^k + \lambda\sum_k^N m_\beta^k - \mathrm{i}q_\alpha + \mathrm{i}q_\beta|^2} \\
&= \left[\prod_{\alpha=1}^{M}\left(2\lambda\sum_k^N m_\alpha^k\right)\right] \det\left[\frac{1}{(\sum_k^N \lambda m_\alpha^k - \mathrm{i}q_\alpha) + (\sum_k^N \lambda m_\beta^k + \mathrm{i}q_\beta)}\right]_{\alpha,\beta=1,\dots,M}. \quad (29)
\end{aligned}
$$

Substituting equation (25)–(28) into equation (23) and using the Airy function relation

$$
\exp\left[\frac{1}{3}\lambda^3\left(\sum_k^N m_\alpha^k\right)^3\right] = \int_{-\infty}^{+\infty} \mathrm{d}y\, \mathrm{Ai}(y)\, \exp\left[\lambda\left(\sum_k^N m_\alpha^k\right) y\right] \quad (30)
$$

we get

$$
\begin{aligned}
W(\mathbf{f};\mathbf{x}) &= 1 + \lim_{\lambda\to\infty}\left(\sum_{M=1}^{\infty}\frac{(-1)^M}{M!}\prod_{\alpha=1}^{M}\left\{\int\int_{-\infty}^{+\infty}\frac{\mathrm{d}q_\alpha \mathrm{d}y_\alpha}{2\pi}\mathrm{Ai}\left(y_\alpha + q_\alpha^2\right)\right.\right. \\
&\quad \times \left.\sum_{\sum_k^N m_\alpha^k \geqslant 1} (-1)^{\sum_k^N m_\alpha^k - 1} \exp\left[\lambda\sum_{k=1}^{N} m_\alpha^k (y_\alpha + f_k + \mathrm{i}x_k q_\alpha) - \frac{1}{2}\lambda^2 \sum_{k,l=1}^{N} m_\alpha^k m_\alpha^l \Delta_{kl}\right]\right\} \\
&\quad \times \left.\det\hat{K}\left[\left(\sum_k^N \lambda m_\alpha^k,\, q_\alpha\right);\left(\sum_k^N \lambda m_\beta^k,\, q_\beta\right)\right]_{\alpha,\beta=1,\dots,M} G_M\left(\frac{\kappa\mathbf{q}}{2\lambda};\{m_\alpha^k\}\right)\right), \quad (31)
\end{aligned}
$$

where

$$
\Delta_{kl} = |x_k - x_l| \quad (32)
$$

and

$$
\hat{K}\left[\left(\sum_k^N \lambda m_\alpha^k,\, q_\alpha\right);\left(\sum_k^N \lambda m_\beta^k,\, q_\beta\right)\right] = \frac{1}{(\sum_k^N \lambda m_\alpha^k - \mathrm{i}q_\alpha) + \left(\sum_k^N \lambda m_\beta^k + \mathrm{i}q_\beta\right)}. \quad (33)
$$

The quadratic in $m_\alpha^k$ term in the exponential of equation (31) can be linearized as follows:

$$
\begin{aligned}
\exp\left\{-\frac{1}{2}\lambda^2\sum_{k,l=1}^{N} m_\alpha^k m_\alpha^l \Delta_{kl}\right\} &= \exp\left\{-\frac{1}{4}\lambda^2\sum_{k,l=1}^{N}\Delta_{kl}\left(m_\alpha^k + m_\alpha^l\right)^2 + \frac{1}{2}\lambda^2\sum_{k=1}^{N}\left(m_\alpha^k\right)^2\sum_{l=1}^{N}\Delta_{kl}\right\} \\
&= \prod_{k,l=1}^{N}\left\{\int_{-\infty}^{+\infty}\frac{\mathrm{d}\xi_{kl}^\alpha}{\sqrt{2\pi}}\exp\left[-\frac{1}{2}(\xi_{kl}^\alpha)^2\right]\right\}\prod_{k=1}^{N}\left\{\int_{-\infty}^{+\infty}\frac{\mathrm{d}\eta_k^\alpha}{\sqrt{2\pi}}\exp\left[-\frac{1}{2}(\eta_k^\alpha)^2\right]\right\} \\
&\quad \times \exp\left\{\lambda\sum_k^N\left[\frac{\mathrm{i}}{\sqrt{2}}\sum_{l=1}^{N}\sqrt{\Delta_{kl}}\,(\xi_{kl}^\alpha + \xi_{lk}^\alpha) - \sqrt{\gamma_k}\eta_k^\alpha\right] m_\alpha^k\right\}, \quad (34)
\end{aligned}
$$

where

$$
\gamma_k = \sum_{l=1}^{N}\Delta_{kl} = \sum_{l=1}^{N}|x_k - x_l|. \quad (35)
$$

Substituting the representation (34) into equation (31) and redefining the integration parameters

$$
\eta_k^\alpha \to \eta_k^\alpha + \frac{\mathrm{i}}{\sqrt{\gamma_k}} q_\alpha x_k + \mathrm{i}\sum_{l=1}^{N}\sqrt{\frac{\Delta_{kl}}{2\gamma_k}}\,(\xi_{kl}^\alpha + \xi_{lk}^\alpha) \quad (36)
$$

we get

$$W(\mathbf{f};\mathbf{x}) = 1 + \sum_{M=1}^{\infty} \frac{(-1)^M}{M!} \prod_{\alpha=1}^{M} \left( \int\int_{-\infty}^{+\infty} \frac{dq_\alpha dy_\alpha}{2\pi} \mathrm{Ai}(y_\alpha + q_\alpha^2) \prod_{k,l=1}^{N} \left( \int_{-\infty}^{+\infty} \frac{d\xi_{kl}^{\alpha}}{\sqrt{2\pi}} \right) \prod_{k=1}^{N} \left( \int_{-\infty}^{+\infty} \frac{d\eta_k^{\alpha}}{\sqrt{2\pi}} \right) \right.$$
$$\left. \times \exp\left\{ -\frac{1}{2} \sum_{k,l=1}^{N} (\xi_{kl}^{\alpha})^2 - \frac{1}{2} \sum_{k=1}^{N} \left[ \eta_k^{\alpha} + \frac{i}{\sqrt{\gamma_k}} q_\alpha x_k + i \sum_{l=1}^{N} \sqrt{\frac{\Delta_{kl}}{2\gamma_k}} (\xi_{kl}^{\alpha} + \xi_{lk}^{\alpha}) \right]^2 \right\} \right) \mathcal{S}(\mathbf{f}, \mathbf{y}, \mathbf{q}, \{\eta_k\}), \tag{37}$$

where

$$\mathcal{S}(\mathbf{f}, \mathbf{y}, \mathbf{q}, \{\eta_k\}) = \lim_{\lambda\to\infty} \prod_{\alpha=1}^{M} \left\{ \sum_{\sum_k^N m_\alpha^k \geq 1} (-1)^{\sum_k^N m_\alpha^k - 1} \exp\left[ \lambda \sum_{k=1}^{N} m_\alpha^k (y_\alpha + f_k - \sqrt{\gamma_k}\eta_k) \right] \right.$$
$$\left. \times \det \hat{K} \left[ \left( \sum_k^N \lambda m_\alpha^k, q_\alpha \right); \left( \sum_k^N \lambda m_\beta^k, q_\beta \right) \right]_{\alpha,\beta=1,\ldots,M} G_M\left( \frac{\kappa \mathbf{q}}{2\lambda}; \{m_\alpha^k\} \right) \right\}. \tag{38}$$

The summations over $m_\alpha^k$ in the above expression can be performed as follows:

$$\mathcal{S}(\mathbf{f}, \mathbf{y}, \mathbf{q}, \{\eta_k\}) = \lim_{\lambda\to\infty} \prod_{\alpha=1}^{M} \left[ \prod_{k=1}^{N} \left( \sum_{m_\alpha^k=0}^{\infty} \delta_{m_\alpha^k,0} \right) - (-1)^N \prod_{k=1}^{N} \left\{ \sum_{m_\alpha^k=0}^{\infty} (-1)^{m_\alpha^k - 1} \exp\left[ \lambda m_\alpha^k (y_\alpha + f_k - \sqrt{\gamma_k}\eta_k) \right] \right\} \right]$$
$$\times \det \hat{K} \left[ \left( \sum_k^N \lambda m_\alpha^k, q_\alpha \right); \left( \sum_k^N \lambda m_\beta^k, q_\beta \right) \right]_{\alpha,\beta=1,\ldots,M} \times G_M\left( \frac{\kappa \mathbf{q}}{2\lambda}; \{m_\alpha^k\} \right)$$
$$= \lim_{\lambda\to\infty} \prod_{\alpha=1}^{M} \left[ \prod_{k=1}^{N} \left( \int_{\mathcal{C}} dz_\alpha^k \delta(z_\alpha^k) \right) - (-1)^N \prod_{k=1}^{N} \left\{ \int_{\mathcal{C}} \frac{dz_\alpha^k}{2i\sin(\pi z_\alpha^k)} \exp\left[ \lambda z_\alpha^k (y_\alpha + f_k - \sqrt{\gamma_k}\eta_k) \right] \right\} \right]$$
$$\times \det \hat{K} \left[ \left( \sum_k^N \lambda z_\alpha^k, q_\alpha \right); \left( \sum_k^N \lambda z_\beta^k, q_\beta \right) \right]_{\alpha,\beta=1,\ldots,M} G_M\left( \frac{\kappa \mathbf{q}}{2\lambda}; \{z_\alpha^k\} \right), \tag{39}$$

where the integration goes over the contour $\mathcal{C}$ shown in figure 1. Redefining $z_\alpha^k \to z_\alpha^k/\lambda$, in the limit

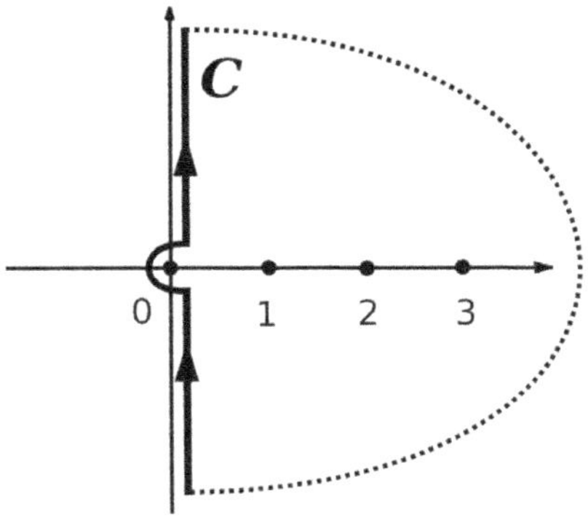


**Figure 1.** The contours of integration in the complex plane used for summing the series equation (39).

$\lambda \to \infty$, we get

$$\mathscr{S}(\mathbf{f},\mathbf{y},\mathbf{q},\{\eta_k\}) = \prod_{\alpha=1}^{M}\left[\prod_{k=1}^{N}\left(\int_{\mathscr{C}} dz_\alpha^k\,\delta(z_\alpha^k)\right) - (-1)^N\prod_{k=1}^{N}\left\{\int_{\mathscr{C}}\frac{dz_\alpha^k}{2\pi i z_\alpha^k}\exp\left[z_\alpha^k\left(y_\alpha + f_k - \sqrt{\gamma_k}\eta_k\right)\right]\right\}\right]$$
$$\times \det\hat{K}\left[\left(\sum_k^N z_\alpha^k,\ q_\alpha\right);\left(\sum_k^N z_\beta^k,\ q_\beta\right)\right]_{\alpha,\beta=1,\dots,M}\lim_{\lambda\to\infty} G_M\left(\frac{\kappa\mathbf{q}}{2\lambda};\{\frac{z_\alpha^k}{\lambda}\}\right). \tag{40}$$

Taking into account the Gamma function property $\lim_{|z|\to 0}\Gamma(1+z) = 1$, one can easily demonstrate (see e.g. [38]) that

$$\lim_{\lambda\to\infty} G_M\left(\frac{\kappa\mathbf{q}}{2\lambda};\left\{\frac{z_\alpha^k}{\lambda}\right\}\right) = 1. \tag{41}$$

Thus, in the limit $\lambda \to \infty$, the expression (37) takes the form of the Fredholm determinant

$$W(\mathbf{f};\mathbf{x}) = 1 + \sum_{M=1}^{\infty}\frac{(-1)^M}{M!}\prod_{\alpha=1}^{M}\left[\int\int_{-\infty}^{+\infty}\frac{dq_\alpha dy_\alpha}{2\pi}\mathrm{Ai}(y_\alpha + q_\alpha^2)\prod_{k,l=1}^{N}\left(\int_{-\infty}^{+\infty}\frac{d\xi_{kl}^\alpha}{\sqrt{2\pi}}\right)\prod_{k=1}^{N}\left(\int_{-\infty}^{+\infty}\frac{d\eta_k^\alpha}{\sqrt{2\pi}}\right)\right.$$
$$\times\exp\left\{-\frac{1}{2}\sum_{k,l=1}^{N}\xi_{kl}^2 - \frac{1}{2}\sum_{k=1}^{N}\left[\eta_k^\alpha + \frac{i}{\sqrt{\gamma_k}}q_\alpha x_k + i\sum_{l=1}^{N}\sqrt{\frac{\Delta_{kl}}{2\gamma_k}}\left(\xi_{kl}^\alpha + \xi_{lk}^\alpha\right)\right]^2\right\}$$
$$\times\left.\prod_{k=1}^{N}\left(\int_{\mathscr{C}} dz_\alpha^k\right)\left\{\prod_{k=1}^{N}\delta(z_\alpha^k) - (-1)^N\prod_{k=1}^{N}\frac{1}{2\pi i z_\alpha^k}\exp\left[z_\alpha^k\left(y_\alpha + f_k - \sqrt{\gamma_k}\eta_k\right)\right]\right\}\right]$$
$$\times\det\hat{K}\left[\left(\sum_k^N z_\alpha^k,\ q_\alpha\right);\left(\sum_k^N z_\beta^k,\ q_\beta\right)\right]_{\alpha,\beta=1,\dots,M} \tag{42}$$
$$\equiv \det\left[\hat{1} - \hat{A}\right] = \exp\left\{-\sum_{M=1}^{\infty}\frac{1}{M}\mathrm{Tr}\hat{A}^M\right\}, \tag{43}$$

where $\hat{A}$ is the integral operator with the kernel

$$A\left[\left(\sum_k^N z^k, q\right);\left(\sum_k^N \tilde{z}^k, \tilde{q}\right)\right] = \int_{-\infty}^{+\infty}\frac{dy}{2\pi}\mathrm{Ai}(y+q^2)\prod_{k,l=1}^{N}\left(\int_{-\infty}^{+\infty}\frac{d\xi_{kl}}{\sqrt{2\pi}}\right)\prod_{k=1}^{N}\left(\int_{-\infty}^{+\infty}\frac{d\eta_k}{\sqrt{2\pi}}\right)$$
$$\times\exp\left\{-\frac{1}{2}\sum_{k,l=1}^{N}\xi_{kl}^2 - \frac{1}{2}\sum_{k=1}^{N}\left[\eta_k + \frac{i}{\sqrt{\gamma_k}}q_\alpha x_k + i\sum_{l=1}^{N}\sqrt{\frac{\Delta_{kl}}{2\gamma_k}}\left(\xi_{kl} + \xi_{lk}\right)\right]^2\right\}$$
$$\times\left\{\prod_{k=1}^{N}\delta(z^k) - (-1)^N\prod_{k=1}^{N}\frac{1}{2\pi i z^k}\exp\left[z^k\left(y + f_k - \sqrt{\gamma_k}\eta_k\right)\right]\right\}$$
$$\times\frac{1}{\sum_k^N z^k - iq + \sum_k^N \tilde{z}^k + i\tilde{q}}. \tag{44}$$

Correspondingly, for the trace of this operator in the $M$-th power [in the exponential representation of the Fredholm determinant, equation (43)] we get

$$\begin{aligned}\mathrm{Tr}\hat{A}^{M} &= \prod_{\alpha=1}^{M}\left[\int\int_{-\infty}^{+\infty}\frac{\mathrm{d}y\mathrm{d}q_\alpha}{2\pi}\mathrm{Ai}(y+q_\alpha^2)\prod_{k,l=1}^{N}\left(\int_{-\infty}^{+\infty}\frac{\mathrm{d}\xi_{kl}}{\sqrt{2\pi}}\right)\prod_{k=1}^{N}\left(\int_{-\infty}^{+\infty}\frac{\mathrm{d}\eta_k}{\sqrt{2\pi}}\right)\right.\\ &\times\exp\left\{-\frac{1}{2}\sum_{k,l=1}^{N}\xi_{kl}^2-\frac{1}{2}\sum_{k=1}^{N}\left[\eta_k+\frac{i}{\sqrt{\gamma_k}}q_\alpha x_k+\mathrm{i}\sum_{l=1}^{N}\sqrt{\frac{\Delta_{kl}}{2\gamma_k}}\,(\xi_{kl}+\xi_{lk})\right]^2\right\}\\ &\times\left.\prod_{k=1}^{N}\left(\int_{\mathscr{C}}\mathrm{d}z_\alpha^k\right)\left\{\prod_{k=1}^{N}\delta(z_\alpha^k)\ -\ (-1)^N\prod_{k=1}^{N}\frac{1}{2\pi\mathrm{i}z_\alpha^k}\exp\left[z_\alpha^k\left(y+f_k-\sqrt{\gamma_k}\eta_k\right)\right]\right\}\right]\\ &\times\prod_{\alpha=1}^{M}\left(\frac{1}{\sum_k^N z_\alpha^k-\mathrm{i}q_\alpha\ +\ \sum_k^N z_{\alpha+1}^k+\mathrm{i}q_{\alpha+1}}\right),\end{aligned} \tag{45}$$

where, by definition, $z_{M+1}^k\equiv z_1^k$ and $q_{M+1}\equiv q_1$.

Substituting

$$\frac{1}{\sum_k^N z_\alpha^k-\mathrm{i}q_\alpha\ +\ \sum_k^N z_{\alpha+1}^k+\mathrm{i}q_{\alpha+1}} = \int_0^\infty \mathrm{d}\omega_\alpha\exp\left\{-\omega_\alpha\left(\sum_k^N z_\alpha^k-\mathrm{i}q_\alpha\ +\ \sum_k^N z_{\alpha+1}^k+\mathrm{i}q_{\alpha+1}\right)\right\} \tag{46}$$

into equation (45) we obtain

$$\mathrm{Tr}\hat{A}^{M} = \int_0^\infty\ldots\int_0^\infty \mathrm{d}\omega_1\ldots\mathrm{d}\omega_M\prod_{\alpha=1}^{M}A(\omega_\alpha;\,\omega_{\alpha+1}), \tag{47}$$

where

$$\begin{aligned}A(\omega;\,\omega') &= \int\int_{-\infty}^{+\infty}\frac{\mathrm{d}y\mathrm{d}q}{2\pi}\mathrm{Ai}(y+q^2+\omega+\omega')\prod_{k,l=1}^{N}\left(\int_{-\infty}^{+\infty}\frac{\mathrm{d}\xi_{kl}}{\sqrt{2\pi}}\right)\prod_{k=1}^{N}\left(\int_{-\infty}^{+\infty}\frac{\mathrm{d}\eta_k}{\sqrt{2\pi}}\right)\\ &\times\exp\left\{-\frac{1}{2}\sum_{k,l=1}^{N}\xi_{kl}^2-\frac{1}{2}\sum_{k=1}^{N}\left[\eta_k+\frac{\mathrm{i}}{\sqrt{\gamma_k}}qx_k+\mathrm{i}\sum_{l=1}^{N}\sqrt{\frac{\Delta_{kl}}{2\gamma_k}}\,(\xi_{kl}+\xi_{lk})\right]^2-\mathrm{i}q(\omega-\omega')\right\}\\ &\times\left\{1\ -\ (-1)^N\prod_{k=1}^{N}\int_{\mathscr{C}}\frac{\mathrm{d}z^k}{2\pi\mathrm{i}z^k}\exp\left[z^k\left(y+f_k-\sqrt{\gamma_k}\eta_k\right)\right]\right\}.\end{aligned} \tag{48}$$

Integrating over $z^1,\ldots,z^N$, we finally get

$$\begin{aligned}A(\omega;\,\omega') &= \int\int_{-\infty}^{+\infty}\frac{\mathrm{d}y\mathrm{d}q}{2\pi}\mathrm{Ai}(y+q^2+\omega+\omega')\prod_{k,l=1}^{N}\left(\int_{-\infty}^{+\infty}\frac{\mathrm{d}\xi_{kl}}{\sqrt{2\pi}}\right)\prod_{k=1}^{N}\left(\int_{-\infty}^{+\infty}\frac{\mathrm{d}\eta_k}{\sqrt{2\pi}}\right)\\ &\times\exp\left\{-\frac{1}{2}\sum_{k,l=1}^{N}\xi_{kl}^2-\frac{1}{2}\sum_{k=1}^{N}\left[\eta_k+\frac{\mathrm{i}}{\sqrt{\gamma_k}}qx_k+\mathrm{i}\sum_{l=1}^{N}\sqrt{\frac{\Delta_{kl}}{2\gamma_k}}\,(\xi_{kl}+\xi_{lk})\right]^2-\mathrm{i}q\left(\omega-\omega'\right)\right\}\\ &\times\left[1\ -\ (-1)^N\prod_{k=1}^{N}\theta\left(-y-f_k+\eta_k\sqrt{\gamma_k}\right)\right],\end{aligned} \tag{49}$$

where $\Delta_{kl}=|x_k-x_l|$ and $\gamma_k=\sum_{l=1}^N\Delta_{kl}$.

Thus, the $N$-point free energy distribution function $W(f_1,\ldots,f_N;\,x_1,\ldots,x_N)$, equation (7), is given by the Fredholm determinant

$$W(\mathbf{f};\,\mathbf{x})\ =\ \det\left[\hat{1}\ -\ \hat{A}\right], \tag{50}$$

where $\hat{A}$ is the integral operator with the kernel $A(\omega;\,\omega')$ (with $\omega,\omega'\geqslant 0$) represented in equation (49).

## 3. Conclusions

In this paper using the method developed in [30] we extended our result to the spatial $N$-point free energy distribution function in the thermodynamic limit $t \to \infty$. It should be noted that following the ideas of the proof [31] for the two-point function, one can easily demonstrate that the result (49)–(50) obtained in this paper is equivalent to that derived earlier by Prolhac and Spohn [32]. It should be stressed, however, that since the obtained result for the kernel $A(\omega; \omega')$, equation (49), has a rather complicated structure, its analytic properties are at present completely unclear and their study would require special efforts.

## Acknowledgements

This work was supported in part by the grant IRSES DCPA PhysBio-269139.

## References

1. Kardar M., Parisi G., Zhang Y.-C., Phys. Rev. Lett., 1986, **56**, 889; doi:10.1103/PhysRevLett.56.889.
2. Halpin-Healy T., Zhang Y.-C., Phys. Rep., 1995, **254**, 215; doi:10.1016/0370-1573(94)00087-J.
3. Burgers J.M., The Nonlinear Diffusion Equation, Reidel, Dordrecht, 1974.
4. Kardar M., Statistical Physics of Fields, Cambridge University Press, Cambridge, 2007.
5. Huse D.A., Henley C.L., Fisher D.S., Phys. Rev. Lett., 1985, **55**, 2924; doi:10.1103/PhysRevLett.55.2924.
6. Huse D.A., Henley C.L., Phys. Rev. Lett., 1985, **54**, 2708; doi:10.1103/PhysRevLett.54.2708.
7. Kardar M., Zhang Y.-C., Phys. Rev. Lett., 1987, **58**, 2087; doi:10.1103/PhysRevLett.58.2087.
8. Kardar M., Nucl. Phys. B, 1987, **290**, 582; doi:10.1016/0550-3213(87)90203-3.
9. Bouchaud J.P., Orland H., J. Stat. Phys., 1990, **61**, 877; doi:10.1007/BF01027306.
10. Brunet E., Derrida B., Phys. Rev. E, 2000, **61**, 6789; doi:10.1103/PhysRevE.61.6789.
11. Johansson K., Commun. Math. Phys., 2000, **209**, 437; doi:10.1007/s002200050027.
12. Prähofer M., Spohn H., J. Stat. Phys., 2002, **108**, 1071; doi:10.1023/A:1019791415147.
13. Ferrari P.L., Spohn H., Commun. Math. Phys., 2006, **265**, 1; doi:10.1007/s00220-006-1549-0.
14. Sasamoto T., Spohn H., Phys. Rev. Lett., 2010, **104**, 230602; doi:10.1103/PhysRevLett.104.230602.
15. Sasamoto T., Spohn H., Nucl. Phys. B, 2010, **834**, 523; doi:10.1016/j.nuclphysb.2010.03.026.
16. Sasamoto T., Spohn H., J. Stat. Phys., 2010, **140**, 209; doi:10.1007/s10955-010-9990-z.
17. Amir G., Corwin I., Quastel J., Commun. Pure Appl. Math., 2011, **64**, 466; doi:10.1002/cpa.20347.
18. Dotsenko V., Klumov B., J. Stat. Mech., 2010, P03022; doi:10.1088/1742-5468/2010/03/P03022.
19. Dotsenko V., Europhys. Lett., 2010, **90**, 20003; doi:10.1209/0295-5075/90/20003.
20. Dotsenko V., J. Stat. Mech., 2010, P07010; doi:10.1088/1742-5468/2010/07/P07010.
21. Calabrese P., Le Doussal P., Rosso A., Europhys. Lett., 2010, **90**, 20002; doi:10.1209/0295-5075/90/20002.
22. Calabrese P., Le Doussal P., Phys. Rev. Lett., 2011, **106**, 250603; doi:10.1103/PhysRevLett.106.250603.
23. Le Doussal P., Calabrese P., J. Stat. Mech., 2012, P06001; doi:10.1088/1742-5468/2012/06/P06001; Preprint arXiv:1204.2607, 2012.
24. Dotsenko V., J. Stat. Mech., 2012, P11014; doi:10.1088/1742-5468/2012/11/P11014.
25. Gueudré T., Le Doussal P., Europhys. Lett., 2012, **100**, 26006; doi:10.1209/0295-5075/100/26006.
26. Corwin I., Random Matrices: Theory Appl., 2012, **1**, 1130001; doi:10.1142/S2010326311300014; Preprint arXiv:1106.1596, 2011.
27. Borodin A., Corwin I., Ferrari P., Preprint arXiv:1204.1024, 2012.
28. Tracy C.A., Widom H., Commun. Math. Phys., 1994, **159**, 151; doi:10.1007/BF02100489.
29. Prolhac S., Spohn H., J. Stat. Mech., 2011, P01031; doi:10.1088/1742-5468/2011/01/P01031.
30. Dotsenko V., J. Phys. A: Math. Theor., 2013, **46**, 355001; doi:10.1088/1751-8113/46/35/355001.
31. Imamura T., Sasamoto T., Spohn H., J. Phys. A: Math. Theor., 2013, **46**, 355002; doi:10.1088/1751-8113/46/35/355002.
32. Prolhac S., Spohn H., J. Stat. Mech., 2011, P03020; doi:10.1088/1742-5468/2011/03/P03020.
33. Lieb E.H., Liniger W., Phys. Rev., 1963, **130**, 1605; doi:10.1103/PhysRev.130.1605.
34. McGuire J.B., J. Math. Phys., 1964, **5**, 622; doi:10.1063/1.1704156.
35. Yang C.N., Phys. Rev., 1968, **168**, 1920; doi:10.1103/PhysRev.168.1920.
36. Calabrese P., Caux J.-S., Phys. Rev. Lett., 2007, **98**, 150403; doi:10.1103/PhysRevLett.98.150403.
37. Dotsenko V.S., Physics-Uspekhi, 2011, **54**, No. 3, 259; doi:10.3367/UFNe.0181.201103b.0269.
38. Dotsenko V., J. Stat. Mech., 2013, P02012; doi:10.1088/1742-5468/2013/02/P02012.

# 17

# Universality and self-similar behaviour of non-equilibrium systems with non-Fickian diffusion*

D.O. Kharchenko, V.O. Kharchenko, S.V. Kokhan

Institute of Applied Physics, Nat. Acad. Sci. of Ukraine, 58 Petropavlivska St., 40000 Sumy, Ukraine

Analytical approaches describing non-Fickian diffusion in complex systems are presented. The corresponding methods are applied to the study of statistical properties of pyramidal islands formation with interacting adsorbate at epitaxial growth. Using the generalized kinetic approach we consider universality, scaling dynamics and fractal properties of pyramidal islands growth. In the framework of generalized kinetics, we propose a theoretical model to examine the numerically obtained data for averaged islands size, the number of islands and the corresponding universal distribution over the island size.

**Key words:** *complex systems, nonlinear diffusion, pattern formation, fractals*

## 1. Introduction

It is well known that complex physical, chemical and/or biological systems manifest strong interactions between their elements (atoms, molecules, individuals/organisms) leading to self-organization processes with anomalous diffusion, nonequilibrium phase transitions, patterning and the formation of self-similar objects such as fractals. The cooperative interactions in such systems result in a reduction of a large number of degrees of freedom binding the subunits of these systems by means of self-organization into synergetic entities. These synergetic systems admit low-dimensional description by the corresponding Fokker-Planck equation (FPE). As was shown previously (see for example, [1]), systems manifesting a collective behavior of interacting species are described in terms of nonlinear FPE (NFPE) capturing the essential dynamics underlying the observed phenomena with anomalous transport and generalized (nonextensive) or $Q$-deformed statistics [2]. The corresponding nonlinearity in NFPE without a drift component, leads to non-Fickian diffusion with density/concetration dependent dispersal resulting in anomalous dynamics as was shown experimentally and theoretically (see, for example, [2–6]).

Nonlinear diffusion equation can be generalized by taking quasi-chemical reactions into account. In such a case, one arrives at nontrivial scenarios for nonequilibrium phase transitions [7], pattern formation [8] and delaying dynamics at phase separation processes [9]. For example, while studying the pattern formation phenomena on surfaces at deposition from a gaseous phase, one describes the corresponding system by reaction-diffusion model with field dependent diffusivity. It was shown that nanosize patterns of adsorbate can emerge due to microscopic interactions of the deposited particles [10–16]. In [17] it was shown that fractal properties of porous-surface condensates are described by the generalized statistics based on NFPE approach and by the corresponding theory of multifractals. In the process of pyramidal islands growth at molecular beam epitaxy, it was found that a structure of pyramids essentially depends on interactions of the elements forming the pattern described by a concentration dependent diffusion coefficient [18–21]. While studying the arrangement of point defects in solids at particle irradiation according to the swelling rate theory [22, 23], it was found that vacancies can arrange into nanosize clusters due to their interactions described by a nonlinear diffusion flux [24–27]. This effect can lead to abnormal grain

*In memory of Prof. A. Olemskoi.

growth dynamics when vacancies segregate on the grain boundaries in a stochastic manner [28]. Nonlinear diffusion was experimentally studied for a large variety of chemical and biological systems [29, 30]. In [31–33] it was shown that nonlinear diffusion leads to anomalous dynamics. From this non-complete literature overview, it follows that in spite of the diversity of this research area, many phenomena in physical, chemical and biological systems have fundamental physical mechanisms in common. A study of interacting mechanisms of elements of complex systems leading to macroscopic self-organization processes with collective behavior of their species remains an urgent problem in modern statistical physics during the last two decades.

In this paper we initially present a generalized kinetic approach permitting to describe the behavior of physical systems with an interaction of their species. Here, we illustrate a self-similar behavior of the main statistical characteristics for the systems described by nonlinear diffusion equation. Next, studying self-organization processes with the surface pattern formation in a system of interacting adsorbate at molecular beam epitaxy, we use the formalism of nonlinear kinetic approach to describe the scaling properties of pyramidal islands growth and the universality of the system behavior.

The work is organized as follows. In section 2 using the generalized master equation we introduce NFPE and discuss the corresponding anomalous dynamics characterized by nonextensive statistics. Here, we consider the nonlinear diffusion equation and study the properties of a related solution. In section 3 we discuss the physical reasons responsible for the emergence of nonlinear diffusion flux in systems with interacting species. Here, (subsection 3.1) we consider a typical model of reaction-diffusion systems with field dependent diffusivity related to the surface pattern formation at molecular beam epitaxy. The original results illustrating the role of particle interactions and scaling dynamics with universal properties of the system described in the framework of NFPE approach are collected in subsection 3.2. We conclude in section 4.

## 2. Generalized kinetic approach and nonlinear diffusion equation

The classical Fickian diffusion assumes that there are no interactions between the random walkers representing individuals in a system. If these walkers interact, the diffusivity or the mobility can change in the presence of walkers and become explicitly dependent functions of the local concentrations of the moving particles. This leads to NFPE as a generalization of ordinary linear FPE where transition probabilities depend on occupation probability densities of initial and arrival states. Following [34], NFPE can be constructed from the generalized master equation written for a probability density function (pdf) $p(\mathbf{r},t)$ to find a particle in the state $\mathbf{r}$ at the time $t$. This equation has the form

$$\frac{\partial p(\mathbf{r},t)}{\partial t} = \int \left[ w(\mathbf{r}',\mathbf{r}-\mathbf{r}')\gamma(p',p) - w(\mathbf{r},\mathbf{r}'-\mathbf{r})\gamma(p,p')\right] \mathrm{d}\mathbf{r}', \tag{1}$$

where $w(\mathbf{r},\mathbf{r}')$ is the transition rate between two states located at $\mathbf{r}$ and at $\mathbf{r}'$ which depends on the nature of interactions between the particle and the bath; it is a function of starting $\mathbf{r}$ and arrival $\mathbf{r}'$ sites. The factor $\gamma(p,p')$ is an arbitrary function of the particle populations of both the initial and the arrival sites. This function satisfies the following conditions: $\gamma(0,p') = 0$ meaning that if the initial site is empty, the transition probability is equal to zero; $\gamma(p,0) \neq 0$ requires that if the arrival site is empty, the transition probability should depend on the population of the initial site. In the case $\gamma(p',p) = p'$ and $\gamma(p,p') = p$, we have the standard liner kinetics when equation (1) reduces to the Chapman–Kolmogorov equation.

In the diffusion limit one can expand slowly varying functions $\gamma$ in equation (1) and obtain NFPE in the form [34]

$$\frac{\partial p}{\partial t} = -\nabla\left[f(\mathbf{r}) - \nabla D(\mathbf{r})\right]\gamma(p) + \nabla\left[D(\mathbf{r})\gamma(p)\frac{\partial \ln\kappa(p)}{\partial p}\nabla p\right], \tag{2}$$

where $\nabla \equiv \partial/\partial\mathbf{r}$. The drift and diffusion terms are: $f(\mathbf{r}) \equiv \int \mathrm{d}\mathbf{r}'\mathbf{r}' w(\mathbf{r},\mathbf{r}')$, $2D(\mathbf{r}) \equiv \int \mathrm{d}\mathbf{r}'\mathbf{r}'^2 w(\mathbf{r},\mathbf{r}')$. Here, $\gamma(p) = \gamma(p,p)$, and the function $\kappa(p) > 0$ is defined through the condition

$$\frac{\partial \ln\kappa(p)}{\partial p} = \left[\frac{\partial}{\partial p}\ln\frac{\gamma(p,p')}{\gamma(p',p)}\right]_{p=p'}. \tag{3}$$

Formally, the functions $\gamma(p)$ and $\kappa(p)$ can be represented through densities of the initial $a(p)$ and arrival $b(p)$ states as follows: $\gamma(p) = a(p)b(p)$, $\kappa(p) = a(p)/b(p)$ [35].

In stationary equilibrium state, the solution $p_s = p(\mathbf{r}, \infty)$ of equation (2) takes the form

$$\ln\kappa(p_s) = U_0 - U_{ef}(\mathbf{r}), \qquad U_{ef}(\mathbf{r}) = -\int^{\mathbf{r}} \frac{f(\mathbf{r}')}{D(\mathbf{r}')}\mathrm{d}\mathbf{r}' + \ln D(\mathbf{r}), \tag{4}$$

where $\gamma(p)$ remains an arbitrary function; $U_0$ takes care of normalization condition $\int p_s \mathrm{d}\mathbf{r} = 1$. A nonlinearity of the Fokker-Planck equation is defined through the form of the function $\kappa(p)$: at $\ln\kappa(p) = \ln p$, we move to the Boltzmann-Gibbs statistics; a $Q$-deformation of the logarithm $\ln\kappa \to \ln_Q\kappa$ promotes Tsallis statistics [2], where $\ln_Q\kappa = (\kappa^{1-Q} - 1)/(1-Q)$.

In the simplest case with $f(\mathbf{r}) = 0$ and $D = \text{const}$, one can consider the time dependent solution of the nonlinear diffusion equation as reduced NFPE. Here, by taking $\gamma = \kappa = p$ one gets

$$\frac{\partial p}{\partial t} = \nabla\left[\mathscr{D}(p)\nabla p\right], \qquad \mathscr{D}(p) \equiv D\gamma(p)\kappa(p)^{-Q}\frac{\mathrm{d}\kappa(p)}{\mathrm{d}p} = Dp^{1-Q}. \tag{5}$$

As was shown in [29, 36, 37], this equation has an exact solution in one dimension at $|r| \leqslant r_0(t/t_0)^H$:

$$p(r,t) = \left(\frac{t_0}{t}\right)^H\left[1 - \left(\frac{r\,t_0^H}{r_0\,t^H}\right)^2\right]^{\frac{1}{1-Q}}, \qquad H = \frac{1}{3-Q}, \qquad r_0 = \frac{N_0\Gamma\left(\frac{1}{1-Q} + \frac{3}{2}\right)}{\pi^{1/2}\Gamma\left(\frac{2-Q}{1-Q}\right)}, \qquad t_0 = \frac{r_0^2(1-Q)}{2D(3-Q)}, \tag{6}$$

where $N_0$ is the initial number of particles at the origin. At $|r| > r_0(t/t_0)^H$, one has $p(r,t) = 0$. It follows that such a nonlinear diffusion equation admits scaling $\langle(r - \langle r\rangle)^2\rangle \propto t^{2H}$, whereas ordinary Brownian diffusion corresponds to $H = 1/2$ with $Q = 1$. Scaling regimes in a self-affine phase space characterized by $D(r,p) = p^{1-Q}r^{\Delta}$ were studied in [5]. An automodel solution of the corresponding nonlinear diffusion equation is characterized by the Hurst exponent $H$ in the form $H = 1/(3-Q-\Delta)$, whereas the related pdf is well-described by the Tsallis form. Therefore, anomalous dynamics can be controlled either by Tsallis parameter $Q$ or by the exponent $\Delta$.

In the case of the constant drift term [$f(\mathbf{r}) = \text{const}$], the corresponding simulations of the time series, satisfying Tsallis statistics, have shown clusterization of time series manifesting a self-similar regime characterized by fractal properties [6, 38]. A study of the stochastic dynamics corresponding to NFPE with a power law dependence $\gamma(p)$ was done in [39]; here, anomalous dynamics was considered using correlation analysis. The relation between NFPE and Levy-type diffusion was discussed in [36, 40]. The application of the generalized statistics followed by NFPE with anomalous diffusion emergent at self-organized criticality regimes was studied in [4], where scaling laws were discussed for avalanche sizes dynamics and the corresponding power-law distributions.

Therefore, the nonlinear kinetic approach is applicable to a description of a universality and scaling behavior of complex systems characterized by field dependent diffusivity related to interactions between system elements. In the next section we study the pattern formation processes by considering the reaction diffusion model that describes epitaxial growth with interacting adsorbate. We use the nonlinear kinetic formalism to analytically describe the scaling regimes and universality of probability density functions of pyramidal islands over their sizes.

## 3. Universality and scaling regimes of pattern formation at epitaxy

In the previous section we have considered the nonlinear diffusion equation, where density dependent diffusion coefficient was defined through occupation probabilities of the initial $a(p)$ and arrival $b(p)$ states. In this section, while studying the pyramidal islands growth at epitaxy, we illustrate that the corresponding nonlinear diffusivity can be immediately obtained considering the interactions of adsorbed particles. Universality and scaling regimes of the arranged pyramidal islands will be studied with the help of NFPE presented in the previous section.

### 3.1. Phase field model of pyramidal islands growth

Considering a system of interacting adsorbate with free diffusion on a surface and the motion of mobile particles caused by their interactions, one can directly obtain the diffusion flux with a field dependent diffusion coefficient. In such a case, instead of the probability density $p$, the relevant quantity that can be used is a coverage field $\rho$ (concentration of atoms/moleculas adsorbed by the surface). An evolution of this quantity is described by a reaction-diffusion equation of the form $\partial_t \rho = R(\rho) - \nabla \cdot \mathbf{J}$, where $R(\rho)$ is the reaction term including adsorption/desorption and/or additional nonequilibrium processes related to the formation of islands, or oxides; $\mathbf{J}$ is the diffusion flux.

At epitaxial growth, the reaction term $R(\rho)$ is defined through a deposition rate characterized by the flux $F_0$ of arriving particles on a surface and by desorption reaction $-\rho/\tau$, where $\tau$ is the relaxation time. Next, we assume that desorbed particles have lateral interactions described by the attractive potential $U(\mathbf{r})$. This leads to renormalization of the desorption rate $\tau^{-1} = \tau_0^{-1} \exp(U/T)$, where $\tau_0^{-1}$ is the constant related to a hopping rate, $T$ is the temperature measured in energetic units.

In general case, the diffusion flux can be written in the standard form $\mathbf{J} = -\mathscr{D}(\rho)\nabla\frac{\delta \mathscr{F}}{\delta \rho}$, where $\mathscr{D}(\rho)$ is a field dependent diffusivity and $\mathscr{F}[\rho]$ is a free energy. Let us define the related diffusivity and the free energy. For the free diffusion of mobile species, one has a standard definition $\mathbf{J}_{\text{dif}} = -D\nabla\rho$, with $D = \text{const}$. A strong local bond induced by the interaction of adsorbed particles can be described by the potential $U[\rho(\mathbf{r})] = -\int u(\mathbf{r}-\mathbf{r}')\rho(\mathbf{r}')d\mathbf{r}'$, where $u(r)$ is a spherically symmetric function depending on the nature of the system. Therefore, the corresponding thermodynamical (chemical) force $\mathbf{f} = -\nabla(U/T)$ induces speed $\mathbf{v} = D\mathbf{f}$ of mobile species (Einstein's relation). Hence, the associated flux $\mathbf{v}\rho$ is possible to $(1-\rho)$ free sites. In such a case, for this flux one has $\mathbf{J}_o = -D\rho(1-\rho)\nabla(U/T)$. Therefore, the total flux $\mathbf{J} = \mathbf{J}_{\text{dif}} + \mathbf{J}_o$ can be written as

$$\mathbf{J} = -D\left[\nabla\rho + \rho(1-\rho)\nabla(U/T)\right]. \tag{7}$$

Assuming that an interaction length $r_0$ between two interacting species is small compared to a diffusion length $\ell = \sqrt{D\tau_0}$, i.e., $r_0 \ll \ell$, one can write $U(\rho(r)) \simeq -\epsilon\rho(r)$ with $\epsilon = \int u(r)dr$. The presented formalism is general and can be applied to a large class of systems: semiconductors and metals, where the quantity $\epsilon$ depends on concrete material properties. Using these assumptions one can rewrite the total diffusion flux (7) through the free energy functional $\mathscr{F}[\rho]$ in the standard form

$$\mathbf{J} = -\mathscr{D}(\rho)\nabla\frac{\delta\mathscr{F}}{\delta\rho}; \qquad \mathscr{D}(\rho) = \frac{D}{T}\rho(1-\rho), \qquad \mathscr{F} = \int d\mathbf{r}\left\{T\left[\rho\ln\rho + (1-\rho)\ln(1-\rho)\right] - \frac{\epsilon}{2}\rho^2\right\}. \tag{8}$$

Therefore, the reaction diffusion model of a system with an interacting adsorbate can be written in the form $\partial_t\rho = F - \rho e^{-\varepsilon\rho/\theta} - \nabla\cdot\mathbf{J}$, where dimensionless quantities $t' = t/\tau_0$, $\mathbf{r}' = \mathbf{r}/\ell$, $F = F_0\tau_0$, $\theta = T/T_0$, $\varepsilon = \epsilon/T_0$ are used; $T_0$ is the bath temperature. Next, we drop the primes for convenience. This approach is widely used to study the pattern formation processes in chemical systems, at condensation under deterministic (see [10–12]) and stochastic conditions (see [13–16]), at epitaxial growth [20, 21], at the formation of point defect clusters due to irradiation effect [24–26, 41] and at other physical systems manifesting interactions between their elements.

The obtained model cannot be used immediately to model the pyramidal islands formation because it does not take into account discrete steps (sharp interfaces) between terraces in pyramids. The problem of sharp interface modelling at the pyramidal islands formation can be solved using the phase field approach proposed by Liu and Metiu [19] and Karma and Plapp [18]. The idea of the phase field approach lies in introducting an order parameter $\phi(\mathbf{r}, t)$ that indicates the phase at a particular position. In our case, the phase field $\phi$ describes the surface height in units of monoatomar layers. According to [19], local stable minima of the order parameter relate to terraces whereas rapid spatial variation of $\phi$ corresponds to the positions of steps. In the framework of the phase field approach [18], one has $\tau_\phi\partial_t\phi = -\frac{\delta\mathscr{H}}{\delta\phi}$. Here, $\tau_\phi$ is the characteristic scale of the time of attachment of adatoms at the step. $\mathscr{H}$ is the effective Hamiltonian, $\mathscr{H} = \int d\mathbf{r}[\varpi^2(\nabla\phi)^2/2 + g(\phi, x)]$, with the density $g(\phi, x) = (2\pi)^{-1}\cos(2\pi[\phi-\phi_s]) - \lambda x[\phi + (2\pi)^{-1}\sin(2\pi[\phi-\phi_s])]$. Here, $\varpi$ stands for the width of the step, $\lambda$ is the dimensionless coupling constant, $\phi_s/2$ is the height of the initial substrate. This model reduces to the solid-on-solid model when the coupling between two fields is a constant supersaturation. The term incorporating $1+\cos(\pi[\phi-\phi_s])$ admits that minima of the free energy $\mathscr{H}$ are possible at $\phi - \phi_s = 2n+1$, independently of the adatom concentration [18]. According to the phase field approach, the corresponding dynamics of the coverage field is described by the equation $\partial_t\rho = R(\rho) - \nabla\cdot\mathbf{J} - \frac{1}{2}\partial_t\phi$ [18]. This model is relevant to the case of a constant adsorbate temperature.

It is known that the temperature of the growing surface can be changed locally at adsorption/desorption processes: when atom becomes adatom, the temperature increases locally, it decreases when the desorption of adatom occurs. Moreover, the temperature can be increased due to the effect of the source of atoms described by $F$. Using the above mechanisms for the temperature variations, one can write the evolution equation for the temperature field in the dimensionless form [21]

$$\mu\partial_t\theta = 1 - \theta + \chi\Delta\theta + r_aF\rho + r_b\partial_t\rho. \tag{9}$$

The relaxation of the temperature to the bath temperature is described by two first terms in the right-hand side with a relaxation time $\tau_T$, where $\mu = \tau_T/\tau_0$; $\chi$ plays the role of thermal diffusivity. The fourth term describes re-heat of the surface with an intensity $r_a$ due to energy exchange with the environment by deposition flux $F$. This is a standard assumption and it is widely used considering temperature instabilities in chemical reactions (see, for example [41, 42]). Such temperature instability can be caused by reorganization of the surface due to annihilation of defects on the surface and their motion to sinks. The fourth term in equation (9) does not directly take into account a change of the temperature for curved steps. Such temperature change can be described by the phase field. In our consideration, this effect is effectively taken into account through the introduction of the last term in equation (9). It also relates to the local heating ($\partial_t\rho > 0$) or cooling ($\partial_t\rho < 0$) during adsorption/desorption processes with intensity $r_b$; generally, it is responsible for the formation of a curved step.

The total system of three equations describing epitaxial growth generalized by introducing fluctuation terms is [21]

$$\begin{cases} \partial_t\rho = & F - \rho e^{-\varepsilon\rho/\theta} - \nabla\cdot\mathbf{J} - \frac{1}{2}\partial_t\phi + \zeta_\rho(\mathbf{r},t), \\ \vartheta\partial_t\phi = & \varpi^2\Delta\phi - \partial_\phi g(\rho,\phi) + \zeta_\phi(\mathbf{r},t), \\ \mu\partial_t\theta = & 1 - \theta + \chi\Delta\theta + r_aF\rho + r_b\partial_t\rho + \zeta_\theta(\mathbf{r},t), \end{cases} \tag{10}$$

where $\vartheta = \tau_\phi/\tau_0$. The last terms in equation (10) are stochastic sources responsible for statistical description of the system dynamics: $\langle\zeta_i\rangle = 0$, $\langle\zeta_i(\mathbf{r},t)\zeta_j(\mathbf{r}',t')\rangle = \delta_{i,j}\delta(t-t')\delta(\mathbf{r}-\mathbf{r}')$, where $i, j \in \{\rho,\phi,\theta\}$.

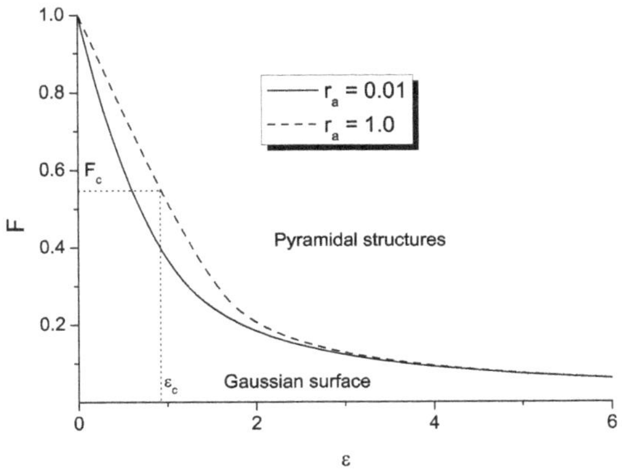


**Figure 1.** Phase diagram for pyramidal structures formation at $\mu = 10$, $\chi = 10$, $\varpi = 2$, $\lambda = 10$, $r_b = 0.05$, $\vartheta = 1$.

Using the linear stability analysis, one can find critical values of main control parameters $F$ and $\varepsilon$ bounding a domain of their values corresponding to the formation of pyramidal islands (see figure 1). At $\varepsilon < \varepsilon_c$ and $F < F_c$, the averaged phase field $\langle\phi\rangle$ does not grow with time meaning the formation of a surface with Gaussian profile related to the effect of fluctuation terms in equation (10). In the opposite case, one has $\partial_t\langle\phi\rangle > 0$, resulting in an increase of the surface height and in the formation of well organized pyramidal structures on the growing surface. In [20, 21], it was shown that the morphology of pyramids essentially depends on the effect of the nonlinear diffusion term in the flux of the adsorbate. Snapshots of a typical evolution of the system are shown in figure 2 at $F > F_c$. In our simulations, we use a square lattice with linear size $L = 256$ sites and periodic boundary conditions. We choose a time step $\Delta t = 2.5 \times 10^{-4}$ and the mesh size $l = 1$. The Gaussian distributed field $\phi(\mathbf{r}, 0)$ with $\langle \phi(\mathbf{r}, 0) \rangle = 0$ and $\langle [\delta\phi(\mathbf{r}, 0)]^2 \rangle = 0.1$ was taken as initial conditions. For initial coverage and temperature fields we use $\rho(\mathbf{r}, 0) = 0$ and $\theta(\mathbf{r}, 0) = 1$. It is seen that at the initial stages, small islands of adsorbate emerge. Here, the temperature field can be

**Figure 2.** (Color online) Evolution of the phase field at $t = 5$, 30, and 60 (first second and third column, respectively). Other parameters are: $\varepsilon = 4$, $F = 4$, $\mu = 10$, $\chi = 10$, $\varpi = 2$, $\lambda = 10$, $r_a = 0.01$, $r_b = 0.05$, $\vartheta = 1$.

locally changed. During the system evolution, such islands become centers of pyramidal patterns, and the corresponding pyramids connect each other by terraces of equivalent heights (see figure 2).

## 3.2. Results and discussions

Let us study the universal dynamics of the islands growth. To this end, we compute the average islands area $\langle s(t) \rangle$ at a half-height of the whole system of pyramids emergent at different times and compute the number of the corresponding islands $N(t)$. It was found that there is a strong relation between $\langle s(t) \rangle$ and $N(t)$: $\langle s(t) \rangle \propto [N(t)]^{-1}$, where $\langle s(t) \rangle \propto t^{\beta}$, $\beta$ is the scaling exponent. This relation means that the total area of all islands $S_0$ calculated at different times should remain constant [see figure 3 (a)]. In other words,

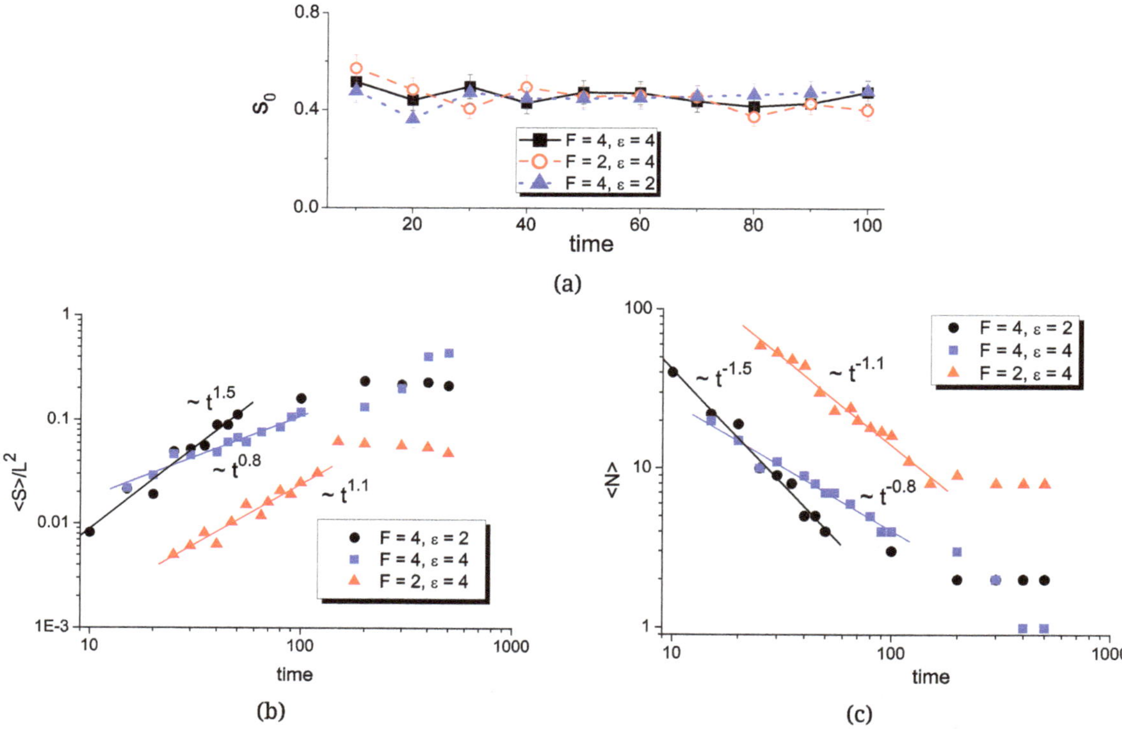


**Figure 3.** (Color online) Time dependencies of the total islands area $S_0$(a), averaged island area (b) and the number of islands (c) at a different set of the system parameters. Other parameters are the same as in figure 2.

pdf of the island area distribution $\mathcal{N}(s,t)$ should satisfy the following two criteria: $\int_0^\infty \mathcal{N}(s,t)\mathrm{d}s = N(t)$ and $\int_0^\infty s\mathcal{N}(s,t)\mathrm{d}s = S_0 \equiv \text{const}$. The first one defines the number of islands, whereas the second one corresponds to the surface conservation law. This law is applicable only at the first stages of the pyramids growth, whereas at late stages one gets only one large constantly growing pyramid. The corresponding dependencies of $\langle s(t)\rangle$ and $N(t)$ are shown in figures 3 (b), (c).

It is seen that the scaling exponent $\beta$ depends on the main system parameters reduced to $F$ and $\varepsilon$, and takes the values smaller or larger than 1 related to the normal growth regime: $\langle s(t)\rangle \propto t$.

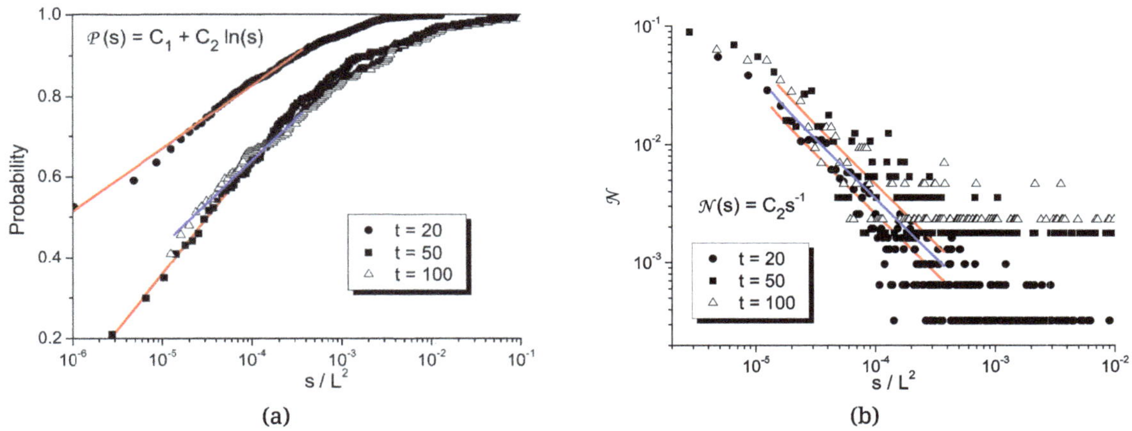


**Figure 4.** (Color online) Integral distribution function (a) and differential distribution function (b) at different times at $F = 4$, $\varepsilon = 4$ (insertion is a typical snapshot of pyramidal islands at half height). Other parameters are the same as in figure 2.

To characterize the distribution of the islands area, we compute the integral distribution function $\mathcal{P}(s,t)$ at first and get the corresponding pdf $\mathcal{N} = \mathrm{d}\mathcal{P}/\mathrm{d}s$ using numerical differentiation (see figure 4). From the obtained dependencies it follows that the universal behavior of both $\mathcal{P}(s,t)$ and $\mathcal{N}(s,t)$ at different times is realized. Moreover, the corresponding fitting procedure gives logarithmic approximation of $\mathcal{P}(s)$ and algebraic dependencies in the form of the Zipf law $\mathcal{N}(s) = C_2 s^{-1}$ at different times, where the normalization parameter is a function of time, i.e., $C_2 = C_2(t)$.

To describe the obtained universal dynamics and the universal behavior of the distribution function over islands area that satisfy the above area conservation condition, we apply the formalism of NFPE to our system. Since the island area grows diffusively (by attachment/deattachment interacting atoms), we start with the nonlinear diffusion equation for $\mathcal{N}(s,t)$ written in the form

$$\partial_t \mathcal{N} = \nabla_s [\mathcal{D}(\mathcal{N})\nabla_s \mathcal{N}], \tag{11}$$

where we assume the form $\mathcal{D}(\mathcal{N}) = D_0\mathcal{N}^{1-Q}$; $D_0 = \text{const}$, $\nabla_s \equiv \partial/\partial s$ for generalized diffusivity. To obtain the corresponding time dependent solution, we consider the system in an automodel regime assuming: $s(t) = a(t)y$, $\mathcal{N}(s,t) = a^{\varrho}\varphi(y)$, where $a(t)$ is the scaling function that measures the diffusion package smearing. Using the normalization condition and the area conservation law, we immediately get $\varrho = -2$. Substituting such constructions into equation (11), we can separate a time dependent part and a part depending on $y$:

$$\dot{a}a^{1+2(1-Q)} = \lambda_0; \quad -\lambda_0\varphi = [\lambda_0 y\varphi + \varphi^{1-Q}\varphi']'. \tag{12}$$

Here, $\lambda_0$ is the constant related to the time dependence. From the first equation we get a relation between Tsallis parameter $Q$ and the scaling exponent $\beta$ in the form

$$\beta = [2(2-Q)]^{-1}. \tag{13}$$

To solve the second equation, we assume a solution in the Tsallis form, i.e., $\varphi(y) = [C_0 - \frac{1-Q}{D_0\delta}y^{\delta}]^{1/(1-Q)}$, where $C_0$, $\delta$ will be defined. Next, using the relation $\varphi' = -D_0^{-1}y^{\delta-1}\varphi^Q$, inserting it into equation for $\varphi$

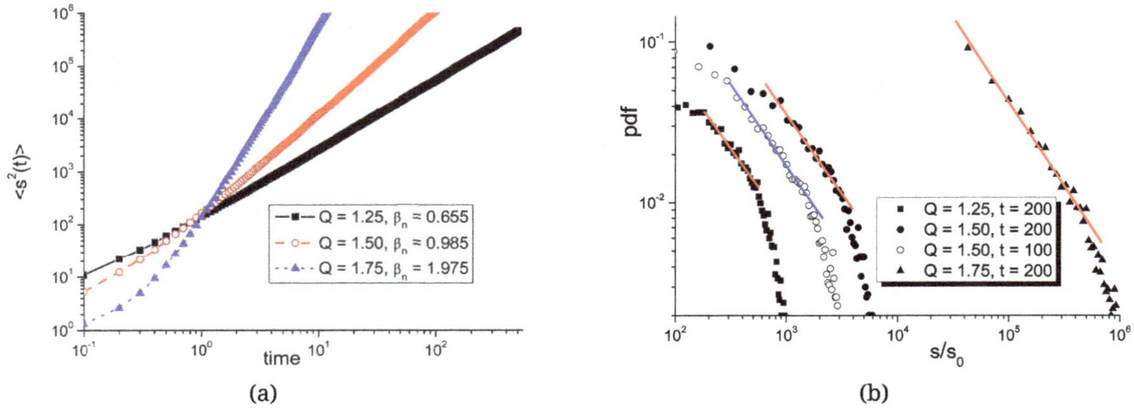


**Figure 5.** (Color online) Protocol for the second moment (a) and the corresponding pdfs (b) at different times (open and filled circles) at fixed $Q$ and at fixed $t = 200$ and different Tsallis exponent $Q$.

we arrive at the algebraic equation having a solution in the form

$$\varphi(y) = \left[\frac{\frac{y^{\delta}}{D_0}\left(1 - \frac{y^{\delta-2}}{D_0\lambda_0}\right)}{2 - \frac{(\delta-1)}{D_0\lambda_0}y^{\delta-2}}\right]^{\frac{1}{1-Q}}. \tag{14}$$

In further consideration, the quantity $1/D_0\lambda_0$ can be considered as a small parameter of the theory. It allows us to obtain a reduced pdf of the form $\varphi \propto y^{-1}$ with $C_0 = 0$ and $\delta = Q - 1$. In such a case, the desired pdf is

$$\mathcal{N}(s,t) \approx \left(\frac{t_0}{t}\right)^{\frac{1}{2(2-Q)}} \frac{s_0}{s}, \tag{15}$$

where $t_0$ and $s_0$ depend on $D_0$ and relate to a normalization condition. It is seen that depending on the parameter $Q$, one gets different dynamics for islands area growth whereas the corresponding area distribution remains universal, independently of the Tsallis parameter $Q$. According to the obtained numerical data [see figures 3 (b), (c)] and equation (13) for the Tsallis parameter, one has $1 \leqslant Q < 2$. For the normal islands growth characterized by $\beta = 1$, we have $Q = 3/2$. Therefore, the delayed dynamics observed at an elevated deposition rate $F$ and at interaction energy $\varepsilon$ is characterized by $Q \in [1, 3/2)$, whereas the enhanced dynamics observed at low $F$ or $\varepsilon$ relates to $Q \in (3/2, 2)$. Using the above formalism for nonlinear diffusion equation for $\mathcal{N}(s,t)$ we can write down the Langevin equation corresponding to equation (11) following [5, 39]. It takes the form

$$\frac{\mathrm{d}s}{\mathrm{d}t} = \sqrt{\mathscr{D}[\mathcal{N}(s,t)]}\xi(t) \equiv \left[t^{\frac{1}{2(2-Q)}} s(t)\right]^{\frac{Q-1}{2}} \xi(t), \tag{16}$$

where $\xi(t)$ is the white noise having standard properties. Using its formal solution together with the correlator $\langle\xi(t)\xi(t')\rangle = \delta(t - t')$ in the automodel regime, we obtain $\langle[s(t) - s(0)]^2\rangle \propto t^{2\beta(Q-1)+1}$. Comparing this time dependence with *a priori* known $\langle[s(t) - s(0)]^2\rangle \propto t^{2\beta}$, one immediately arrives at the relation (13). Testing numerical solutions of the Langevin equation (16) we have found a good agreement between the analytical results and numerical data for time dependencies of $\langle s^2(t)\rangle \propto t^{\beta_n}$, where $\beta_n$ is the fitting exponent, and the corresponding pdfs with hyperbolic approximation $s^{-1}$ at fixed times (see figure 5). In our simulations, we took fixed $Q$ and obtained $\beta(Q) \simeq \beta_n(Q)$ with errors up to 5% [see protocols in figure 5 (a)]. The corresponding data for pdfs shown in figure 5 (b) allow one to elucidate that there is a domain for $s$ where universal behavior of pdf is realized at different times and at different exponent $Q$. It follows that with an increase in $Q$ one gets a growing interval for $s/s_0$ where universal $Q$-independent behavior of pdf is realized.

The jagging of time series $s(t)$ can be studied using multifractal detrended fluctuation analysis (see [43]) as a generalization of the standard multifractal theory [38]. Following [43] we can obtain a set

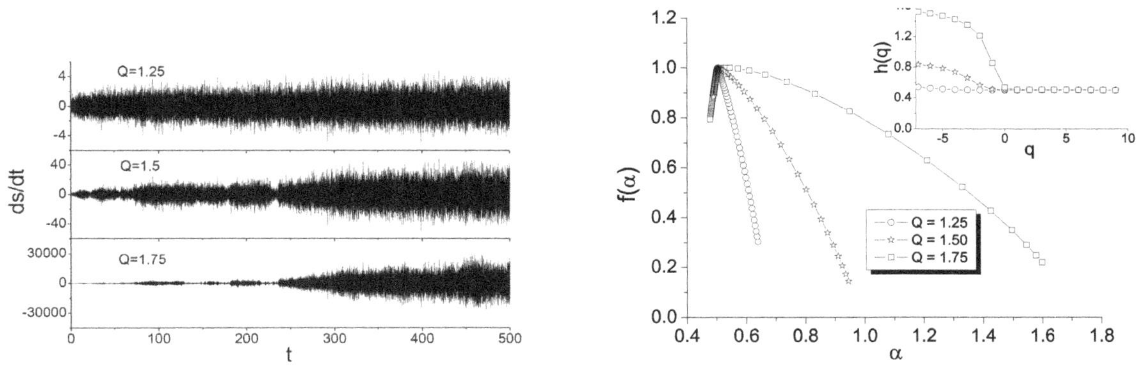


(a) (b)

**Figure 6.** (a) Protocols for the fluctuations in time series $s(t)$ at different $Q$ and (b) the singularity spectrum $f(\alpha)$ with the dependencies $h(q)$ as insertion .

of fractal exponents $h(q)$ and a singularity spectrum $f(\alpha) = q[\alpha - h(q)] + 1$ for multifractals, where $\alpha = h(q) + qh'(q)$ is the singularity strength, $q$ is the index for multifractality. It is known that the quantity $h(q = 2)$ coincides with the Hurst exponent $0 \leqslant H \leqslant 1$ measuring the smoothness of the time series [44] and defining the fractal dimension of time series as $D_\mathrm{f} = 2 - H$ [38]. The corresponding protocols of fluctuations $\mathrm{d}s(t)/\mathrm{d}t$ at different exponent $Q$ used in detrended fluctuation analysis are shown in figure 6 (a). The related dependencies $f(\alpha)$ and $h(q)$ are shown in figure 6 (b). It follows that at $Q \gtrsim 1$, the corresponding fluctuations are characterized by small variations in $h(q) \simeq 1/2$ and by narrow spectrum $f(\alpha)$ which means a weak multifractality of the time series. In other words, at $Q \gtrsim 1$ one gets Gaussianly distributed fluctuations. At elevated $Q$, the time series manifests multifractal properties with wide spectrum $f(\alpha)$. The physical reason for multifractal properties emergence lies in time correlations of the time series. This result comes from the analysis of shuffled time series leading to the fact that $h(q)$ for this series remains constant, 1/2. [43].

## 4. Conclusions

Using a nonlinear kinetic approach we have described universality and scaling properties of pyramidal islands formation at epitaxial growth. In the framework of the phase-field model for pyramidal islands growth in systems with interacting adsorbate, we have analyzed the dynamics of the number of islands, their size and the corresponding distribution functions. We proposed a generalized theoretical model for the island size dynamics manifesting a self-similar behavior and universality of the probability density over the island size. The corresponding dynamics is studied using multifractal time series analysis. It is shown that multifractality of the corresponding time series is related to time correlations.

## References

1. Frank T.D., Nonlinear Fokker-Planck equations, Springer-Verlag, Berlin, Heidelberg, 2005.
2. Tsallis C., In: Nonextensive Statistical Mechanics and its Applications, Lecture Notes in Physics, Abe S., Okamoto Y. (Eds.), Springer, Berlin, 2000.
3. Kumar N., Horsthemke W., Physica A, 2010, **389**, 1812; doi:10.1016/j.physa.2009.12.052.
4. Olemskoi A.I., Khomenko A.V., Kharchenko D.O., Physica A, 2003, **323**, 263; doi:10.1016/S0378-4371(02)01991-X.
5. Kharchenko D.O., Kharchenko V.O., Physica A, 2005, **354**, 262; doi:10.1016/j.physa.2005.01.057.
6. Olemskoi A., Kokhan S., Physica A, 2006, **360**, 37; doi:10.1016/j.physa.2005.06.048.
7. Kharchenko V.O., Physica A, 2009, **388**, 268; doi:10.1016/j.physa.2008.10.016.
8. Kharchenko D.O., Kokhan S.V., Dvornichenko A.V., Physica D, 2009, **238**, No. 23–24, 2251; doi:10.1016/j.physd.2008.12.005.
9. Kharchenko D.O., Dvornichenko A.V., Physica A, 2008, **387**, No. 22, 5342; doi:10.1016/j.physa.2008.05.041.
10. Battogtokh D., Hildebrant M., Krischer F., Mikhailov A., Phys. Rep., 1997, **288**, 435; doi:10.1016/S0370-1573(97)00036-7.

11. Hildebrand M., Mikhailov A.S., Ertl G., Phys. Rev. Lett., 1998, **81**, 2602(4); doi:10.1103/PhysRevLett.81.2602.
12. Hildebrand M., Mikhailov A.S., Ertl G., Phys. Rev. E, 1998, **58**, 5483(11); doi:10.1103/PhysRevE.58.5483.
13. Mangioni S.E., Wio H.S., Phys. Rev. E, 2005, **71**, 056203; doi:10.1103/PhysRevE.71.056203.
14. Mangioni S.E., Physica A, 2010, **389**, 1799; doi:10.1016/j.physa.2010.01.011.
15. Kharchenko V.O., Kharchenko D.O., Phys. Rev. E, 2012, **86**, 041143; doi:10.1103/PhysRevE.86.041143.
16. Kharchenko V.O., Kharchenko D.O., Kokhan S.V., Vernyhora I.V., Yanovsky V.V., Phys. Scr., 2012, **86**, 055401; doi:10.1088/0031-8949/86/05/055401.
17. Olemskoi A., Shuda I., Borisyuk V., EPL, 2010, **89**, 50007; doi:10.1209/0295-5075/89/50007.
18. Karma A., Plapp M., Phys. Rev. Lett., 1998, **81**, 4444; doi:10.1103/PhysRevLett.81.4444.
19. Liu F., Metiu H., Phys. Rev. E, 1994, **49**, 2601; doi:10.1103/PhysRevE.49.2601.
20. Kharchenko D.O., Kharchenko V.O., Lysenko I.O., Phys. Scr., 2011, **83**, 045802; doi:10.1088/0031-8949/83/04/045802.
21. Kharchenko D.O., Kharchenko V.O., Zhylenko T., Dvornichenko A.V., Eur. Phys. J. B, 2012, **86**, 175; doi:10.1140/epjb/e2013-31053-1.
22. Walgraef D., Lauzeral J., Ghoniem N.M., Phys. Rev. B, 1996, **53**, 14782; doi:10.1103/PhysRevB.53.14782.
23. Walgraef D., Spatio-Temporal Pattren Formation, Springer-Verlag, New York, Berlin, Heidelberg, 1996.
24. Kharchenko V.O., Kharchenko D.O., Eur. Phys. J. B, 2012, **85**, 383; doi:10.1140/epjb/e2012-30522-3.
25. Kharchenko D.O., Kharchenko V.O., Bashtova A.I., Ukr. J. Phys., 2013, **58**, No. 10, 993.
26. Kharchenko V.O., Kharchenko D.O., Condens. Matter Phys., 2013, **16**, No. 3, 33001; doi:10.5488/CMP.16.33001.
27. Kharchenko D.O., Kharchenko V.O., Bashtova A.I., Radiat. Eff. Defects Solids, 2014, **169**, 5, 418; doi:10.1080/10420150.2014.905577.
28. Kharchenko V.O., Kharchenko D.O., Phys. Rev. E, 2014, **89**, 042133; doi:10.1103/PhysRevE.89.042133.
29. Murray J.D., Mathematical Biology, Springer-Verlag, Berlin, 1989.
30. Witelski T.P., J. Math. Biol., 1997, **35**, 695; doi:10.1007/s002850050072.
31. Kumar N., Viswanathan G.M., Kenkre V.M., Physica A, 2009, **388**, 3687; doi:10.1016/j.physa.2009.05.015.
32. Ghosh R.N., Webb W.W., Biophys. J., 1994, **68**, 766.
33. Berkowitz B., Klafter J., Metzler R., Scher H., Water Resour. Res., 2002, **38**, No. 10, 1191; doi:10.1029/2001WR001030.
34. Kaniadakis G., Phys. Lett. A, 2001, **288**, 283; doi:10.1016/S0375-9601(01)00543-6.
35. Chavanis P.H., Eur. Phys. J. B, 2006, **54**, 525; doi:10.1140/epjb/e2007-00021-y.
36. Tsallis C., Bukman D.J., Phys. Rev. E, 1996, **54**, R2197; doi:10.1103/PhysRevE.54.R2197.
37. Stariolo D.A., Phys. Rev. E, 1997, **55**, 4806; doi:10.1103/PhysRevE.55.4806.
38. Feder J., Fractals, Plenum Press, New York, London, 1988.
39. Borland L., Phys. Rev. E, 1998, **57**, 6634; doi:10.1103/PhysRevE.57.6634.
40. Tsallis C., Levy S.V.F., Souza A.M.C., Maynard R., Phys. Rev. Lett., 1995, **75**, 3589; doi:10.1103/PhysRevLett.75.3589.
41. Mirzoev F.Kh., Panchenko V.Ya., Shelepin L.A., Phys.-Usp. (Physics-Uspekhi), 1996, **39**, 1; doi:10.1070/PU1996v039n01ABEH000125.
42. Horsthemke W., Lefever R., Noise induced Transitions, Springer–Verlag, 1984.
43. Kantelhardt J.W., Zschiegner S.A., Koscielny-Bunde E., Havlind S., Bundea A., Stanley H.E., Physica A, 2002, **316**, 87; doi:10.1016/S0378-4371(02)01383-3.
44. Hurst H.E., Trans. Am. Soc. Civ. Eng., 1951, **116**, 770.

# 18

# Zubarev nonequilibrium statistical operator method in Renyi statistics. Reaction-diffusion processes*

P. Kostrobij[1], R. Tokarchuk[1], M. Tokarchuk[1,2], B. Markiv[2]

[1] National University "Lviv Polytechnic", 12 Bandera St., 79013 Lviv, Ukraine

[2] Institute for Condensed Matter Physics of the National Academy of Sciences of Ukraine, 1 Svientsitskii St., 79011 Lviv, Ukraine

The Zubarev nonequilibrium statistical operator (NSO) method in Renyi statistics is discussed. The solution of $q$-parametrized Liouville equation within the NSO method is obtained. A statistical approach for a consistent description of reaction-diffusion processes in "gas-adsorbate-metal" system is proposed using the NSO method in Renyi statistics.

**Key words:** *Renyi entropy, nonequilibrium statistical operator, chemical reaction*

## 1. Introduction

Nowadays, investigating complex, self-organizing, fractal structures and various physical phenomena, such as subdiffusion, turbulence and chemical reactions, as well as various economical, social and biological systems, Tsallis [1], Renyi [2, 3], Sharma-Mittal [4, 5] statistics as well as superstatistics [6, 7] are extensively used along with the Gibbs one. A significant contribution to these investigations was made by A.I. Olemskoi [8–12] whose scientific activity we really lack. In particular, problems in synergetic description of self-organizing systems, description of the dynamics of phase transitions within the synergetic approach, the theory of stochastic systems with singular multiplicative noise are elegantly presented in the original work [9].

The Tsallis entropy is widely used in various directions of nonextensive statistical mechanics (for example, see [13–15] and references therein). Some examples are the phenomena of subdiffusion [16, 17] and turbulence [18, 19], and the investigations of transport coefficients in gases and plasma [20], as well as quantum dissipative systems [21]. The energy fluctuations [22], kinetics of nonequilibrium plasma [23], problems of self-gravitating systems [24] and complex systems [25, 26] were investigated within the Tsallis formalism. In references [27–29], Tsallis statistics was applied to a description of chemical reactions, in particular, nonlinear equations of reaction-diffusion processes were obtained in reference [27]. Despite the wide application of Tsallis entropy as a generalization of Gibbs-Shannon entropy, the Renyi entropy is of great interest as well [16, 30–37]. In particular, in this case it is possible to determine a connection between the parameter $q$ and the heat capacity of the system [32]. It is important to note the papers [38, 39] by Luzzi et al., where the nonequilibrium statistical operator method and the Renyi entropy are used in describing the systems far from equilibrium. In particular, the nonequilibrium $q$-dependent Renyi ensemble as well as the generalized distribution functions of bosons and fermions were obtained in reference [38]. Therein the experiments on anomalous luminescence at nanometer quantum dots in semiconductor heterostructures were also described in this approach. A statistical approach for a description of fractal physical-chemical systems based on non-Fickian diffusion processes was proposed

* In memory of Prof. A. Olemskoi.

in reference [39]. Therein the investigations of anomalous diffusion in fractal-like electrodes in microbatteries were carried out. The nonextensive approach [40] as well as other ones [41, 42] leading to Lindblad equation was used to describe a decoherence in quantum mechanics. The references [43–45] are devoted to the investigation of nonlinear kinetics based on the Kramers, Boltzmann and Fokker-Planck equations within the framework of generalized statistics.

In the present paper, an approach to the formulation of extensive statistical mechanics of nonequilibrium processes [34] is considered, based on the Zubarev NSO method [46–48] and the maximum entropy principle for the Renyi entropy. This statistical approach is applied to a consistent description of reaction-diffusion processes in the "gas-adsorbate-metal" system. Reaction-diffusion and adsorption-desorption processes on the metal surface are nonlinear. They manifest an oscillation character, possess memory effects and are actual in terms of nanostructure formation on the surfaces occurring in catalytic phenomena [49–56].

## 2. Renyi entropy and nonequilibrium statistical operator method

The nonequilibrium state of a classical or quantum system of interacting particles is completely described by the nonequilibrium statistical operator $\varrho(x^N;t)$, which satisfies the classical or quantum Liouville (von Neumann) equation:

$$\frac{\partial}{\partial t}\varrho(x^N;t)+\mathrm{i}L_N\varrho(x^N;t)=0. \tag{2.1}$$

Here, $\mathrm{i}L_N$ is the Liouville operator of a system.

Within the NSO method framework, we will be looking for solutions of equation (2.1) which are independent of the initial conditions. The solutions will explicitly depend only on the observable quantities

$$\int \mathrm{d}\Gamma_N \hat{P}_n \varrho(x^N;t)=\langle\hat{P}_n\rangle^t. \tag{2.2}$$

The nonequilibrium statistical operator has the following form:

$$\varrho(x^N;t)=\varrho_{\mathrm{rel}}(x^N;t)-\int_{-\infty}^{t}\mathrm{e}^{\varepsilon(t'-t)}T(t,t')\left[1-P_{\mathrm{rel}}(t')\right]\mathrm{i}L_N\varrho_{\mathrm{rel}}(x^N;t')\mathrm{d}t', \tag{2.3}$$

where

$$T(t,t')=\exp_+\left\{-\int_{t'}^{t}\left[1-P_{\mathrm{rel}}(t')\right]\mathrm{i}L_N\mathrm{d}t'\right\}$$

is the evolution operator containing the projection ($\exp_+$ denotes ordered exponential). The relevant statistical operator (distribution function) $\varrho_{\mathrm{rel}}(x^N;t)$ will be determined using the maximum entropy principle for the Renyi entropy

$$S^{\mathrm{R}}(\varrho)=\frac{1}{1-q}\ln\int \mathrm{d}\Gamma_N\varrho^q(t) \tag{2.4}$$

at fixed parameters of a reduced description, taking into account the normalization condition.

The relevant statistical operator corresponding to the Renyi entropy maximum has the following form:

$$\varrho_{\mathrm{rel}}(t)=\frac{1}{Z_{\mathrm{R}}}\left[1-\frac{q-1}{q}\sum_n F_n(t)\delta\hat{P}_n\right]^{\frac{1}{q-1}}, \tag{2.5}$$

$$Z_{\mathrm{R}}(t)=\int \mathrm{d}\Gamma_N\left[1-\frac{q-1}{q}\sum_n F_n(t)\delta\hat{P}_n\right]^{\frac{1}{q-1}}. \tag{2.6}$$

$Z_{\mathrm{R}}(t)$ is the partition function of the relevant statistical operator, $\delta\hat{P}_n=\hat{P}_n-\langle\hat{P}_n\rangle^t$. The Lagrange multipliers $F_n(t)$ are defined from the self-consistency conditions:

$$\langle\hat{P}_n\rangle^t=\langle\hat{P}_n\rangle^t_{\mathrm{rel}}, \tag{2.7}$$

where $\langle\ldots\rangle^t_{\rm rel} = \int {\rm d}\Gamma_N \ldots \varrho_{\rm rel}(x^N;t)$.

In order to determine the generalized Kawasaki-Gunton projection operator entering (2.3)

$$P_{\rm rel}(t)\varrho' = \left(\varrho_{\rm rel}(t) - \sum_n \frac{\delta\varrho_{\rm rel}(t)}{\delta\langle\hat{P}_n\rangle^t}\langle\hat{P}_n\rangle^t\right)\int d\Gamma_N\varrho' + \sum_n \frac{\delta\varrho_{\rm rel}(t)}{\delta\langle\hat{P}_n\rangle^t}\int d\Gamma_N\hat{P}_n\varrho' \tag{2.8}$$

it is convenient to present the relevant statistical operator in a slightly different form:

$$\varrho^*_{\rm rel}(t) = \frac{1}{Z^*_{\rm R}}\left[1 - \frac{q-1}{q}\sum_n F^*_n(t)\hat{P}_n\right]^{\frac{1}{q-1}}, \tag{2.9}$$

with the partition function

$$Z^*_{\rm R}(t) = \int {\rm d}\Gamma_N\left[1 - \frac{q-1}{q}\sum_n F^*_n(t)\hat{P}_n\right]^{\frac{1}{q-1}} \tag{2.10}$$

and

$$F^*_n(t) = \frac{F_n(t)}{1 + \frac{q-1}{q}\sum_l F_l(t)\langle\hat{P}_l\rangle^t}. \tag{2.11}$$

An action of the operators $P_{\rm rel}(t){\rm i}L_N$ on the relevant statistical operator can be presented as follows, by means of generalized projection which now acts on the dynamic variables, $P_{\rm rel}(t){\rm i}L_N\varrho_{\rm rel}(t) = P_{\rm rel}(t)A(t)\varrho_{\rm rel}(t) = [P(t)A(t)]\varrho_{\rm rel}(t)$, where

$$P(t)\ldots = \sum_{m,n}\langle\ldots\hat{P}_m\rangle_{\rm rel}\langle\hat{P}_m\delta\{[q\psi(t)]^{-1}\hat{P}_n\}\rangle^{-1}_{\rm rel}\delta\{[q\psi(t)]^{-1}\hat{P}_n\}, \tag{2.12}$$

$$\psi(t) = 1 - \frac{q-1}{q}\sum_n F^*_n(t)\hat{P}_n.$$

Taking into account that $[1 - P_{\rm rel}(t)]{\rm i}L_N\varrho_{\rm rel}(t) = -\sum_n I_n(t)F_n(t)\varrho_{\rm rel}(t)$, where

$$I_n(t) = [1 - P(t)]\frac{1}{q}\psi^{-1}(t)\dot{\hat{P}}_n \tag{2.13}$$

are the generalized flows, we can now write down an explicit expression for the *nonequilibrium statistical operator*

$$\varrho(x^N;t) = \varrho_{\rm rel}(x^N;t) + \sum_n\int_{-\infty}^{t}{\rm e}^{\varepsilon(t'-t)}T(t,t')I_n(t')F_n(t')\varrho_{\rm rel}(x^N;t'){\rm d}t'. \tag{2.14}$$

This allows us to obtain generalized *transport equations* for the reduced-description parameters. They can be presented in the form

$$\frac{\partial}{\partial t}\langle\hat{P}_m\rangle^t = \langle\dot{\hat{P}}_m\rangle^t_{\rm rel} + \sum_n\int_{-\infty}^{t}{\rm e}^{\varepsilon(t'-t)}\varphi_{mn}(t,t')F_n(t'){\rm d}t', \tag{2.15}$$

with the generalized *transport kernels* (memory functions)

$$\varphi_{mn}(t,t') = \int {\rm d}\Gamma_N\left\{\dot{\hat{P}}_m T(t,t')I_n(t')\varrho_{\rm rel}(t')\right\} \tag{2.16}$$

which describe the dissipative processes in the system.

## 3. $q$-generalization of Liouville equation

An interesting generalization of Liouville equation was proposed in [40], where the $q$-parametrized Liouville equation was obtained:

$$\frac{\partial}{\partial t}\varrho(x^N;t) + \mathrm{i}\tilde{L}_N(t)\varrho(x^N;t) = 0, \tag{3.1}$$

where

$$\mathrm{i}\tilde{L}_N(t) = \frac{\mathrm{i}L_N}{1+(1-q)t\mathrm{i}L_N} \tag{3.2}$$

is the $q$-parametrized Liouville operator. When $q=1$ we have $\mathrm{i}\tilde{L}_N(t) = \mathrm{i}L_N$. For $|1-q|\Omega t \ll 1$, where $\Omega$ is the characteristic frequency of the physical system considered, we may write [40] for (3.1):

$$\frac{\partial}{\partial t}\varrho(x^N;t) + \left[\mathrm{i}L_N - (1-q)t(\mathrm{i}L_N)^2\right]\varrho(x^N;t) = 0. \tag{3.3}$$

This is the Lindblad type equation for nonequilibrium statistical operator $\varrho(x^N;t)$. The Lindblad type equation within Renyi statistics was obtained in [21].

The solution of $q$-parametrized Liouville equation within the NSO method can be presented as follows:

$$\begin{aligned}
\varrho(x^N;t) &= -\varepsilon\int_{-\infty}^{t} \mathrm{e}^{\varepsilon(t'-t)}T_q(t,t')\varrho_{\mathrm{rel}}(x^N;t')\mathrm{d}t' \\
&= \varrho_{\mathrm{rel}}(x^N;t) - \int_{-\infty}^{t}\mathrm{e}^{\varepsilon(t'-t)}\Bigg\{T_q(t,t')\frac{\partial}{\partial t'} \\
&\quad - \int_0^1 T_q^{\tau}(t,t')\frac{\mathrm{i}L_N}{1+(1-q)(t'-t)\mathrm{i}L_N}T_q^{1-\tau}(t,t')d\tau\Bigg\}\varrho_{\mathrm{rel}}(x^N;t')\mathrm{d}t',
\end{aligned} \tag{3.4}$$

where

$$T_q(t,t') = \exp_+\left\{-\int_{t'}^{t}\frac{\mathrm{i}L_N}{1+(1-q)t''\mathrm{i}L_N}\mathrm{d}t''\right\} \tag{3.5}$$

is the parametrized evolution operator. For $|1-q|\Omega t \ll 1$ from (3.4) we obtain the solution of Lindblad type equation for $\varrho(x^N;t)$.

It is important to note that, at $q \to t$, from (2.14) and (2.15) we reproduce the nonequilibrium statistical operator and the generalized transport equations for the reduced-description parameters within Gibbs statistics [46–48]. In the following section we apply the discussed approach to a description of reaction-diffusion processes, in particular, in catalytic processes.

## 4. Reaction-diffusion processes

Let us start with the Hamiltonian of the system of "gas-adsorbate-metal" in the following form $H = H' + H_{\mathrm{reac}}$, $H' = H_a + H_a^{\mathrm{int}}$. Here, $H_a$ is the Hamiltonian of the gas subsystem considered according to the classical approach; $H_a^{\mathrm{int}}$ is the Hamiltonian describing the interaction between the gas atoms and the atoms adsorbed on the metal surface; $H_{\mathrm{reac}}$ is the Hamiltonian of interaction for chemical reactions between atoms or molecules adsorbed on the metal surface [49]

$$H_{\mathrm{reac}} = \sum_{\bar{a},\bar{b},\bar{a}',\bar{b}'}\left(\langle\bar{a}',\bar{b}'|\Phi_{\mathrm{reac}}|\bar{a},\bar{b}\rangle\hat{q}^+_{\bar{a}'}\hat{q}^+_{\bar{b}'}\hat{q}_{\bar{a}}\hat{q}_{\bar{b}} + \langle\bar{a}',\bar{b}'|\Phi_{\mathrm{reac}}|\bar{a},\bar{b}\rangle^*\hat{q}^+_{\bar{a}}\hat{q}^+_{\bar{b}}\hat{q}_{\bar{a}'}\hat{q}_{\bar{b}'}\right), \tag{4.1}$$

where $\langle\bar{a}',\bar{b}'|\Phi_{\text{reac}}|\bar{a},\bar{b}\rangle = \langle\bar{a},\bar{b}|\Phi_{\text{reac}}|\bar{a}',\bar{b}'\rangle$ are the amplitudes of reaction between reagents $A$ and $B$ (supposed to be known from quantum mechanics). We introduce the notation $\bar{a},\bar{b}$ and $\bar{a}',\bar{b}'$ for state of the reagents $A$, $B$ (atoms or molecules) and for the state of atoms in the reaction products $AB$. Here, $\hat{q}^{+}_{\bar{a}'}$, $\hat{q}^{+}_{\bar{b}'}$, $\hat{q}^{+}_{\bar{a}}$, $\hat{q}^{+}_{\bar{b}}$ and $\hat{q}_{\bar{a}'}$, $\hat{q}_{\bar{b}'}$, $\hat{q}_{\bar{a}}$, $\hat{q}_{\bar{b}}$ are the operators of creation and annihilation of atomic states $\bar{a}',\bar{b}'$ for molecule $AB$, and $\bar{a},\bar{b}$ for $A$ and $B$, respectively.

Parameters of the reduced description are the averaged densities of gas atoms absorbed and not absorbed on the metal surface $\langle\hat{n}_{\bar{a}}(\vec{R})\rangle^{t} = \mathrm{Sp}\{\hat{n}_{\bar{a}}(\vec{R})\varrho(t)\}$, $\langle\hat{n}_{a}(\vec{R})\rangle^{t} = \mathrm{Sp}\{\hat{n}_{a}(\vec{R})\varrho(t)\}$. $\hat{n}_{\bar{a}}(\vec{R})$ is the density operator for gas atoms adsorbed on the metal surface in the state $\nu$; $\hat{n}_{\bar{a}}(\vec{R}) = \sum_{j}^{N_a^{\text{ad}}} \hat{\psi}^{+}_{\nu j}(\vec{R})\hat{\psi}_{\nu j}(\vec{R})$; $\hat{\psi}^{+}_{\nu j}(\vec{R})$, $\hat{\psi}_{\nu j}(\vec{R})$ are the creation and annihilation operators of gas atoms absorbed in the state $\nu$ on the metal surface which satisfy the Bose-type commutation relations. Since we do not consider a catalyst surface explicitly in this model, the states $\nu$ and $\mu$ mean the adsorption centers, where atoms can be located. $\hat{n}_{a}(\vec{r}) = \sum_{j}^{N_a^{\text{ad}}} \delta(\vec{r}-\vec{r}_j)$ is the microscopic density of gas atoms, $\langle\hat{G}^{\nu\mu}_{\bar{a}\bar{b}}(\vec{R},\vec{R}')\rangle^{t} = \mathrm{Sp}\{\hat{G}^{\nu\mu}_{\bar{a}\bar{b}}(\vec{R},\vec{R}')\varrho(t)\}$ is the nonequilibrium pair distribution function of the atoms or molecules adsorbed on the metal surface, and $\hat{G}^{\nu\mu}_{\bar{a}\bar{b}}(\vec{R},\vec{R}') = \hat{n}^{\nu}_{\bar{a}}(\vec{R})\hat{n}^{\mu}_{\bar{b}}(\vec{R}')$.

The relevant statistical operator has the form:

$$\varrho_{\text{rel}} = \frac{1}{Z_{\text{R}}(t)}\left\{1-\frac{q-1}{q}\beta\left[\delta H(t)-\sum_{a}\int \mathrm{d}\vec{r}\mu_{a}(\vec{r};t)\delta\hat{n}_{a}(\vec{r};t)\right.\right.$$
$$\left.\left.-\sum_{\bar{a}}\sum_{\nu}\int \mathrm{d}\vec{R}\mu^{\nu}_{\bar{a}}(\vec{R};t)\delta\hat{n}^{\nu}_{\bar{a}}(\vec{R};t)-\sum_{\bar{a}\bar{b}}\sum_{\nu\mu}\int \mathrm{d}\vec{R}\mathrm{d}\vec{R}' M^{\nu\mu}_{\bar{a}\bar{b}}(\vec{R},\vec{R}';t)\hat{G}^{\nu\mu}_{\bar{a}\bar{b}}(\vec{R},\vec{R}';t)\right]\right\}^{\frac{1}{q-1}}. \tag{4.2}$$

The partition function of the relevant statistical operator is as follows

$$Z_{\text{R}}(t) = \mathrm{Sp}\left\{1-\frac{q-1}{q}\beta\left[\delta H(t)-\sum_{a}\int \mathrm{d}\vec{r}\mu_{a}(\vec{r};t)\delta\hat{n}_{a}(\vec{r};t)\right.\right.$$
$$\left.\left.-\sum_{\bar{a}}\sum_{\nu}\int \mathrm{d}\vec{R}\mu^{\nu}_{\bar{a}}(\vec{R};t)\delta\hat{n}^{\nu}_{\bar{a}}(\vec{R};t)-\sum_{\bar{a}\bar{b}}\sum_{\nu\mu}\int \mathrm{d}\vec{R}\mathrm{d}\vec{R}' M^{\nu\mu}_{\bar{a}\bar{b}}(\vec{R},\vec{R}';t)\hat{G}^{\nu\mu}_{\bar{a}\bar{b}}(\vec{R},\vec{R}';t)\right]\right\}^{\frac{1}{q-1}}. \tag{4.3}$$

Here,

$$\mathrm{Sp}\{\dots\} = \prod_{\alpha}\int \frac{(\mathrm{d}\vec{r}\mathrm{d}\vec{p})^{N_\alpha}}{N_\alpha!(2\pi\hbar)^{3N_\alpha}}\mathrm{Sp}_{(\nu,\xi,\sigma)}\{\dots\},$$

$N_\alpha = \{N_a, N_{\bar{a}}\}$, $\mathrm{Sp}_{(\nu,\xi,\sigma)}$ means the averaged summation over all values of spin and quantum numbers. The parameters $\mu_{a}(\vec{r};t)$, $\mu^{\nu}_{\bar{a}}(\vec{R};t)$, $M^{\nu\mu}_{\bar{a}\bar{b}}(\vec{R},\vec{R}';t)$ should be determined from the corresponding self-consistency conditions

$$\langle\hat{n}_{a}(\vec{r})\rangle^{t} = \langle\hat{n}_{a}(\vec{r})\rangle^{t}_{\text{rel}},\qquad \langle\hat{n}^{\nu}_{\bar{a}}(\vec{R})\rangle^{t} = \langle\hat{n}^{\nu}_{\bar{a}}(\vec{R})\rangle^{t}_{\text{rel}},\qquad \langle\hat{G}^{\nu\mu}_{\bar{a}\bar{b}}(\vec{R},\vec{R}')\rangle^{t} = \langle\hat{G}^{\nu\mu}_{\bar{a}\bar{b}}(\vec{R},\vec{R}')\rangle^{t}_{\text{rel}},$$

hence, we found that $\mu_{a}(\vec{r};t)$ defines a local chemical potential of gas atoms; $\mu^{\nu}_{\bar{a}}(\vec{R};t)$ is a local chemical potential of an atom adsorbed in a state $\nu$ on the metal surface. $\delta H(t) = H-\langle H\rangle^{t}$, $\delta\hat{n}_{a}(\vec{r};t) = \hat{n}_{a}(\vec{r})-\langle\hat{n}_{a}(\vec{r})\rangle^{t}$, $\delta\hat{n}^{\nu}_{\bar{a}}(\vec{R};t) = \hat{n}^{\nu}_{\bar{a}}(\vec{R})-\langle\hat{n}^{\nu}_{\bar{a}}(\vec{R})\rangle^{t}$, $\delta\hat{G}^{\nu\mu}_{\bar{a}\bar{b}}(\vec{R},\vec{R}';t) = \hat{G}^{\nu\mu}_{\bar{a}\bar{b}}(\vec{R},\vec{R}')-\langle\hat{G}^{\nu\mu}_{\bar{a}\bar{b}}(\vec{R},\vec{R}')\rangle^{t}$.

According to (2.14), NSO of the system "gas-adsorbate-metal" has the following form

$$\varrho(t) = \varrho(t)_{\text{rel}}+\sum_{a}\int \mathrm{d}\vec{r}\int_{-\infty}^{t} \mathrm{e}^{\varepsilon(t'-t)}T(t,t')\left[\int_{0}^{1}\mathrm{d}\tau\varrho^{\tau}_{\text{rel}}(t')I_{a}(\vec{r};t)\varrho^{1-\tau}_{\text{rel}}(t')\beta\mu_{a}(\vec{r};t')\right]\mathrm{d}t'$$
$$+\sum_{\bar{a}}\sum_{\nu}\int \mathrm{d}\vec{R}\int_{-\infty}^{t} \mathrm{e}^{\varepsilon(t'-t)}T(t,t')\left[\int_{0}^{1}\mathrm{d}\tau\varrho^{\tau}_{\text{rel}}(t')I^{\nu}_{\bar{a}}(\vec{R};t)\varrho^{1-\tau}_{\text{rel}}(t')\beta\mu^{\nu}_{\bar{a}}(\vec{R};t')\right]\mathrm{d}t'$$
$$+\sum_{\bar{a},\bar{b}}\sum_{\nu,\mu}\int \mathrm{d}\vec{R}\mathrm{d}\vec{R}'\int_{-\infty}^{t} \mathrm{e}^{\varepsilon(t'-t)}T(t,t')\left[\int_{0}^{1}\mathrm{d}\tau\varrho^{\tau}_{\text{rel}}(t')I^{\nu\mu}_{G_{\bar{a}\bar{b}}}(\vec{R},\vec{R}';t)\varrho^{1-\tau}_{\text{rel}}(t')\beta M^{\nu\mu}_{\bar{a}\bar{b}}(\vec{R},\vec{R}';t')\right]\mathrm{d}t'. \tag{4.4}$$

$$I_a(\vec{r};t) = [1-P(t)]\frac{1}{q}\psi^{-1}(t)\dot{\hat{n}}_a(\vec{r}), \qquad I^{\nu}_{\bar{a}}(\vec{R};t) = [1-P(t)]\frac{1}{q}\psi^{-1}(t)\dot{\hat{n}}^{\nu}_{\bar{a}}(\vec{R}),$$
$$I^{\nu\mu}_{G_{\bar{a}\bar{b}}}(\vec{R},\vec{R}';t) = [1-P(t)]\frac{1}{q}\psi^{-1}(t)\dot{\hat{G}}^{\nu\mu}_{\bar{a}\bar{b}}(\vec{R},\vec{R}') \tag{4.5}$$

are the generalized flows describing reaction-diffusion processes. The function $\psi(t)$ is defined by the relation

$$\psi(t) = 1 - \frac{q-1}{q}\beta\Bigg[\delta H(t) - \sum_a \int \mathrm{d}\vec{r}\mu_a(\vec{r};t)\delta\hat{n}_a(\vec{r};t)$$
$$-\sum_{\bar{a}}\sum_{\nu}\int \mathrm{d}\vec{R}\mu^{\nu}_{\bar{a}}(\vec{R};t)\delta\hat{n}^{\nu}_{\bar{a}}(\vec{R};t) - \sum_{\bar{a}\bar{b}}\sum_{\nu\mu}\int \mathrm{d}\vec{R}\mathrm{d}\vec{R}' M^{\nu\mu}_{\bar{a}\bar{b}}(\vec{R},\vec{R}';t)\hat{G}^{\nu\mu}_{\bar{a}\bar{b}}(\vec{R},\vec{R}';t)\Bigg]. \tag{4.6}$$

By means of NSO (4.4), we obtain the set of self-consistent generalized transport equations for averaged densities of adsorbed and non-absorbed atoms along with nonequilibrium pair distribution function of absorbed atoms (or molecules). It has the following form

$$\frac{\partial}{\partial t}\langle\hat{n}_a(\vec{r})\rangle^t = \langle\dot{\hat{n}}_a(\vec{r})\rangle^t_{\mathrm{rel}} + \sum_b \int \mathrm{d}\vec{r}' \int_{-\infty}^{t} \mathrm{e}^{\varepsilon(t'-t)}\varphi_{n_a n_b}(\vec{r},\vec{r}';t,t')\beta\mu_b(\vec{r}';t')\mathrm{d}t'$$
$$+\sum_{\bar{b}}\sum_{\nu'}\int \mathrm{d}\vec{R}' \int_{-\infty}^{t} \mathrm{e}^{\varepsilon(t'-t)}\varphi^{\nu'}_{n_a n_{\bar{b}}}(\vec{r},\vec{R}';t,t')\beta\mu^{\nu'}_{\bar{b}}(\vec{R}';t')\mathrm{d}t'$$
$$+\sum_{\bar{a}'\bar{b}}\sum_{\nu'\mu'}\int \mathrm{d}\vec{R}'\mathrm{d}\vec{R}'' \int_{-\infty}^{t} \mathrm{e}^{\varepsilon(t'-t)}\varphi^{\nu'\mu'}_{n_a G_{\bar{a}'\bar{b}}}(\vec{r},\vec{R}',\vec{R}'';t,t')\beta M^{\nu'\mu'}_{\bar{a}'\bar{b}}(\vec{R}',\vec{R}'';t')\mathrm{d}t', \tag{4.7}$$

$$\frac{\partial}{\partial t}\langle\hat{n}^{\nu}_{\bar{a}}(\vec{R})\rangle^t = \langle\frac{1}{\mathrm{i}\hbar}[\hat{n}^{\nu}_{\bar{a}}(\vec{R}),H']\rangle^t_{\mathrm{rel}} + \langle\frac{1}{\mathrm{i}\hbar}[\hat{n}^{\nu}_{\bar{a}}(\vec{R}),H_{\mathrm{reac}}]\rangle^t_{\mathrm{rel}}$$
$$+\sum_b \int \mathrm{d}\vec{r}' \int_{-\infty}^{t} \mathrm{e}^{\varepsilon(t'-t)}\varphi^{\nu}_{n_{\bar{a}} n_b}(\vec{R},\vec{r}';t,t')\beta\mu_b(\vec{r}';t')\mathrm{d}t'$$
$$+\sum_{\bar{b}}\sum_{\nu'}\int \mathrm{d}\vec{R}' \int_{-\infty}^{t} \mathrm{e}^{\varepsilon(t'-t)}\varphi^{\nu\nu'}_{n_{\bar{a}} n_{\bar{b}}}(\vec{R},\vec{R}';t,t')\beta\mu^{\nu'}_{\bar{b}}(\vec{R}';t')\mathrm{d}t'$$
$$+\sum_{\bar{a}'\bar{b}}\sum_{\nu'\mu'}\int \mathrm{d}\vec{R}'\mathrm{d}\vec{R}'' \int_{-\infty}^{t} \mathrm{e}^{\varepsilon(t'-t)}\varphi^{\nu'\mu'}_{n_{\bar{a}} G_{\bar{a}'\bar{b}}}(\vec{r},\vec{R}',\vec{R}'';t,t')\beta M^{\nu'\mu'}_{\bar{a}'\bar{b}}(\vec{R}',\vec{R}'';t')\mathrm{d}t', \tag{4.8}$$

$$\frac{\partial}{\partial t}\langle\hat{G}^{\nu\mu}_{\bar{a}\bar{b}}(\vec{R},\vec{R}')\rangle^t = \langle\frac{1}{\mathrm{i}\hbar}[\hat{G}^{\nu\mu}_{\bar{a}\bar{b}}(\vec{R},\vec{R}'),H']\rangle^t_{\mathrm{rel}} + \langle\frac{1}{\mathrm{i}\hbar}[\hat{G}^{\nu\mu}_{\bar{a}\bar{b}}(\vec{R},\vec{R}'),H_{\mathrm{reac}}]\rangle^t_{\mathrm{rel}}$$
$$+\sum_{b'}\int \mathrm{d}\vec{r}' \int_{-\infty}^{t} \mathrm{e}^{\varepsilon(t'-t)}\varphi^{\nu\mu}_{G_{\bar{a}\bar{b}} n_{b'}}(\vec{R},\vec{R}';t,t')\beta\mu_{b'}(\vec{r}';t')\mathrm{d}t'$$
$$+\sum_{\bar{b}}\sum_{\nu'}\int \mathrm{d}\vec{R}'' \int_{-\infty}^{t} \mathrm{e}^{\varepsilon(t'-t)}\varphi^{\nu\mu\nu'}_{G_{\bar{a}\bar{b}} n_{\bar{b}'}}(\vec{R},\vec{R}',\vec{R}'';t,t')\beta\mu^{\nu'}_{\bar{b}}(\vec{R}';t')\mathrm{d}t'$$
$$+\sum_{\bar{a}'\bar{b}'}\sum_{\nu'\mu'}\int \mathrm{d}\vec{R}''\mathrm{d}\vec{R}''' \int_{-\infty}^{t} \mathrm{e}^{\varepsilon(t'-t)}\varphi^{\nu\mu\nu'\mu'}_{G_{\bar{a}\bar{b}} G_{\bar{a}'\bar{b}}}(\vec{r},\vec{R}',\vec{R}'',\vec{R}''';t,t')\beta M^{\nu'\mu'}_{\bar{a}'\bar{b}'}(\vec{R}'',\vec{R}''';t')\mathrm{d}t', \tag{4.9}$$

where $\varphi_{n_a n_b}, \varphi^{\nu\nu'}_{n_{\bar{a}} n_{\bar{b}}}, \varphi^{\nu}_{n_a n_{\bar{b}}}, \varphi^{\nu\mu\nu'\mu'}_{G_{\bar{a}\bar{b}} G_{\bar{a}'\bar{b}}}$ are the generalized transport kernels. The second term in the right-hand side of (4.8), $\langle\frac{1}{\mathrm{i}\hbar}[\hat{n}^{\nu}_{\bar{a}}(\vec{R}),H_{\mathrm{reac}}]\rangle^t_{\mathrm{rel}}$, defines an averaged value of the operator of the rate of the reaction between adsorbed atoms on the metal surface. Transport kernels are built on the generalized flows (4.5) taking into account the contributions of amplitudes of chemical reactions in the flows $I^{\nu}_{\bar{a}}(\vec{R};t')$ and $I^{\nu\mu}_{G_{\bar{a}\bar{b}}}(\vec{R},\vec{R}';t')$, and have the following form

$$\varphi_{BB'} = \mathrm{Sp}\left\{ \dot{\hat{P}}_B T(t,t') \int_0^1 \mathrm{d}\tau \varrho^{\tau}_{\mathrm{rel}}(t') I_{B'}(t') \varrho^{1-\tau}_{\mathrm{rel}}(t') \right\}. \tag{4.10}$$

In particular, $\varphi_{n_a n_b}(\vec{r},\vec{r}';t,t')$ describes dynamical correlations of diffusive flows of gas atoms and is connected to the inhomogeneous diffusion coefficient of atoms (or molecules) $D_{ab}(\vec{r},\vec{r}';t)$. Similarly, the transport kernel $\varphi^{\nu\nu'}_{n_{\bar{a}}n_{\bar{b}}}(\vec{R},\vec{R}';t,t')$ describes dynamical dissipative correlations of diffusive flows of atoms adsorbed in the states $\nu$ and $\nu'$ on the metal surface and determines an inhomogeneous diffusion coefficient of the atoms adsorbed on the metal surface $D^{\nu\nu'}_{\bar{a}\bar{b}}(\vec{R},\vec{R}';t)$. Transport kernel $\varphi^{\nu}_{n_{\bar{a}}n_b}(\vec{R},\vec{r}';t,t')$, $\varphi^{\nu'}_{n_a n_{\bar{b}}}(\vec{r},\vec{R}';t,t')$ describes dynamical dissipative correlations between the flows of gas atoms and the atoms adsorbed on the metal surface, and determines the inhomogeneous coefficient of mutual diffusion "gas atom-adsorbed atom" $D^{\nu'}_{a\bar{b}}(\vec{r},\vec{R}';t)$. It is very important to investigate these diffusion coefficients. The transport kernel $\varphi^{\nu\mu}_{G_{\bar{a}\bar{b}}p}$ $(p = n, \bar{n})$ describes dissipative correlations of the flows and densities of adsorbed atoms with the flows of atoms, molecules and adsorbed atoms. Memory function $\varphi^{\nu\mu\nu'\mu'}_{G_{\bar{a}\bar{b}}G_{\bar{a}'\bar{b}}}(\vec{R},\vec{R}',\vec{R}'',\vec{R}''';t,t')$ describes reaction-diffusion processes between the atoms adsorbed on the metal surface. They are higher memory functions with respect to dynamical variables $G^{\nu\mu}_{\bar{a}\bar{b}}$. We note here that at $q \to 1$, the generalized equations of reaction-diffusion processes correspond to the ones within the Gibbs statistics.

## 5. Conclusions

Summarizing, we proposed an approach to the formulation of extensive statistical mechanics of nonequilibrium processes based on the Zubarev NSO method and maximum entropy principle for Renyi entropy. We consider a $q$-parametrized Liouville equation which leads to Lindblad type equation at $|1-q|\Omega t \ll 1$. The solution of this $q$-parametrized Liouville equation is obtained within the NSO method. The proposed approach is used to describe the reaction-diffusion processes which are relevant in catalytic nanotechnologies. Consequently, we obtain generalized transport equations (4.7)–(4.9) for the nonequilibrium averaged densities of adsorbed and non-adsorbed atoms of consistent description of the reaction-diffusion processes in the system "gas-adsorbate-metal" within Renyi statistics. At $q = 1$, these equations coincide with the equations of reaction-diffusion processes within Gibbs statistics [49]. As we can see, these equations are nonlinear and spatially inhomogeneous. They can describe strong as well as weak nonequilibrium processes in a system. The application of the obtained results to the consideration of a particular physical model will be done in our forthcoming works.

## References


1. Tsallis C., J. Stat. Phys., 1988, **52**, No. 1–2, 479; doi:10.1007/BF01016429.
2. Renyi A., Probability theory, North-Holland, Amsterdam, 1970.
3. Selected papers by Alfred Renyi, Vol. 2, Turan P. (Ed.), Akademiai Kiado, Budapest, 1976.
4. Sharma B.D., Mittal D.P., J. Math. Sci., 1975, **1**, 28.
5. Akturk E., Bagci G.B., Sever R., Preprint arXiv:cond-mat/0703277, 2007.
6. Beck C., Continuum Mech. Thermodyn., 2004, **16**, 293; doi:10.1007/s00161-003-0145-1.
7. Beck C., In: Anomalous Transport: Foundations and Applications, Klages R., Radons G., Sokolov I.M. (Eds.), Wiley-VCH, New York, 2008, chap. 15; doi:10.1002/9783527622979.ch15.
8. Olemskoi A.I., Katsnelson A.A., Synergetics of Condensed Matter, Editorial URSS, Moscow, 2003.
9. Olemskoi A.I., Theory of Structure Transformations in Non-Equilibrium Condensed Matter, Horizons in World Physics Series, Vol. 231, NOVA Science Publishers, New York, 1999,
10. Olemskoi A.I., Sklyar I.A., Sov. Phys. Usp., 1992, **35**, No. 6, 455; doi:10.1070/PU1992v035n06ABEH002241.
11. Olemskoi A.I., Savelyev A., Phys. Rep., 2005, **416**, No. 4–5, 145; doi:10.1016/j.physrep.2005.08.003.
12. Olemskoi A.I., Koplyk I.V., Phys. Usp., 1995, **38**, No. 10, 1061; doi:10.1070/PU1995v038n10ABEH000112.

13. Nonextensive Statistical Mechanics and its Applications, Abe S., Okamoto Y. (Eds.), Spriger-Verlag, Heidelberg, 2001.
14. Nonextensive Entropy — Interdisciplinary Applications, Gell-Mann M., Tsallis C. (Eds.), Oxford University Press, New York, 2004.
15. Introduction to Nonextensive Statistical Mechanics (Approaching a Complex World), Tsallis C. (Ed.), Springer, New York, 2009.
16. Essex C., Schulzky C., Franz A., Hoffmann K.H., Physica A, 2000, **284**, 299; doi:10.1016/S0378-4371(00)00174-6.
17. Boon J.P., Lutsko J.F., Physica A, 2006, **368**, 55; doi:10.1016/j.physa.2005.11.054.
18. Arimitsu T., Arimitsu N., Phys. Rev. E, 2000, **61**, 3237; doi:10.1103/PhysRevE.61.3237.
19. Ramos F.M., Rosa R.R., Neto C.R., Bolzan M.J.A., Abreu Sa L.D., Campos Velho H.F., Physica A, 2001, **295**, 250; doi:10.1016/S0378-4371(01)00083-8.
20. Bezerra J.R., Silva R., Lima J.A.S., Physica A, 2003, **322**, 256; doi:10.1016/S0378-4371(02)01813-7.
21. Kirchanov V.S., Theor. Math. Phys., 2008, **156**, No. 3, 1347; doi:10.1007/s11232-008-0111-y.
22. Feng Z-H., Liu L-Y., Physica A, 2010, **389**, 237; doi:10.1016/j.physa.2009.09.005.
23. Du J.L., Phys. Lett. A, 2004, **329**, 262; doi:10.1016/j.physleta.2004.07.010.
24. Du J.L., Physica A, 2004, **335**, 107; doi:10.1016/j.physa.2003.11.027.
25. Yulmetyev R.M., Gafarov F.M., Yulmetyeva D.G., Emeljanova N.A., Physica A, 2002, **303**, 427; doi:10.1016/S0378-4371(01)00561-1.
26. Yulmetyev R.M., Emeljanova N.A., Gafarov F.M., Physica A, 2004, **341**, 649; doi:10.1016/j.physa.2004.03.094.
27. Plastino A.R., Casas M., Plastino A., Physica A, 2000, **280**, 289; doi:10.1016/S0378-4371(00)00006-6.
28. Niven R.K., Chem. Eng. Sci., 2006, **61**, 3785; doi:10.1016/j.ces.2005.12.004.
29. Bagci G.B., Physica A, 2007, **386**, 79; doi:10.1016/j.physa.2007.06.045.
30. Bashkirov A.G., Vityazev A.V., Physica A, 2000, **277**, 136; doi:10.1016/S0378-4371(99)00449-5.
31. Jizba P., Arimitsu T., Ann. Phys., 2004, **312**, 17; doi:10.1016/j.aop.2004.01.002.
32. Bashkirov A.G., Sukhanov A.D., J. Exp. Theor. Phys., 2002, **95**, No. 3, 440; doi:10.1134/1.1513816.
33. Bashkirov A.G., Physica A, 2004, **340**, 153; doi:10.1016/j.physa.2004.04.002.
34. Markiv B., Tokarchuk R., Kostrobii P., Tokarchuk M., Physica A, 2011, **390**, 785; doi:10.1016/j.physa.2010.11.009.
35. Parvan A.S., Biro T.S., Phys. Lett. A, 2005, **340**, 375; doi:10.1016/j.physleta.2005.04.036.
36. Parvan A.S., Biro T.S., Phys. Lett. A, 2010, **374**, 1952; doi:10.1016/j.physleta.2010.03.007.
37. Figueiredo A., Amato M.A., Filho M.T.R., Physica A, 2006, **367**, 191; doi:10.1016/j.physa.2005.11.036.
38. Luzzi R., Vasconcellos A.R., Ramos J.G., Riv. Nuovo Cimento, 2007, **30**, No. 3, 95; doi:10.1393/ncr/i2007-10018-6.
39. Vasconcellos A.R., Ramos J.G., Gorenstein A., Kleinke M.U., Souza Cruz T.G., Luzzi R., Int. J. Mod. Phys. B, 2006, **20**, 4821; doi:10.1142/S0217979206035667.
40. Vidiella-Barranco A., Moya-Cessa H., Phys. Lett. A, 2001, **279**, 56; doi:10.1016/S0375-9601(00)00820-3.
41. Milburn G.J., Phys. Rev. A, 1991, **44**, 5401; doi:10.1103/PhysRevA.44.5401.
42. Bonifacio R., Olivares S., Tombesi P., Vitali D., Phys. Rev. A, 2000, **61**, 053802; doi:10.1103/PhysRevA.61.053802.
43. Kaniadakis G., Physica A, 2001, **296**, 405; doi:10.1016/S0378-4371(01)00184-4.
44. Borland L., Phys. Rev. E, 1998, **57**, 6634; doi:10.1103/PhysRevE.57.6634.
45. Frank T.D., Nonlinear Fokker-Planck Equations. Fundamentals and Applications, Springer, New York, 2004.
46. Zubarev D.N., Nonequilibrium Statistical Thermodynamics, Consultant Bureau, New-York, 1974.
47. Zubarev D.N., Morozov V.G., Röpke G. Statistical Mechanics of Nonequilibrium Processes, Vol. 1, Moscow, Fizmatlit, 2002 (in Russian).
48. Zubarev D.N., Morozov V.G., Röpke G. Statistical Mechanics of Nonequilibrium Processes, Vol. 2, Moscow, Fizmatlit, 2002 (in Russian).
49. Kostrobii P.P., Tokarchuk M.V., Markovych B.M., Ignatyuk V.V., Gnativ B.V., Reaction-Diffusion Processes in the "Metal-Gas" Systems, Publishing House of National University "Lviv Polytechnic", Lviv, 2009 (in Ukrainian).
50. Ala-Nissila T., Ferrando R., Ying S.C., Adv. Phys., 2002, **51**, 949; doi:10.1080/00018730110107902.
51. Ignatyuk V.V., Phys. Rev. E, 2009, **80**, 041133; doi:10.1103/PhysRevE.80.041133.
52. Ignatyuk V.V., Phys. Rev. E, 2011, **84**, 021111; doi:10.1103/PhysRevE.84.021111.
53. Ignatyuk V.V., J. Chem. Phys., 2012, **136**, 184104; doi:10.1063/1.4711863.
54. Galenko P.K., Kharchenko D., Lysenko I., Physica A, 2010, **389**, 3443; doi:10.1016/j.physa.2010.05.002.
55. Sen Sh., Ghosh P., Riaz S.Sh., Ray D.Sh., Phys. Rev. E, 2010, **81**, 017101; doi:10.1103/PhysRevE.81.017101.
56. Kharchenko V.O., Kharchenko D.O., Kokhan S.V., Vernigora I.V., Yanovsky V.V., 2012, Phys. Scr., **86**, 055401; doi:10.1088/0031-8949/86/05/055401.

# 19

# Analytical calculation of the Stokes drag of the spherical particle in a nematic liquid crystal

M.V. Kozachok[1,2], B.I. Lev[1]

[1] Bogolyubov Institute for Theoretical Physics of the National Academy of Sciences of Ukraine, 14-b Metrolohichna St., 03680 Kyiv, Ukraine

[2] Open International University of Human Development "Ukraine", 23 Lvivska St., Kyiv, Ukraine

As an approach to the motion of particles in an anisotropic liquid, we analytically study the Stokes drag of spherical particles in a nematic liquid crystal. The Stokes drag of spherical particles for a general anisotropic case is derived in terms of multipoles. In the case of weak anchoring, we use the well-known distribution of the elastic director field around the spherical particle. In the case of strong anchoring, the multipole expansion may be also used by modifying the size of a particle to the size of the deformation coating. For the case of zero anchoring (uniform director field) we found that the viscosities along the director $\eta_{\parallel}$ and perpendicular direction $\eta_{\perp}$ are almost the same, which is quite reasonable because in this case the liquid behaves as isotropic. In the case of non-zero anchoring, the general ratio $\eta_{\parallel}/\eta_{\perp}$ is about 2 which is satisfied by experimental observations.

**Key words:** *Stokes drag, liquid crystal, diffusion, viscosity*

## 1. Introduction

Colloidal particles in liquid crystals (LC) have attracted a great research interest during the recent years. Anisotropic properties of the host fluid-liquid crystal give rise to a new class of colloidal anisotropic interactions that never occur in isotropic hosts. Liquid crystal colloidal systems have shown much recent interest as the models for diverse phenomena in condense matter physics. Particles suspended in a fluid are under the effect of the hits from the surrounding particles and perform Brownian motion. They perform random walk whose diffusion constant obeys the famous Stokes-Einstein relation. A simple Langevine approach predicts that the velocity autocorrelation function of random walkers decays exponentially [1]. The drag force can be derived from the Navier-Stokes equations with an additional assumption on the character of the random force. The Navier-Stokes equations, which describe the hydrodynamic behavior of fluids, assume that molecules are point particles or smooth spheres and, as a consequence, do not exert a torque on one another. These equations originate from the conservation of mass, linear momentum and energy during the collision processes. If the particles in a fluid are of non-spherical shape, they can induce rotation to each other during the collisions and the energy can be transferred from the translational motion to the rotational motion. During these collisions, the total angular momentum of colliding particles should be conserved. The requirement that the angular momentum should be conserved together with the Navier-Stokes equations leads to a complete hydrodynamic description of the fluid. Such a complete hydrodynamic description was applied to the fluid composed of finite-sized spherical particles with internal rotational degrees of freedom and it is shown that the friction force becomes memory dependent even for this simple liquid [2].

In anisotropic liquids, the rod-like organic molecules align, on average, along a common direction indicated by a unit vector $\vec{n}$ called director. For this case, to find the drag force we need to solve the dynamic equations of a nematic liquid crystal LC, i.e., the Ericksen-Leslie equations. In these equations, the independent variables — the director and the fluid velocity — are coupled and this fact causes the complexity of these equations. Thus, only a few examples with analytical solution exist, e.g., the flow between two parallel plates which defines the different Miesowicz viscosities [3], the Couette flow [4, 5], the Poiseuille flow [6] which was first measured by Cladis et al. [7], or back flow [8]. It is expected that the knowledge of the more or less general solutions of these equations will shed light upon some effects. The solutions of the Ericksen-Leslie equations are also of technological interest since they are indispensable for determining the switching times of liquid-crystal displays.

Every particle immersed in a liquid crystal produces a deformation director field around the particle if the LC molecules are specifically anchored to the closed surface. In the case of a weak anchoring, the area of deformation of the director field around the particle is small and every deformation of the director field can be presented as a small deformation of the ground state, which represents the orientation of all the molecules in one direction.

In the case of a strong anchoring, we have a distortion director field around the immersed particle, which can be called a dipole or quadrupole configuration [9] (figure 1). This configuration directly depends on the strength of coupling with the surface and on the size of particles. In this case, it is necessary to describe the possible configurations and to note that in the long-range distance we have a configuration which shows the same behavior as in the case of week anchoring represented by a multiple expansion. There exist two approaches to describe the distribution of the director field at a short and long distance from the immersed particle. The first theoretical approach was developed in [9] combining the ansatz functions for the director and the use of the multiple expansion in the far field area. The authors investigated spherical particles with hyperbolic hedgehog and found dipole and quadruple elastic interactions between such particles. Another approach [10] made it possible to find approximate solutions in terms of the geometrical shape of particles for the case of small anchoring strength and has provided the way to connect the type of the interaction potential with the local symmetry of the director field around the particles [11]. The concept of coating has been introduced that contains all the topological defects located inside and carries the symmetry of the director, and enables us to qualitatively determine the type of the interaction potential. However, the coating is not quantitatively exactly defined. The configuration of director distribution plays a crucial role when the particle moves through a liquid crystal.

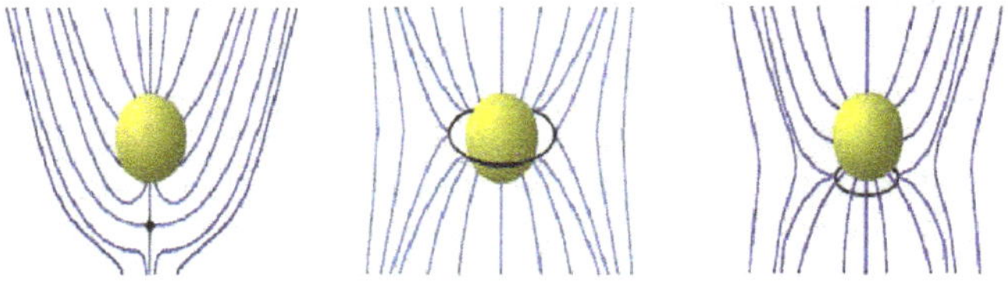

**Figure 1.** (Color online) The distortion of the molecules around the spherical particle in the case of the strong anchoring. We can see that the change of the distortion of the director near the particle is very strong. The form of the distortion of the director field in the case of the strong anchoring was theoretically obtained in article [31].

The hydrodynamic solution for the flow of a nematic liquid crystal around a particle at rest, which is equivalent to the problem of a moving particle, still requires its full result. The experiment with the inverted nematic emulsion [12, 13] and investigations by Ruhwandl and Terentjev [14] urged Stark and Ventzki [15–17] to perform Stokes drag calculations for a particle in a nematic environment, especially for the particle-defect dipole. They concentrated on the low Eriksen numbers, where the director field is not affected by the velocity field. The authors presented streamline patterns, interpreted them, calculated Stokes drags for motions parallel and perpendicular to the overall symmetry axis, and compared the results to the Saturn-ring configuration and a uniform director-field. Heuer et al. presented analytical and numerical solutions for both the velocity field and the Stokes drag where the director field was kept uniform [18, 19]. They were the first to investigate a cylinder of infinite length [20]. Diogo [21] put the velocity field to be the same as the one for an isotropic fluid and calculated the drag force for simple director configurations. He investigated the case where the viscous forces largely exceed the elastic forces from director distortions, i.e., Ericksen numbers much larger than one, as it was explained in the [15]. Roman and Terentjev, have focused on the opposite case. They obtained an analytical solution for the flow velocity in a spatially uniform director field by an expansion in the anisotropy of the viscosities [22]. Chono and Tsuji performed a numerical solution of the Ericksen-Leslie equations around a cylinder determining both the velocity and director field [23]. They found that the director field strongly depends on the Eriksen number, but for homeotropic anchoring their director fields did not exhibit any topological defects required by the boundary conditions signaling about some shortcomings in the exploration. Billeter and Pelcovits used molecular-dynamic simulations to determine the Stokes drag of very small particles [24]. They observed that the Saturn ring is strongly deformed due to the motion of the particles. Ruhwandl and Terentjev have investigated a nonuniform but fixed director configuration, and numerically calculated the velocity field and Stokes drag of a cylinder [25] or spherical particle [14]. The particle was surrounded by the Saturn-ring configuration, and the cylinder was accompanied by two disclination lines. It is known when a particle is surrounded by a disclination ring, the Stokes drag strongly depends on the presence of line defects. There are a few studies that determine both experimentally [26] and theoretically [27–29] the drag force of a moving disclination.

We cannot fully describe all the effects associated with the possible configurations of the director field around the immersed particle, but we attempt to find a general motive of the change dissipation energy of the moving particle in a liquid crystal. First of all, we focus on increasing the effective mass of the immersed colloidal particle and analytically calculate the Stokes drag for colloidal particles in a nematic liquid crystal.

## 2. Theory and details of calculations

The essence of this paper is the calculation of the Stokes drag of a spherical particle in a nematic liquid crystal when the angle between the director and particle velocity is arbitrary. In other words, we have calculated the Stokes drag of the spherical particle in a nematic liquid crystal for the fully anisotropic case. The drag force is caused by the interaction of the particles of the fluid and a foreign body immersed in it. As we mentioned in the introduction, every particle immersed in a liquid crystal is dressed in a deformation coating with the region of deformation of the director field at the distance of the correlation length. The efficacy of the coating was investigated in [30]. To describe this phenomena we can also use the results on the inertial characteristic and viscosity, which present a different approach to the motion of the immersed particles. The inertial characteristic is the effective mass which is an analogue of the hydrodynamical mass in the usual hydrodynamics. Every moving particle immersed in a liquid crystal has two principal different characteristics. One is an inertial characteristic as an effective mass and another characteristic determines the dissipative part. When the particle moves, the region of deformation — i.e., coating, moves too. This causes an increase of its inertia mass. Under these conditions, the effective mass becomes the anisotropic value and can be expressed via formulas [30]:

$$m_{\text{eff}}^{\perp} = m + I\int d\vec{r}\left\{\left(\frac{\partial \vec{n}}{\partial x}\right)^2 + \left(\frac{\partial \vec{n}}{\partial y}\right)^2\right\}, \tag{2.1}$$

$$m_{\text{eff}}^{\parallel} = m + I\int d\vec{r}\left(\frac{\partial \vec{n}}{\partial z}\right)^2, \tag{2.2}$$

where $I$ is density of the moment inertia of the liquid crystal. To determine the inertial characteristic, we can use the distribution of the elastic director around the particles. As was shown in [31], in the case of a weak anchoring, when only small deviations of the director for homeotropic boundary conditions on the surface of a particle are expected, the problem can be linearized, and to describe the director field one can use the two principal angles of a spherical coordinate system $n_z = \cos\beta(\vec{r}), n_x = \sin\beta(\vec{r})\cos\phi$, and $n_y = \sin\beta(\vec{r})\sin\phi$, where $\phi$ is the azimuthal angle, thus respecting an obvious azimuthal symmetry of the problem. At a small anchoring $\beta \ll 1$, the director rotation angle takes the form

$$\beta = \frac{WR}{4K}\left(\frac{R}{r}\right)^3 \sin 2\theta. \tag{2.3}$$

If we substitute the known director field distribution for weak anchoring, we get the value of the inertial effective mass:

$$m_{\text{eff}} = m + \frac{4I}{3}\left(\frac{W}{4K}\right)^2 R^3, \tag{2.4}$$

which can be by an order higher than the mass of the immersed particle [30]. It is analogue of the hydrodynamic mass for the moving particle in an ordinary liquid.

The friction force for a spherical region of a radius $R/\varepsilon$ with a centered particle within is expressed via formula [15].

The same arguments relate to the viscosity coefficient. The theoretical calculations [16, 17] revealed that the viscosity coefficient depends on the configuration of the director distribution and is much bigger than in an ordinary viscose liquid. The essence of this phenomena can be understood from simple considerations. Every particle that moves in the viscose environment undergoes the action of the additional friction force, which is described by Stokes formula $f = 6\pi\eta R$, where $\eta$ is the friction coefficient, which is associated with a diffusion coefficient of the Brownian particle via the relation $D = (kT)/(6\pi\eta R)$. The friction force for the spherical region of radius $R/\varepsilon$ with a centered particle within is expressed via formula [15]

$$f = 6\pi\eta R \frac{1 - \frac{3\varepsilon}{2} + \varepsilon^3 - \frac{\varepsilon^5}{2}}{\left(1 - \frac{3\varepsilon}{2} + \varepsilon^3\right)^2}. \tag{2.5}$$

From this formula it is easy to see that the friction force increases if the particle is inside the shell. It can be a solvate shell and in the case of the liquid crystal this is the region of a strong change of a director deformation. If we now take into account the configuration of the director distribution around the spherical inclusion, then the diffusion of this inclusion will depend on the direction of the motion with regard to equilibrium director distribution. This leads to the anisotropy of the diffusion coefficient and to a dependence of these coefficients on the conditions of anchoring on the surface of the inclusion. The results of numerical calculations of these phenomena can be found in [16, 17].

To determine the character and the value of the Stokes drag of the spherical particle in a nematic liquid crystal, we can use different distributions of elastic director field around it. The stress tensor $\sigma_{ik}$ is used to calculate the Stokes drag force [32]. From the known stress tensor $\sigma_{ik}$, the drag force can be calculated by the following formula [2]:

$$F_i = \int \sigma_{ij} \mathrm{d}s_j. \tag{2.6}$$

The expression for the stress tensor $\sigma_{ik}$ in a nematic environment is well known and can be found in the literature [32]

$$\sigma_{ik} = -p\delta_{ik} + \sigma_{ik}^{(r)} + \sigma'_{ik}. \tag{2.7}$$

Here, $p$ is macroscopic pressure, $\sigma_{ik}^{(r)}$ is "reactive" part of stress tensor and $\sigma'_{ik}$ is a dissipative part of stress tensor. The expressions for "reactive" and dissipative parts of stress tensor can be found in [32]

$$\sigma_{ik}^{(r)} = -\pi_{kl}\partial_i n_l - \frac{\lambda}{2}(n_i h_k + n_k h_i) + \frac{1}{2}(n_i h_k - n_k h_i), \tag{2.8}$$

$$\begin{aligned} \sigma'_{ik} = {} & 2\eta_1 v_{ik} + (\eta_2 - \eta_1)\delta_{ik} v_{ll} \\ & + (\eta_4 + \eta_1 - \eta_2)(\delta_{ik} n_l n_m v_{lm} + n_i n_k v_{ll}) \\ & + (\eta_3 - 2\eta_1)(n_i n_l v_{kl} + n_k n_i v_{il}) \\ & + (\eta_5 + \eta_1 + \eta_2 - 2\eta_3 - 2\eta_4) n_i n_k n_l n_m v_{lm}. \end{aligned} \tag{2.9}$$

To find the stress tensor we need to know the solution of the Eriksen-Leslie equations that link the director field and the fluid velocity. The general solution of these equations is a challenge to a theorist.

Here we suggest an approach for finding the stress tensor. As the first step we use the director structure around a colloid particle suspended in a nematic liquid crystal, found in [31]. We assume here the situation when a spherical particle moves slowly and the nematic liquid crystal environment has enough time to relax to the equilibrium state during the motion of a spherical particle. We consider a smooth hard sphere which moves through the fluid with the velocity $\vec{u}(t) = u(t)\vec{e}_z$. The fact that it is smooth means that no torques and no force directed tangent to its surface can be exerted on it. Under these conditions, only the component $\sigma_{rr}$ of the stress tensor contributes to the drag force. Since we use the director field for equilibrium state, the "reactive" part of stress tensor will not contribute to the drag force, but only a dissipative part. It is obvious that the drag force $\vec{F}$ has the same direction as the velocity of a spherical particle $\vec{u}$, and formula (2.6) will reduce to the following:

$$F = \int \sigma'_{rr} \cos\theta ds. \tag{2.10}$$

Substituting the components of director field from [31] and the components for velocity field which are the same as the one for an isotropic fluid [33] in the stress tensor and keeping terms up to the first order of small parameter $\beta$, we have obtained the Stokes drag of spherical particle in a nematic environment at weak anchoring

$$\begin{aligned} F &= 4\pi R\eta_1 u + 3\pi R u(\eta_1 + \eta_2 - \eta_3 - \eta_4)\left(-0.27 - 0.02\frac{WR}{K}\right) \\ &\quad + 3\pi R u(\eta_5 + \eta_1 + \eta_2 - 2\eta_3 - 2\eta_4)\left(0.11 + \frac{WR}{K}\right). \end{aligned} \tag{2.11}$$

The presented director field structure contains configurations of the director field at a small anchoring. We would like to determine the Stokes drag in the case of strong anchoring. The task of finding a director distribution around a spherical particle consists in minimizing the Frank free-energy functional with boundary conditions provided by it and by the surface energy. Generally, this class of problems is not solvable analytically due to its nonlinearity brought in by the unit-vector constraint $|\vec{n}(\vec{r})|^2 = 1$. In [34], in particular, the director distribution was obtained in the one-constant approximation in terms of the multipole expansion. However, the expansion coefficients were not associated with the physical and geometrical parameters of macroparticles. In [10], the director distribution is derived for the general case of different elastic Frank constants and, moreover, the multipole expansion parameters are found in terms of geometric and physical characteristics of macroparticles. Thus, both the behavior and the value of the pair interaction energy are described with no additional restrictions. However, only in [9] there was proposed a theoretical approach combining the ansatz functions for the director field and the use of the multiple expansion in the far field area that was a satisfactory solution to many problems. Thus, the use of the director field in terms of the multipole expansion becomes particularly relevant. In the framework of this approach we have

$$n_x = p_z\frac{x}{r^3} + 2c\frac{zx}{r^5}, \qquad n_y = p_z\frac{y}{r^3} + 2c\frac{zy}{r^5}, \qquad n_z = 1 - \frac{1}{2}\left(n_x^2 + n_y^2\right), \tag{2.12}$$

where the vector $\vec{p}$ is the dipole moment of the droplet-defect configuration and the parameter $c$ is the amplitude of the quadrupole moment tensor $c_{ij}$ of the particle-defect configuration. Assuming that, at small anchoring, the director deviates from its uniform orientation $\vec{n}_0 \| \vec{e}_z$ by only a small amount, we can consider $n_x$ and $n_y$ as small parameters. Repeating all the steps as in the first approach, except that now the small parameters are $n_x$ and $n_y$, we obtain the Stokes drag of a spherical particle in a nematic environment at a weak anchoring in the framework of the present approach

$$\begin{aligned} F &= 4\pi R\eta_1 u + (\eta_4 + \eta_3 - \eta_2 - \eta_1)\left[\cos^2 2\Omega \cdot \frac{0.06u\pi c}{R^2} + \cos 2\Omega \cdot \left(\frac{0.18u\pi c}{R^2} + 0.6u\pi R\right) + 0.2u\pi R\right] \\ &\quad + (\eta_5 + \eta_1 + \eta_2 - 2\eta_3 - \eta_4)\left[(1 + \sin 2\Omega)\left(-0.08\cos^4\Omega + 0.44\cos^2\Omega - 0.12\sin\Omega - 0.24\right)\left(\frac{u\pi c}{R^2}\right)\right. \\ &\quad + \left(\frac{u\pi c}{R^2}\right)\left(0.17\cos 2\Omega + 0.6\cos^6\Omega - 0.12\cos 4\Omega - 0.06\cos 6\Omega\right) + 3u\pi R\left(0.08\cos^2\Omega \cdot \sin^2\Omega\right. \\ &\quad \left.\left. - 0.06\sin^4\Omega\right) + 0.06u\pi R\sin^2 2\Omega + 0.165u\pi R\left(3\cos^4\Omega - \cos^2\Omega\right)\right]. \end{aligned} \tag{2.13}$$

**Table 1.** The viscosities for zero and planar anchoring. The middle column is obtained from equations (2.14) and (2.15) with the Leslie coefficients of 5CB and MBBA [38, 39]; the last one is derived from Stark's numerical calculations [36].

| Uniform director | | Present work | Numerically exact |
|---|---|---|---|
| 5CB | $\eta_{\parallel}$ | 0.27 | 0.38 |
| | $\eta_{\perp}$ | 0.24 | 0.75 |
| MBBA | $\eta_{\parallel}$ | 0.28 | 0.38 |
| | $\eta_{\perp}$ | 0.27 | 0.68 |
| Planar anchoring | | | |
| 5CB | $\eta_{\parallel}$ | 0.16 | – |
| | $\eta_{\perp}$ | 0.34 | – |
| MBBA | $\eta_{\parallel}$ | 0.18 | – |
| | $\eta_{\perp}$ | 0.36 | – |

Here, $\Omega$ is the angle between the director $\vec{n}_0$ and the velocity of a spherical particle $\vec{u}$. This is a general expression for the drag force which includes anisotropy. We can note that it is a very good approximation for a spherical particle with the coating size $R$. Outside this region, there are only small deformations and we should take into account the multiple explanation. If we assume that the director field is equal to zero we see that expressions (2.11) and (2.13) become the same, which confirms the rightness of our calculations. In practice, only two directions are measured i.e., along and perpendicular to the director direction. We rewrite the expression (2.13) for these two directions

$$\begin{aligned} F_{\parallel} &= 4\pi\eta_1 u + (\eta_4+\eta_3-\eta_2-\eta_1)\left(\frac{0.24u\pi c}{R^2}+0.8u\pi R\right) \\ &\quad + (\eta_5+\eta_1+\eta_2-2\eta_3-2\eta_4)\left(\frac{0.71u\pi c}{R^2}+0.33u\pi R\right), \end{aligned} \tag{2.14}$$

$$\begin{aligned} F_{\perp} &= 4\pi\eta_1 u + (\eta_4+\eta_3-\eta_2-\eta_1)\left(\frac{-0.12u\pi c}{R^2}-0.4u\pi R\right) \\ &\quad + (\eta_5+\eta_1+\eta_2-2\eta_3-2\eta_4)\left(\frac{-0.59u\pi c}{R^2}-0.18u\pi R\right). \end{aligned} \tag{2.15}$$

Alternative investigation was recently conducted by [35]. In [35] the authors have developed a perturbative approach to the Leslie-Ericksen equations and related the diffusion coefficients to the Miesovicz viscosity parameters $\eta_i$. We present our results in the same order as in [35] in table 1. The value of quadrupole moment is used as in [9]. Two cases are considered, i.e., uniform director field and planar anchoring.

## 3. Conclusion

For uniform director field, our results differ from the results in [35, 36], but we believe that our results are reasonable. Our arguments are as follows: in case of a uniform director field, the field of the director is the same in space. The difference in the description of the dynamics of the usual liquid and the liquid crystal is in the expression of the free energy [32]. The expression for the free energy for the liquid crystal contains, in comparison with the expression for the free energy for an ordinary liquid, an additional term, i.e., deformation free energy which depends only on the derivatives of the director regarding the position [32]. In the case of a uniform director field, this term becomes zero and the liquid crystal behaves as an ordinary liquid. Thus, the viscosities along the director $\eta_{\parallel}$ and perpendicular direction $\eta_{\perp}$ should be the same. Our results confirm this fact in contradiction to the results of [35, 36]. The possible explanation of this discrepancy might be in the fact that the director evolution depends on the molecular field and the gradient of the velocities [32]. A situation is possible when the molecular field is zero but the gradient of velocities is big enough (the particle moves quickly) and, consequently, the deviation of the director is not as small as in our approach. If the anchoring is not zero, we find that for both kinds of liquid crystals the ratio $\eta_{\parallel}/\eta_{\perp}$ is about 2 which agrees well with the general tendency [35]. The approach used by us is valid for a weak anchoring when $n_x$ and $n_y$ are small parameters. However, it may be applied to the case of strong anchoring as well. As was shown in [11, 30, 37], under strong anchoring, the effective mass of an ion increases due to the formation of a polarization coating, moving together with the ion. Thus, we can consider the particle with the coating to be a new single moving particle. The anchoring for this "new particle" is weak [11, 30, 37] and the above approach can be applicable too. It should be noted that while extracting the viscosity, the size of the polarization coating should be taken into account. We can conclude that our approach works well for the two limiting cases, i.e., weak and strong anchoring, and it does not include the case of the "middle" anchoring. In [35] there was measured the ratio $D_{\parallel}/D_{\perp} \approx 4$. This might be the case of the "middle" anchoring. We cannot claim a complete theory of motion of immersed particles in a nematic liquid crystal, but we suggest the approach which makes the analytical calculation of the diffusion process of a particle in this viscose media possible. This process should take into account the change as an inertial effect and the Stokes drag of a particle in a liquid crystal which are linked with the deformation of the elastic director field.

## References

1. Russel W.B., Saville D.A., Schowalter W.R., Colloidal Dispersions, Cambridge University Press, Cambridge, 1995.
2. Reichl L.E., Phys. Rev. A, 1980, **24**, 1609; doi:10.1103/PhysRevA.24.1609.
3. Currie P.K., J. Phys. (France), 1979, **40**, 501; doi:10.1051/jphys:01979004005050100.
4. Atkin R.J., Leslie F.M., Q. J. Mech. Appl. Math., 1970, **23**, S3; doi:10.1093/qjmam/23.2.3.
5. Currie P.K., Arch. Ration. Mech. An., 1970, **37**, 222; doi:10.1007/BF00281478.
6. Atkin R.J., Arch. Ration. Mech. An., 1970, **38**, 224; doi:10.1007/BF00251660.
7. White A.E., Cladis P.E., Torza S., Mol. Cryst. Liq. Cryst., 1977, **43**, 13; doi:10.1080/00268947708084931.
8. Pieransky P., Brochard F., Guyon E., J. Phys. (France), 1973, **34**, 35; doi:10.1051/jphys:0197300340103500.
9. Lubensky T.C., Pettey D., Currier N., Stark H., Phys. Rev. E, 1997, **57**, 610; doi:10.1103/PhysRevE.62.711.
10. Lev B.I., Tomchuk P.M., Phys. Rev. E, 1998, **59**, 591; doi:10.1103/PhysRevE.59.591.
11. Chernyshuk S.B., Lev B.I., Yokoyama H., Phys. Rev. E, 2005, **71**, 062701; doi:10.1103/PhysRevE.71.062701.
12. Poulin P., Stark H., Lubensky T.C., Weitz D.A., Science, 1997, **275**, 1770; doi:10.1126/science.275.5307.1770.
13. Poulin P., Cabuil V., Weitz D.A., Phys. Rev. Lett., 1997, **79**, 4862; doi:10.1103/PhysRevLett.79.4862.
14. Ruhwandl R.W., Terentjev E.M., Phys. Rev. E, 1996, **54**, 5204; doi:10.1103/PhysRevLett.79.4862.
15. Stark H., Ventzki D., Phys. Rev. E, 2001, **64**, 031711; doi:10.1103/PhysRevE.64.031711.
16. Fukuda J., Stark H., Yoneya M., Yokoyama H., J. Phys.: Condens. Matter, 2004, **16**, S1957; doi:10.1088/0953-8984/16/19/008.
17. Fukuda J., Stark H., Yokoyama H., Phys. Rev. E, 2005, **72**, 021701; doi:10.1103/PhysRevE.72.021701.
18. Kneppe H., Schneider F., Schwesinger B., Mol. Cryst. Liq. Cryst., 1991, **205**, 9; doi:10.1080/00268949108032075.
19. Heuer H., Kneppe H., Schneider F., Mol. Cryst. Liq. Cryst., 1992, **214**, 43; doi:10.1080/10587259208037281.
20. Heuer H., Kneppe H., Schneider F., Mol. Cryst. Liq. Cryst., 1991, **200**, 51; doi:10.1080/00268949108044231.
21. Diogo A.C., Mol. Cryst. Liq. Cryst., 1983, **100**, 153; doi:10.1080/00268948308073729.
22. Roman V.G., Terentjev E.M., Colloid J. USSR, 1989, **51**, 435.
23. Chono S., Tsuji T., Mol. Cryst. Liq. Cryst., 1998, **309**, 217; doi:10.1080/10587259808045530.
24. Billeter J.L., Pelcovits R.A., Phys. Rev. E, 2000, **62**, 711; doi:10.1103/PhysRevE.62.711.
25. Ruhwandl R.W., Terentjev E.M., Z. Naturfors. A, 1995, **50**, 1023.
26. Cladis P.E., Saarloos W., Finn P.L., Kortan A.R., Phys. Rev. Lett., 1987, **58**, 222; doi:10.1103/PhysRevLett.58.222.
27. Imura H., Okano K., Phys. Lett. A, 1973, **42**, 403; doi:10.1016/0375-9601(73)90728-7.
28. de Gennes P.G., Molecular Fluids, Gordon and Breach, London, 1976, 373–400.
29. Ryskin G., Kremenetsky M., Phys. Rev. Lett., 1991, **67**, 1574; doi:10.1103/PhysRevLett.67.1574.
30. Lev B.I., Chernyshuk S.B., Tomchuk P.M., Yokoyama H., Phys. Rev. E, 2002, **65**, 021709; doi:10.1103/PhysRevE.65.021709.
31. Kuksenok O.V., Ruhwandl R.W., Shiyanovskii S.V., Terentjev E.M., Phys. Rev. E, 1996, **54**, 5198; doi:10.1103/PhysRevE.54.5198.
32. Landau L.D., Lifhitz E.M., Theory of Elasticity, Pergamon Press, Oxford, 1986.
33. Landau L.D., Lifhitz E.M., Mechanics of Fluids, Pergamon Press, Oxford, 1987.
34. Brochar F., de Gennes P.G., J. Phys. (France), 1970, **31**, 691; doi:10.1051/jphys:01970003107069100.

35. Mondiot F., Loudet J-C., Mondain-Monval O., Snabre P., Vilquin A., Wurger A., Phys. Rev. E, 2012, **86**, 010401; doi:10.1103/PhysRevE.86.010401.
36. Stark H., Phys. Rep., 2001, **353**, 387; doi:10.1016/S0370-1573(00)00144-7.
37. Belotskii E.D., Lev B.I., Tomchuk P.M., Mol. Cryst. Liq. Cryst., 1992, **213**, 99; doi:10.1080/10587259208028721.
38. de Gennes P.G., Prost J., The Physics of Liquid Crystals, Clarendon Press, Oxford, 1993.
39. Oswald P., Pieranski P., Nematic and Cholesteric Liquid Crystals, Taylor Francis Group, Boca Raton, FL 33487-2742, 2005.

# 20

# Long-range interaction between dust grains in plasma

D.Yu. Mishagli*

[1] Department of Theoretical Physics, Faculty of Physics, Mechnicov National University, 2 Dvoryanska St., 65026 Odessa, Ukraine

[2] Institut für Theoretische Physik, Universität Leipzig, 10/11 Augustusplatz, 04109 Leipzig, Germany†

The nature of long-range interactions between dust grains in plasma is discussed. The dust grain interaction potential within a cell model of dusty plasma is introduced. The attractive part of intergrain potential is described by multipole interaction between two electro-neutral cells. This allowed us to draw an analogy with molecular liquids where the attraction between molecules is determined by dispersion forces. The main ideas of the fluctuation theory for electrostatic field in a cell model are formulated, and the dominating contribution to the attractive part of intergrain potential is obtained.

**Key words:** *dusty plasma, dust crystals, cell model, interaction potential, fluctuations*

## 1. Introduction

Dusty (complex) plasma consists of a weakly ionized gas (plasma) and charged submicron- and micron-sized particles (grains). It represents a new type of soft matter. It is an interdisciplinary field of research: geophysics, geology, meteorology, ecology, planetary science, different applications in technology — dusty plasmas can bring new results to all these directions. Progress in the research of dusty plasma properties is documented in resent monographs [1–9] and review articles [10–16]. However, there are still problems to be solved.

Dust grains being highly charged (up to $10^6$ elementary charges per grain) substantially affect the properties of the whole system. Dust grains can be either positively or negatively charged. The processes of charging are very important in the theory of dusty plasma. However, we will not consider them in the present paper (for this purpose see the above mentioned references and the recent article by the author [17]). Under certain conditions, dust grains can form various condensed ("plasma crystal" and "plasma liquid") or gaseous phases depending on the relative strengths of the intergrain interaction. In a theoretical research, the problem of describing an intergrain interaction and phase transition in a subsystem of dust grains holds an important position.

The forces that govern the dust grains do not correspond to a direct Coulomb interaction and are long-range ones (see e.g., work [18]). Thus, in the present paper the model intergrain potential is proposed that predicts both the repulsion (at small distances) and the attraction (at large distances) between the dust grains. The cell approach to dusty plasma discussed in work [17] plays an essential role in our construction. It allowed us: (i) to create a simple theory of charge fluctuations and (ii) to study a long-range interaction between the dust grains. The approach presented can be used to describe the interaction effects in similar systems such as the mixture of ionic and nonionic liquids [19], where spherical clusters occur.

The article is built as follows. In section 2, basic statements of the cell model and some results obtained in [17] are reviewed. In section 3, we present a model intergrain interaction potential. In order

* E-mail: mishagli@onu.edu.ua

† Present address

to establish the main contribution to the attractive part of intergrain potential, we consider an electro-neutral cell in the external electric field in section 4, where the dipole polarizability of a cell is obtained. Then, in section 5 we present the principles of construction of the fluctuation electrostatic field inside and outside a cell and obtain fluctuation multipole moments of a cell. Moreover, the fluctuation dipole moment is obtained and the main contribution to the attractive part of the intergrain potential is presented in the explicit form. A brief discussion of the obtained results is presented in the concluding section 6. In appendix A, we get an expression for the potential of the fluctuation electrostatic field inside a cell. In appendix B, the energy of the fluctuation electrostatic field is obtained.

## 2. The cell model of complex plasma

Let us consider a system of charged dust grains of the same radii $r_{\rm p}$ and the emitted electrons. The dust grains have an average charge $Ze$ ($e$ is the electron charge). The system is in a thermal equilibrium (thermal plasma). We assume that the complex plasma can be represented as the collection of electro-neutral cells due to its electro-neutrality. Each cell contains only one dust grain. In the mean field approximation, the cells should have a spherical form of the radius

$$r_{\rm c} = \frac{1}{2}\left(\frac{3}{4\pi n_{\rm p}}\right)^{1/3}, \tag{2.1}$$

where $n_{\rm p}$ is an average dust grain density. Note, that such an approach is applicable only for describing the equilibrium thermodynamic properties of a dusty plasma.

The electro-neutrality condition for a cell is as follows:

$$Ze + \int \rho(\mathbf{r})\mathrm{d}\mathbf{r} = 0, \tag{2.2}$$

where $\rho(\mathbf{r})$ is the volume-charge density, the distance $r$ is reckoned from the center of a grain. The integration occurs over the area occupied by electrons inside a cell. Note, that in the absence of an external electric field, the distributions of an electric charge $\rho(\mathbf{r})$ and electric potential $\phi(\mathbf{r})$ have spherical symmetry: $\rho(\mathbf{r}) \Rightarrow \rho(r)$, $\phi(\mathbf{r}) \Rightarrow \phi_0(r)$. The electrostatic field distribution is described in the self-consistent field approximation: the potential $\phi(\mathbf{r})$ satisfies the Poisson equation, in which the charge density $\rho(\mathbf{r})$ is determined by the Boltzmann distribution.

Its use is justified by the inequality

$$\tau_{\rm c} \ll \tau_*, \tag{2.3}$$

where $\tau_{\rm c}$ is the time required for the formation of an electro-neutral cell around the grain, $\tau_*$ is the characteristic macroscopic relaxation time for a system. This inequality expresses the fact that the electron mobility substantially exceeds the mobility of other plasma components.

Reference [17] discusses the system of identical dust grains of the radius $r_{\rm p}$ with the mean charge $Ze$, which are in equilibrium with the emitted electrons, without an external electric field. The problem of proper boundary conditions for such a model is also considered therein. It is shown that (i) setting the electrostatic potential $\phi$ equal to zero on the surface of a cell and (ii) connecting the electrostatic field strength on the surface of a grain with its average charge are sufficient conditions for a full description of a system:

$$\begin{cases} \phi_0(r_{\rm c}) = 0, \\ \left.\dfrac{\partial\phi_0(r)}{\partial r}\right|_{r=r_{\rm p}} = -4\pi\sigma, \end{cases} \tag{2.4}$$

where $\sigma$ is the average surface charge density on a dust grain. Note, that there is no electric field outside a cell due to the Gauss law.

For a further analysis in [17], the dimensionless variables

$$\tilde{r} = \frac{r}{r_{\rm p}}, \qquad \varsigma = \frac{r_{\rm c}}{r_{\rm p}}, \qquad \psi_0(\tilde{r}) = 1 + \frac{e\phi_0(\tilde{r})}{kT} \tag{2.5}$$

were used. Here, $\varsigma$ and $\psi_0(\tilde{r})$ are, respectively, the dimensionless radius of a cell and the dimensionless potential of the electrostatic field inside a cell, and $k$ is the Boltzmann constant. Note, that $\tilde{r} \in [1, \varsigma]$. The solution of the linearized Poisson equation ($e\phi_0(r) \ll kT$) satisfying the boundary conditions (2.4) is as follows:

$$\psi_0(\tilde{r}) = \frac{1}{\tilde{r}} \frac{(Z/Z_0)\lambda \sinh \frac{\varsigma-\tilde{r}}{\lambda} + \varsigma\left(\lambda \sinh \frac{\tilde{r}-1}{\lambda} + \cosh \frac{\tilde{r}-1}{\lambda}\right)}{\lambda \sinh \frac{\varsigma-1}{\lambda} + \cosh \frac{\varsigma-1}{\lambda}}. \tag{2.6}$$

Here, $Z_0 = kTr_p/e^2$ (typical values of $Z/Z_0 \sim 2–4$ [15]) and $\lambda$ is the dimensionless Debye radius defined as

$$\lambda = \frac{r_D}{r_p}, \qquad r_D = \sqrt{\frac{kT}{4\pi e^2 n_{e0}}}, \tag{2.7}$$

where $n_{e0}$ is the mean density of the emitted electrons for $\phi_0(r) = 0$, i.e., on the boundary of a cell (for $r = r_c$ or, in dimensionless form, $\tilde{r} = \varsigma$). The buffer gas is not taken into account here. Note, that in the presented cell model dust grains do not take part in the screening.

The equation (2.2) allows one to establish the dependence of Debye radius $\lambda$ on the average dust grain charge $Z$, as it is also shown in [17]. The dependence $\lambda = \lambda(Z)$ is set as a root of the equation

$$\frac{Z}{Z_0} = \frac{1}{\lambda}\left[(\varsigma - \lambda^2)\sinh \frac{\varsigma-1}{\lambda} + \lambda(\varsigma - 1)\cosh \frac{\varsigma-1}{\lambda}\right]. \tag{2.8}$$

In the mean-field approximation described above, there is no interaction between the cells. However, as it is shown below, the charge fluctuations (beyond the mean-field cell approach) can lead to the interaction between two electro-neutral cells.

## 3. Averaged potential of multipole interaction between dust grains

The equilibrium value of the dust grain charge is mainly violated by the thermal electron motion. Therefore, the fluctuation electric multipole moments of cells occur and generate long-range electric fields.

The electric field of one cell acts on its neighbours and generates interaction effects, which are similar to a dispersion interaction between neutral atoms (molecules). In particular, there are two interaction mechanisms:

1. The fluctuation field of one cell polarizes the neighbour one. Therefore, the second cell gains a certain value of induced multipole moments. The latter interact with multipole moments of the first cell. The average value of such an interaction is an analogue of a dispersion interaction between neutral atoms (molecules):

   $$\Phi_1 = \langle W_{\mathrm{mim}} \rangle, \tag{3.1}$$

   where index "mim" denotes the interaction between multipole and the induced multipole.

2. The interaction of fluctuation multipole moments of both cells. The average value of such an interaction is determined in a different way:

   $$\Phi_2 = -\beta \langle W_{\mathrm{mm}}^2 \rangle, \tag{3.2}$$

   where $\beta = 1/kT$.

It is easy to ascertain that both components lead to the attraction between "plasma atoms" (i.e., electroneutral spherical cells). The dominating contribution in both cases is determined by "dipole–dipole" interactions, decreasing as $1/R^6$ ($R$ is an average distance between the centers of two grains). At the same time, since the grains have charges of the same sign, repulsion forces arise at distances $R < 2r_D$, where Coulomb repulsion between dust grains is not screened. This allows us to qualitatively model the interaction between two cells (i.e., grains) with the potential

$$U(R) = U_r(R) + U_a(R),$$

where $U_r(R)$ and $U_a(R)$ are the repulsive and the attractive parts of $U(R)$, respectively. The repulsive part of $U(R)$ is modelled by the combination of the hard-core potential (the radius of hard-core coincides with the radius of a grain $r_p$) and the potential of Coulomb repulsion at distances $2r_p < R < 2r_D$:

$$U_r(R) = \begin{cases} \infty, & R \leqslant 2r_p\,; \\ \dfrac{(Ze)^2}{R}, & 2r_p < R < 2r_D\,. \end{cases}$$

The attractive part $U_a(R)$ is conditioned by the forces of an electric multipole interaction:

$$U_a(R) = \Phi_1(R) + \Phi_2(R).$$

The proposed interaction potential can lead to the formation of ordered structures for certain values of temperature and dust grain density. It should be noted that the random distribution of dust grains in the volume is not taken into account. Thus, the proposed approach could be applied to the regular homogeneous grain distribution (like in the case of dust crystals).

Let us determine the dominating contributions to $\Phi_1(R)$ and $\Phi_2(R)$ according to (3.1) and (3.2).

### 3.1. "Dispersion" intergrain interaction

The average energy of electrostatic "multipole–induced multipole" interaction is determined according to (3.1) as follows:

$$\Phi_1(R) = \langle W_{\rm did}\rangle + \langle W_{\rm diq}\rangle + \langle W_{\rm qiq}\rangle + \ldots, \tag{3.3}$$

where contributions $\langle W_{\rm did}\rangle$, $\langle W_{\rm diq}\rangle$ and $\langle W_{\rm qiq}\rangle$ describe the averaged "dipole–induced dipole", "dipole–induced quadrupole" and "quadrupole–induced quadrupole" interactions, respectively.

The main contribution $\langle W_{\rm did}\rangle$ can be estimated from the following simple considerations. It was noted above that the fluctuation field of one cell polarizes the other one. Therefore, the second cell acquires the induced dipole moment $\mathbf{d}^{(\rm ind)} = \alpha\mathbf{d}/R^3$, where $\alpha$ is the polarizability of a cell and $\mathbf{d}$ is the fluctuating dipole moment of the first cell. Thus, the energy of "dipole–induced dipole" interaction is $W_{\rm did} = -\mathbf{d}\cdot\mathbf{E}^{(\rm ind)}$ (here $\mathbf{E}^{(\rm ind)} = \mathbf{d}^{(\rm ind)}/R^3$ is the induced field of the second cell). The careful analysis of this problem yields:

$$\Phi_1(R) = \langle W_{\rm did}\rangle + \ldots \simeq -4\alpha\frac{\langle \mathbf{d}^2\rangle}{R^6}\,. \tag{3.4}$$

### 3.2. Electrostatic multipole intergrain interaction

The average value of the direct electrostatic multipole interaction given by (3.2) can be represented as follows:

$$\Phi_2(R) = -kT\left(\frac{A_6}{R^6} + \frac{A_8}{R^8} + \frac{A_{10}}{R^{10}} + \ldots\right). \tag{3.5}$$

Here, coefficients $A_i$ ($i = 6, 8, 10, \ldots$) are expressed through the average values of square multipole moments of a cell. To get the "dipole–dipole" contribution, we can follow the authors of work [20] where the dipole fluid is considered. Thus,

$$\Phi_2(R) \simeq -\frac{2}{3}\beta\frac{\langle \mathbf{d}^2\rangle^2}{R^6}\,. \tag{3.6}$$

The rest contributions to (3.3) and (3.5) require a more detailed analysis and will be considered in a further paper.

## 4. Dipole polarizability of a cell

In the previous section, the model potential of intergrain interaction $U(R)$ has been introduced. This section is devoted to the dipole polarizability $\alpha$ of a cell. For this purpose, let us consider the reaction of electro-neutral cell to an external electric field $\mathbf{E}_0$. The latter polarizes a cell and it gains the polarization vector $\mathbf{P} = \alpha\mathbf{E}_0$. On the other hand, the polarization vector is a dipole moment of the unit cell volume:

$$\mathbf{P} = \int \mathbf{r}\cdot\rho(\mathbf{r})d\mathbf{r}, \tag{4.1}$$

where the volume-charge density $\rho(\mathbf{r})$, as it was noted in section 2, is connected with the electrostatic potential $\phi(\mathbf{r})$ by the Boltzmann distribution. In the presence of an external electric field (as opposed to the case considered above) the distributions of $\rho(\mathbf{r})$ and $\phi(\mathbf{r})$ lose their spherical symmetry: $\rho(\mathbf{r}) \Rightarrow \rho(r,\vartheta)$, $\phi(\mathbf{r}) \Rightarrow \phi(r,\vartheta)$.

The boundary conditions (2.4) for the Poisson equation have the following form:

$$\begin{cases} \phi(r_c,\vartheta) = -E_0 r_c \cos\vartheta, \\ \left.\dfrac{\partial\phi(r,\vartheta)}{\partial r}\right|_{r=r_p} = -\dfrac{Ze}{r_p^2} + E_0\cos\vartheta. \end{cases} \tag{4.2}$$

Corresponding to (4.2), the solution of the linearized Poisson equation in dimensionless variables (2.5) for the renormalized dimensionless potential $\psi(\tilde{r},\vartheta) = 1 + e\phi(r,\vartheta)/kT$ has the form

$$\psi(\tilde{r},\vartheta) = \psi_0(\tilde{r}) + \psi_f(\tilde{r},\vartheta). \tag{4.3}$$

Here, $\psi_0$ is the isotropic part of $\psi(\tilde{r},\vartheta)$, determined by equation (2.6). The angular part $\psi_f$ (index "f" expresses the fact that this part of the potential is due to an external electric field) is proportional to $\cos\vartheta$ and is equal to

$$\psi_f(\tilde{r},\vartheta) = -\tilde{E}_0\cos\vartheta\left\{\frac{1}{\tilde{r}}\frac{\lambda\left(\varsigma\sinh\frac{\varsigma-\tilde{r}}{\lambda} - \lambda\cosh\frac{\varsigma-\tilde{r}}{\lambda}\right) + \varsigma^3\left[2\lambda\sinh\frac{\tilde{r}-1}{\lambda} + (2\lambda^2+1)\cosh\frac{\tilde{r}-1}{\lambda}\right]}{\lambda(2\varsigma-2\lambda^2-1)\sinh\frac{\varsigma-1}{\lambda} + \left[2\lambda^2(\varsigma-1)+\varsigma\right]\cosh\frac{\varsigma-1}{\lambda}}\right.$$
$$\left. + \frac{1}{\tilde{r}^2}\frac{\lambda^2\left(-\lambda\sinh\frac{\varsigma-\tilde{r}}{\lambda} + \varsigma\cosh\frac{\varsigma-\tilde{r}}{\lambda}\right) - \varsigma^3\lambda\left[(2\lambda^2+1)\sinh\frac{\tilde{r}-1}{\lambda} + 2\lambda\cosh\frac{\tilde{r}-1}{\lambda}\right]}{\lambda(2\varsigma-2\lambda^2-1)\sinh\frac{\varsigma-1}{\lambda} + \left[2\lambda^2(\varsigma-1)+\varsigma\right]\cosh\frac{\varsigma-1}{\lambda}}\right\},$$

where $\tilde{E}_0 = eE_0 r_p/kT$.

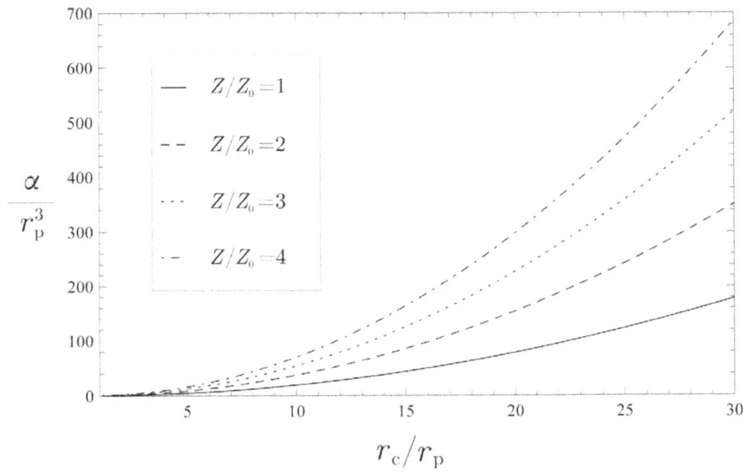


**Figure 1.** Dimensionless polarizability $\alpha/r_p^3$ as the function of $r_c/r_p$.

The volume-charge density in (4.1) obviously connected with the potential $\psi(\tilde{r},\vartheta)$ by the relation

$$\rho(r,\vartheta) = -\frac{kT}{4\pi e r_p^2}\frac{1}{\lambda^2}\psi(\tilde{r},\vartheta).$$

Therefore, after integration and comparing the result with $P = \alpha E_0$, we obtain

$$\alpha = \frac{r_\mathrm{p}^3}{3\lambda}\left\{\frac{\varsigma^5+\{\varsigma^3[2(\varsigma-3)\varsigma+3]+3\varsigma\}\lambda^2+3\left(2\varsigma^3-1\right)\lambda^4}{\lambda\left(2\varsigma-2\lambda^2-1\right)\sinh\frac{\varsigma-1}{\lambda}+\left[2(\varsigma-1)\lambda^2+\varsigma\right]\cosh\frac{\varsigma-1}{\lambda}}\sinh\frac{\varsigma-1}{\lambda}\right.$$
$$\left.+\frac{(\varsigma-1)\left[\varsigma^2(\varsigma-1)(2\varsigma+1)-\varsigma\right]\lambda-3(2\varsigma^3-1)\lambda^2}{\lambda\left(2\varsigma-2\lambda^2-1\right)\sinh\frac{\varsigma-1}{\lambda}+\left[2(\varsigma-1)\lambda^2+\varsigma\right]\cosh\frac{\varsigma-1}{\lambda}}\cosh\frac{\varsigma-1}{\lambda}\right\}. \quad (4.4)$$

The dependence of $\alpha/r_\mathrm{p}^3$ on $r_\mathrm{c}/r_\mathrm{p}$ is shown in figure 1. Note, that the value of Debye radius $\lambda$ for a corresponding $Z$ is determined from equation (2.8). Therefore, polarizability $\alpha$ also depends on the average dust grain charge $Z$. Indeed, one can see that substitution of the potential (4.3) into (2.2) instead of (2.6) also leads to (2.8), so far as the integration over the angular part gives zero.

## 5. Fluctuation multipole moments of a cell

In this section we introduce the method which allows one to construct the fluctuation electrostatic field inside and outside an electro-neutral cell. Moreover, a fluctuation dipole moment is obtained and the main contribution to the attractive part of inter-grain potential is presented in explicit form. The fluctuation theory of electrostatic field for the case of ellipsoidal cells will be the subject of a separate report.

### 5.1. Principles of construction of the fluctuation electrostatic field

The electrostatic field deviates from its equilibrium value described in section 2 due to fluctuations. Thus, let us consider two areas: inside (interior) and outside (exterior) a cell. We assume that the potentials of the fluctuation electrostatic field in the interior and exterior, $\phi'_\mathrm{in}(\mathbf{r})$ and $\phi'_\mathrm{ex}(\mathbf{r})$, respectively, satisfy the linear equations of the same type as the averaged potential $\phi_0(r)$:

$$\Delta\phi'_\mathrm{in}(\mathbf{r})-\frac{1}{r_\mathrm{D}^2}\phi'_\mathrm{in}(\mathbf{r})=0 \quad (5.1)$$

and

$$\Delta\phi'_\mathrm{ex}(\mathbf{r})=0. \quad (5.2)$$

It is supposed that the Laplace equation is proper in the exterior, as far as the whole dust grain charge is in a cell.

According to the form of equations (5.1) and (5.2), we claim that potentials $\psi'_\mathrm{in}\equiv e\phi'_\mathrm{in}/kT$ and $\psi'_\mathrm{ex}\equiv e\phi'_\mathrm{ex}/kT$ have the following structure:

$$\psi'_\mathrm{in}(\tilde{r},\vartheta,\varphi)=\sqrt{\frac{\pi\lambda}{2\tilde{r}}}\sum_{n=0}^{\infty}\sum_{m=-n}^{n}Y_{nm}(\vartheta,\varphi)\left[C^{(1)}_{nm}I_{n+\frac{1}{2}}(\tilde{r}/\lambda)+C^{(2)}_{nm}I_{-n-\frac{1}{2}}(\tilde{r}/\lambda)\right] \quad (5.3)$$

(here, $C^{(1)}_{nm}$ and $C^{(2)}_{nm}$ are the unknown coefficients, $I_{n+\frac{1}{2}}(x)$ and $I_{-n-\frac{1}{2}}(x)$ are the Modified Spherical Bessel functions of the first and the second kind, respectively) and

$$\psi'_\mathrm{ex}(\tilde{r},\vartheta,\varphi)=\sum_{n=0}^{\infty}\sum_{m=-n}^{n}\frac{\tilde{D}_{nm}}{\tilde{r}^{n+1}}Y_{nm}(\vartheta,\varphi) \quad (5.4)$$

(here, coefficients $\tilde{D}_{nm}=eD_{nm}/kTr_\mathrm{p}^{n+1}$ are the required multipole moments of a cell). Here, we have used the dimensionless variables (2.5).

The interconnection between the coefficients $C^{(1)}_{nm}$, $C^{(2)}_{nm}$ and $\tilde{D}_{nm}$ is determined from the conditions for continuity of potential and strength of the fluctuation electric field on the surface of a cell (see appendix A):

$$\begin{cases}\psi'_\mathrm{in}(\varsigma,\vartheta,\varphi)=\psi'_\mathrm{ex}(\varsigma,\vartheta,\varphi),\\ \left.\dfrac{\partial\psi'_\mathrm{in}(\tilde{r},\vartheta,\varphi)}{\partial\tilde{r}}\right|_{\tilde{r}=\varsigma}=\left.\dfrac{\partial\psi'_\mathrm{ex}(\tilde{r},\vartheta,\varphi)}{\partial\tilde{r}}\right|_{\tilde{r}=\varsigma}.\end{cases} \quad (5.5)$$

According to the thermodynamic fluctuation theory [21], the average value of square multipole moments of a cell is

$$\langle |D_{n0}|^2 \rangle \simeq \frac{kT}{\chi_n}, \tag{5.6}$$

where $\chi_n$ are the coefficients in the expansion of fluctuation electrostatic field energy $W'_{\mathrm{el}}$:

$$W'_{\mathrm{el}} = \sum_{n=0}^{\infty} \chi_n |D_{n0}|^2. \tag{5.7}$$

The explicit form of $W'_{\mathrm{el}}$ is discussed in appendix B. Thus, the comparison of equations (5.7) and (B.1) immediately gives us

$$\chi_n = -\frac{1}{8\pi}\frac{1}{r_{\mathrm{p}}^{2n+1}}\left\{[L_i(n)a_i(1,n)]\cdot[L_j(n)b_j(1,n)] + \frac{1}{\lambda^2}\int_1^{\varsigma}[L_i(n)a_i(\tilde{r},n)]^2\,\tilde{r}^2\mathrm{d}\tilde{r}\right\}, \tag{5.8}$$

where the coefficients $a_i(1,n)$, $b_j(1,n)$, $L_i(n)$ and $L_j(n)$ $(i,j=1,2)$ are determined by equations (A.1) and (A.2).

## 5.2. Fluctuation dipole moment of a cell

The dipole moment **d** of a cell required to determine the dominating contributions to the intergrain potential (see section 3) is obtained after setting in the expression for multipole moments $n=1$, i.e., $|\mathbf{d}| = D_{10}$. Therefore, the expressions (A.1) for coefficients $a_1(\tilde{r},n)$, $a_2(\tilde{r},n)$, $b_1(\tilde{r},n)$ and $b_2(\tilde{r},n)$ read as follows:

$$a_1(\tilde{r},1) = -\frac{\sinh\tilde{r}/\lambda}{(\tilde{r}/\lambda)^2} + \frac{\cosh\tilde{r}/\lambda}{\tilde{r}/\lambda}, \qquad a_2(\tilde{r},1) = \frac{\sinh\tilde{r}/\lambda}{\tilde{r}/\lambda} - \frac{\cosh\tilde{r}/\lambda}{(\tilde{r}/\lambda)^2},$$

$$b_1(\tilde{r},1) = \left(\frac{2\lambda^2}{\tilde{r}^3} + \frac{1}{\tilde{r}}\right)\sinh\tilde{r}/\lambda - \frac{2\lambda}{\tilde{r}^2}\cosh\tilde{r}/\lambda,$$

$$b_2(\tilde{r},1) = -\frac{2\lambda}{\tilde{r}^2}\sinh\tilde{r}/\lambda + \left(\frac{2\lambda^2}{\tilde{r}^3} + \frac{1}{\tilde{r}}\right)\cosh\tilde{r}/\lambda,$$

and for the mean-square value of a fluctuation dipole moment of a cell, after integration in (5.8) and

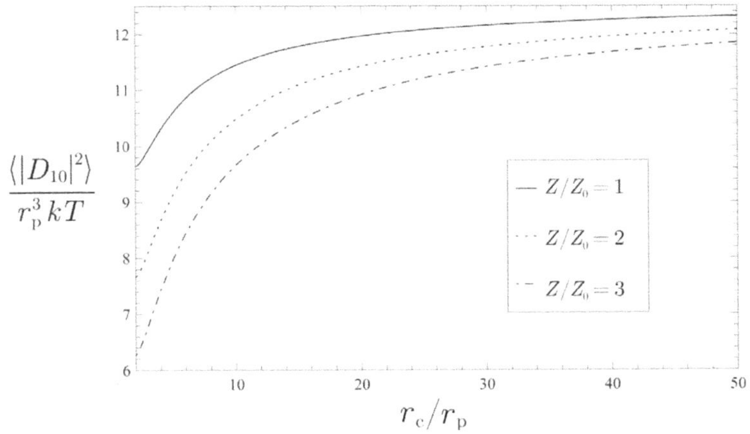


**Figure 2.** The dependence of the fluctuation dipole momentum on the dimensionless radius of a cell $r_{\mathrm{c}}/r_{\mathrm{p}}$.

after a simple calculation, we have

$$\langle |D_{10}|^2 \rangle = 16\pi\varsigma^2\lambda^2 r_{\mathrm{p}}^3 kT\left[\frac{1}{2}\lambda(8\lambda^2+1)\sinh 2\frac{\varsigma-1}{\lambda} + 2\lambda^2(\lambda^2+1)\cosh 2\frac{\varsigma-1}{\lambda} - 2\lambda^2(\lambda^2-1) - \varsigma + 1\right]^{-1}. \tag{5.9}$$

Let us consider the result obtained.

The mean-square value of a fluctuating dipole moment decreases when the dust grain charge increases (see figure 2). Note, that this result is correct for charges $Z/Z_0$ that are not very small. This limit is due to a general restriction on our approach: $r_D < r_c$. The case $r_D > r_c$ corresponds to a rarefied gas, and thus the distribution functions for low-density plasma should be used.

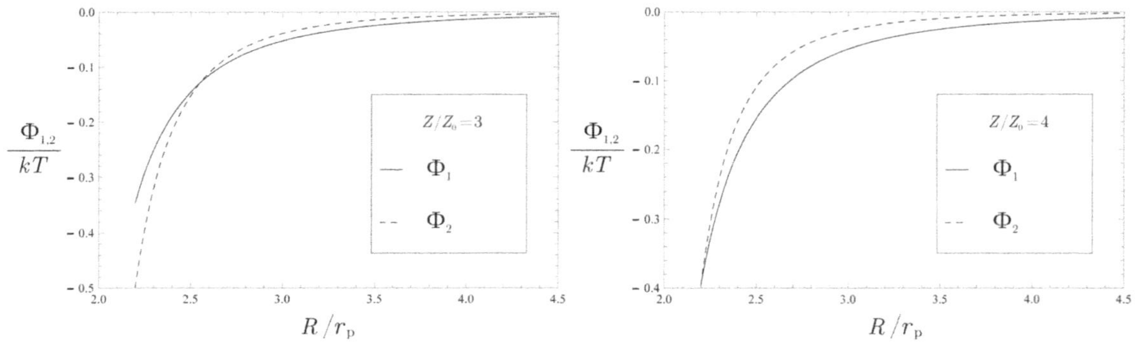


**Figure 3.** The dependencies of $\Phi_1/kT$ and $\Phi_2/kT$ on the average intergrain distance $R/r_p$ (in units of the radius of a grain). *Left*: the case of $Z/Z_0 = 3$. *Right*: the case of $Z/Z_0 = 4$.

The dipole-dipole contributions to the attractive part of interaction potential [equations (3.6) and (3.4)] are presented in figure 3. One can see that for small charges, the contribution $\Phi_2$ exceeds the contribution $\Phi_1$ at small intergrain distances. With an increase of the dust grain charge, the polarizability of a cell (4.4) increases (see figure 1), and the contributions $\Phi_1$ and $\Phi_2$ become comparable. These contributions together, $\Phi_1 + \Phi_2$, lead to the formation of a potential well having the depth $\sim kT$.

## 6. Conclusion

The main results and conclusions of this paper are as follows.

(i) The nature of long-range interaction between dust grains in complex plasma is discussed. It is supposed that plasma is a combination of electro-neutral cells of equal radius (mean-field approximation).

(ii) The main ideas of the fluctuation theory for electrostatic field in a cell model are formulated. The general expressions for fluctuation multipole moments are obtained. The average value of square fluctuation dipole moment is presented.

(iii) It is shown that the contributions of direct "dipole–dipole" interaction and "dipole–induced dipole" interaction form the potential well having the depth $\sim kT$. Multipole contributions of higher order lead to a further deepening of this well.

In molecular liquids, the attraction is determined by dispersion forces which decrease as $1/R^6$. The interaction between electro-neutral cells is similar to that in liquids. Thus, we expect that the grains in complex plasma form quasi-crystal structures like the molecules of argon in the transition to a solid state. The mechanism of ordering for solid argon is scrupulously considered in work [22]. A detailed analysis of the proposed potential will be the subject of a further work.

It is necessary to note that "plasma atom" is not fully similar to electroneutral molecules: it is more "soft". Thus, the use of a mean field approximation is incorrect for some problems. The characteristic example here is a problem of charge oscillations in dusty plasma. To determine the interaction potential between "plasma atoms", we consider charge fluctuations which is also out of the mean field approximation. However, these fluctuations are small compared to the mean field profile of a charge as for the case of charge fluctuations generating disperse forces in molecular liquids.

## Acknowledgements

The general statement of the problem and idea of its solution belong to Prof. N. P. Malomuzh. I thank him for that and for the help in preparing the manuscript. I am also grateful to S. A. Lantratov for numerous discussions and to B. I. Lev for the discussion of the obtained results during "XII Workshop and Awards for young researchers in the field of statistical physics and condensed matter theory" (Lviv, 2012). Finally, I highly appreciate the anonymous Referees for stimulating remarks.

## A. Fluctuation electrostatic field inside a cell

By substituting the expressions (5.3) and (5.4) in the continuity conditions (5.5), we get the following combined equations:

$$\begin{cases} a_1(\varsigma,n)C^{(1)}_{nm} + a_2(\varsigma,n)C^{(2)}_{nm} = \dfrac{\tilde{D}_{nm}}{\varsigma^{n+1}}, \\ b_1(\varsigma,n)C^{(1)}_{nm} + b_2(\varsigma,n)C^{(2)}_{nm} = -(n+1)\dfrac{\tilde{D}_{nm}}{\varsigma^{n+2}}, \end{cases}$$

where coefficients $a_1(\tilde{r},n)$, $a_2(\tilde{r},n)$, $b_1(\tilde{r},n)$ and $b_2(\tilde{r},n)$ read

$$\begin{cases} a_1(\tilde{r},n) = \sqrt{\dfrac{\pi\lambda}{2\tilde{r}}} I_{n+\frac{1}{2}}(\tilde{r}/\lambda), \qquad a_2(\tilde{r},n) = \sqrt{\dfrac{\pi\lambda}{2\tilde{r}}} I_{-n-\frac{1}{2}}(\tilde{r}/\lambda), \\ b_1(\tilde{r},n) = \sqrt{\dfrac{\pi\lambda}{2\tilde{r}}} \left[\dfrac{1}{\lambda} I_{n+\frac{3}{2}}(\tilde{r}/\lambda) + \dfrac{n}{\tilde{r}} I_{n+\frac{1}{2}}(\tilde{r}/\lambda)\right], \\ b_2(\tilde{r},n) = \sqrt{\dfrac{\pi\lambda}{2\tilde{r}}} \left[\dfrac{1}{\lambda} I_{-n-\frac{3}{2}}(\tilde{r}/\lambda) + \dfrac{n}{\tilde{r}} I_{-n-\frac{1}{2}}(\tilde{r}/\lambda)\right]. \end{cases} \tag{A.1}$$

For coefficients $C^{(1)}_{nm}$ and $C^{(2)}_{nm}$, we obtain

$$\begin{cases} C^{(1)}_{nm} = L_1(n)\tilde{D}_{nm}, \quad L_1(n) = \dfrac{1}{\varsigma^{n+2}} \cdot \dfrac{\varsigma b_2(\varsigma,n) + (n+1)a_2(\varsigma,n)}{a_1(\varsigma,n)b_2(\varsigma,n) - a_2(\varsigma,n)b_1(\varsigma,n)}, \\ C^{(2)}_{nm} = L_2(n)\tilde{D}_{nm}, \quad L_2(n) = -\dfrac{1}{\varsigma^{n+2}} \cdot \dfrac{\varsigma b_1(\varsigma,n) + (n+1)a_1(\varsigma,n)}{a_1(\varsigma,n)b_2(\varsigma,n) - a_2(\varsigma,n)b_1(\varsigma,n)}. \end{cases} \tag{A.2}$$

Thus, the fluctuation potential in the interior (5.3) can be represented as follows:

$$\psi'_{\text{in}}(\tilde{r},\vartheta,\varphi) = \sum_{n=0}^{\infty} \sum_{m=-n}^{n} Y_{nm}(\vartheta,\varphi) \sum_{i=1}^{2} L_i(n) a_i(\tilde{r},n) \tilde{D}_{nm}. \tag{A.3}$$

## B. Energy of fluctuation electrostatic field

The energy of fluctuation electrostatic field is determined either by a field in the interior or by a field in the exterior:

$$W'_{\text{el}} = \frac{1}{8\pi} \int_{V_{\text{in}}} [\nabla\phi'_{\text{in}}(\mathbf{r})]^2 \, \mathrm{d}\mathbf{r} + \frac{1}{8\pi} \int_{V_{\text{ex}}} [\nabla\phi'_{\text{ex}}(\mathbf{r})]^2 \, \mathrm{d}\mathbf{r},$$

where $V_{\text{in}}$ and $V_{\text{ex}}$ are, respectively, the volume occupied by the electrons inside a cell and by the volume outside a cell.

Sequentially using the transformation $(\nabla\phi)^2 = \nabla(\phi\nabla\phi) - \phi\Delta\phi$, equation (5.2) and continuity conditions (5.5), we obtain

$$W'_{\text{el}} = -\frac{1}{8\pi} \left[ \oint_{r=r_{\text{p}}} \phi'_{\text{in}} \frac{\partial\phi'_{\text{in}}}{\partial r} r^2 \mathrm{d}\Omega + \frac{1}{r_{\text{D}}^2} \int_{r_{\text{p}}}^{r_{\text{c}}} (\phi'_{\text{in}})^2 r^2 \mathrm{d}r \right],$$

or, in dimensionless variables (2.5) and using the explicit form (A.3),

$$W'_{\rm el} = -\frac{1}{8\pi}\sum_{n=0}^{\infty}\left\{[L_i(n)a_i(1,n)]\cdot[L_j(n)b_j(1,n)] + \frac{1}{\lambda^2}\int_1^{\varsigma}[L_i(n)a_i(\tilde{r},n)]^2\,\tilde{r}^2{\rm d}\tilde{r}\right\}\frac{|D_{n0}|^2}{r_{\rm p}^{2n+1}}. \qquad \text{(B.1)}$$

Here, the summation over indexes $i, j = 1, 2$ occurs.

## References

1. Dusty Plasmas: Physics, Chemisty and Technological Impacts in Plasma Processing, Bouchoule A. (Ed.), Wiley, Chichester, 1999.
2. Shukla P.K., Mamun A.A., Introduction to Dusty Plasma Physics, IOP Publishing, London, 2002.
3. Vladimirov S.V., Ostrikov K., Samarian A.A., Physics and Applications of Complex Plasmas, Imperial College, London, 2005.
4. Lieberman M.A., Lichtenberg A. J., Principles of Plasma Discharges and Materials Processing, 2nd Edition, Wiley-Interscience, New Jersey, 2005.
5. Fortov V., Iakubov I., Khrapak A., Physics of Strongly Coupled Plasma, Oxford University, New York, 2007.
6. Lightning: Principles, Instruments and Applications, Betz H.D., Schumann U., Laroche P. (Eds.), Springer, New York, 2008.
7. Tsytovich V.N., Morfill G.E., Vladimirov S.V., Thomas H.M., Elementary Physics of Complex Plasmas, Lecture Notes in Physics, Springer, Berlin Heidelberg, 2008; doi:10.1007/978-3-540-29003-2.
8. Complex and Dusty Plasmas: From Laboratory to Space, Series in Plasma Physics, Fortov V.E.., Morfill G.E. (Eds.), CRC Press, New York, 2009;
9. Introduction to Complex Plasmas, Springer Series on Atomic, Optical, and Plasma Physics, Bonitz M., Horing N., Ludwig P. (Eds.), Springer, Berlin Heidelberg, 2010; doi:10.1007/978-3-642-10592-0.
10. Tsytovich V.N., Phys. Usp., 1997, **40**, No. 1, 53; doi:10.1070/PU1997v040n01ABEH000201 [Usp. Fiz. Nauk, 1997, **167**, No. 1, 57; doi:10.3367/UFNr.0167.199701e.0057].
11. Smirnov B.M., Phys. Usp., 2000 **43**, No. 5, 453; doi:10.1070/PU2000v043n05ABEH000722 [Usp. Fiz. Nauk, 2000, **170**, No. 5, 495 (in Russian); doi:10.3367/UFNr.0170.200005b.0495].
12. Shukla P.K., Phys. Plasmas, 2001, **8**, No. 5, 1791; doi:10.1063/1.1343087.
13. Fortov V.E., Khrapak A.G., Khrapak S.A., Molotkov V.I., Petrov O.Ph., Phys. Usp., 2004 **47**, No. 5, 447; doi:10.1070/PU2004v047n05ABEH001689 [Usp. Fiz. Nauk, 2004, **174**, No. 5, 495 (in Russian); doi:10.3367/UFNr.0174.200405b.0495].
14. Shukla P.K., Eliasson B., Rev. Mod. Phys., 2009, **81**, No. 1, 25; doi:10.1103/RevModPhys.81.25.
15. Morfill G.E., Ivlev A.V., Rev. Mod. Phys., 2009, **81**, No. 4, 1353; doi:10.1103/RevModPhys.81.1353.
16. Chaudhuri M., Ivlev A.V., Khrapak S.A., Thomasa H.M., Morfilla G.E., Soft Matter, 2011, 7, No. 4, 1287; doi:10.1039/c0sm00813c.
17. Mishagli D.Yu., Ukr. J. Phys., 2012, **57**, No. 8, 824; arXiv:1212.0143.
18. Tsytovich V.N., Phys. Usp., 2007, **50**, No. 4, 409; doi:10.1070/PU2007v050n04ABEH006290 [Usp. Fiz. Nauk, 2007, **177**, No. 4, 427 (in Russian); doi:10.3367/UFNr.0177.200704l.0427].
19. Aerov A.A., Khokhlov A.R., Potemkin I.I., J. Chem. Phys., 2012, **136**, 014504; doi:10.1063/1.3670016.
20. Kulinskii V.L., Malomuzh N.P., Phys. Rev. E, 2003, **67**, 011501; doi:10.1103/PhysRevE.67.011501.
21. Landau L.D., Lifshits E.M., Statistical Physics, Pergamon, Oxford, 1980.
22. Bondarev V.N., Phys. Rev. E, 2005, **71**, 051102; doi:10.1103/PhysRevE.71.051102.

# 21

# Application of Levin's transformations to virial series

C.C.F. Florindo, A.B.M.S. Bassi*

Institute of Chemistry, University of Campinas — UNICAMP, 13083–970 Campinas, Brazil

A new method of estimating high-order virial coefficients for fluids composed of equal three-dimensional rigid spheres is proposed. The predicted $B_{11}$ and $B_{12}$ values are in good agreement with reliable estimates previously reported. A new application of the Levin's transformations is developed, as well as a new way of using Levin's transformations is suggested. For the virial series of packing factor powers, this method estimates the $B_{13}$ value near 173.

**Key words:** *Levin's transformations, virial series, rigid spheres*

## 1. Introduction

The virial series is extremely important for obtaining accurate state equations, because it expands the compressibility factor of a fluid through a power series of an adequate variable. Using the packing factor, $\eta$,

$$Z = 1 + \sum_{i=2}^{\infty} B_i \eta^{(i-1)}, \tag{1.1}$$

where $Z$ is the compressibility factor and $B_i$ is the virial coefficient of order $i$. In the late nineteenth century, van der Waals [1], Boltzmann [2], and van Laar [3], analytically calculated the virial coefficients $B_2$, $B_3$, and $B_4$, of a gas formed by equal three-dimensional spherical rigid particles. So far, there are no analytical expressions to calculate the coefficients after $B_4$, even considering such simple particles. Thus, the best values for higher-order coefficients are obtained by numerical calculations of the Mayer's functions [4].

Using Mayer's functions, in 1953 the coefficient $B_5$ was obtained by Rosenbluth and Rosenbluth [5]. Subsequently, the coefficients $B_6$ and $B_7$ were calculated by Ree and Hoover [6], and $B_8$ by van Rensburg [7], and by Vlasov and You [8]. $B_9$ was obtained by Labík and collaborators [9] in 2005, and $B_{10}$ by Clisby and McCoy in 2006 [10]. Calculations of coefficients after $B_{10}$ were not performed, on account of the huge increase in the number of Mayer's diagrams and integrals to be analyzed. Thus, the coefficients subsequent to $B_{10}$ are estimated (see estimates in [11]).

There are some methods reported in the literature for extrapolating the values of virial coefficients to orders higher than the tenth. Among them, stand out the Padé approximants, the maximum entropy approximation, the density functional method, the series of continuous exponential, molecular dynamics and the differential approximation method. All these methods are considered to be very plausible. Nevertheless, unproved assumptions on the mathematical behavior of the series are imposed in their applications. Indeed, it is not even proved whether the 3D rigid sphere virial series does converge for all physically significant $\eta$ values, or does not. Then, to avoid such assumptions is a desirable aim.

Slow convergent or even divergent series frequently appear in problems involving the evaluation of integrals, solutions of differential equations, perturbation theory and others [12]. Moreover, in many scientific problems, the series permits the computation of a small number of terms, which is not sufficient

* E-mail: bassi@iqm.unicamp.br

to obtain the required accuracy. In this context, the sequence transformations play an essential role, since they accelerate the convergence of the series without the need to compute higher-order terms [13].

In 1973, David Levin [14] introduced new sequence transformations, which improved the convergence of slowly convergent series. Moreover, this method is particularly suitable for the summation of strongly divergent series. According to Smith and Ford [15], who compared the performance of several linear and nonlinear series transformations, the Levin-type ones are probably the most powerful and versatile convergence accelerators ever known. Baram and Luban [16] were the first to demonstrate the applicability of the Levin's transformations to the virial expansions of hard discs and rigid spheres through estimates for $B_7$ [17]. In recent years, many applications of Levin-type transformations have been reported in the literature, though the focus has been mainly in the field of quantum physics (see, for example, [18–23]).

In this work, the Levin's transformations are used to estimate the $B_{11}$, $B_{12}$, and $B_{13}$, virial coefficients for gases composed of 3D equal rigid spheres, assuming the known values of the coefficients up to $B_{10}$. That is, a completely new methodology for estimating virial coefficients is proposed, and an unused way of using Levin's transformations is suggested. This work is organized into four more sections. In section 2, the Levin's transformations are briefly described, highlighting their mathematical structure. In section 3, the methodology used to estimate the virial coefficients is presented. In section 4, the obtained estimates are indicated and compared to those reported in the literature. Finally, in section 5 the results are commented.

## 2. Levin's transformations

In this section, the mathematical background of the Levin's acceleration method is summarized. However, for a more detailed mathematical description of the method, references [24] and [25] are suggested. The Levin's sequence transformations are applicable to the model sequence

$$s_r = s + \omega_r \sum_{j=1}^{k} c_j/(r+\gamma)^{j-1}, \qquad k, r \in \mathbb{N}, \tag{2.1}$$

where $k$ represents the order of the transformation, $\gamma$ is an arbitrary parameter which may not be a negative integer, $\omega_r$ is the remainder estimate and $s$ is the limit of the sequence when it converges, or the antilimit if it diverges. The convergence or divergence of the sequence depends on the behavior of $\omega_r$, for $r \to \infty$.

In equation (2.1) there are $k+1$ unknown quantities, that is, the limit or antilimit $s$ and the $k$ linear coefficients $c_1, \dots, c_k$. Thus, $k+1$ sequence elements $s_r, \dots, s_{r+k}$, and the corresponding remainder estimates $w_r, \dots, w_{r+k}$, are required for determining $s$. Evidently, imposing some kind of remainder estimate, as the three ones considered by Levin, most of the sequences do not follow this model sequence. However, in many cases for which $n \geqslant N$, where $N \in \mathbb{N}$, the sequence can be considered of $k$th order, namely

$$s_r = s_{kn} + \omega_r \sum_{j=1}^{k} c_{jn}/(r+\gamma)^{j-1}, \qquad n \leqslant r \leqslant n+k, \qquad n \geqslant N, \tag{2.2}$$

where $s = \lim_{n\to\infty} s_{kn}$. Thus, $s$ is both the limit of the sequences $\{s_r\}_{r=1}^{\infty}$ and $\{s_{kn}\}_{n=N}^{\infty}$, whose convergence was accelerated relatively to $\{s_r\}_{r=1}^{\infty}$, or even created if $s$ is an antilimit of $\{s_r\}_{r=1}^{\infty}$.

According to Cramer's rule, the general Levin's transformation is

$$\mathfrak{L}_k^{(n)}(\gamma, s_r, \omega_r) = \frac{\begin{vmatrix} s_n & \dots & s_{n+k} \\ \omega_n & \dots & \omega_{n+k} \\ \vdots & \ddots & \vdots \\ \omega_n/(\gamma+n)^{k-1} & \dots & \omega_{n+k}/(\gamma+n+k)^{k-1} \end{vmatrix}}{\begin{vmatrix} 1 & \dots & 1 \\ \omega_n & \dots & \omega_{n+k} \\ \vdots & \ddots & \vdots \\ \omega_n/(\gamma+n)^{k-1} & \dots & \omega_{n+k}/(\gamma+n+k)^{k-1} \end{vmatrix}}. \tag{2.3}$$

If the sequence elements satisfy equation (2.1), then the Levin's general sequence transformation is exact, i.e. $\mathfrak{L}_k^{(n)}(\gamma, s_r, \omega_r) = s$. But if they satisfy equation (2.2), then

$$\mathfrak{L}_k^{(n)}(\gamma, s_r, \omega_r) = s_{kn}. \tag{2.4}$$

As a ratio of two determinants, the Levin's transformation is unsuitable for practical applications involving reliable evaluations of large order determinants. Therefore, alternative expressions are commonly employed. For example, considering the Vandermonde determinant, the equation (2.3) can be rewritten

$$\mathfrak{L}_k^{(n)}(\gamma, s_r, \omega_r) = \frac{\sum_{j=0}^{k}(-1)^j \binom{k}{j} \left(\frac{\gamma+n+j}{\gamma+n+k}\right)^{k-1} \frac{s_{(n+j)}}{\omega_{(n+j)}}}{\sum_{j=0}^{k}(-1)^j \binom{k}{j} \left(\frac{\gamma+n+j}{\gamma+n+k}\right)^{k-1} \frac{1}{\omega_{(n+j)}}}, \qquad k, r, n \in \mathbb{N}. \tag{2.5}$$

According to [24], the Levin's transformation should work very well for a given sequence $\{s_r\}$ if the sequence $\{\omega_r\}$ of the remainder estimates is chosen in such a way that $\omega_r$ is proportional to the dominant term of an asymptotic expansion of the remainder

$$\varpi_r = s_r - s = \omega_r \left[c + O\left(r^{-1}\right)\right], \qquad r \to \infty. \tag{2.6}$$

However, $\omega_r$ is not determined by this asymptotic condition, so that it is possible to find a variety of sequences $\{\omega_r\}$ of the remainder estimates for a given sequence $\{s_r\}$. Thus, the practical problem that arises is how to find the sequence of the remainder estimates.

Based on purely heuristic arguments, Levin suggested three kinds of the remainder estimates, $\omega_r$, for sequences of partial sums

$$s_r = \sum_{i=1}^{r} a_i, \qquad r \in \mathbb{N}. \tag{2.7}$$

In the case of alternating partial sums, $s_r$, Levin suggested

$$\omega_r = a_r, \qquad r \in \mathbb{N}. \tag{2.8}$$

Substituting this relationship in equation (2.5), the Levin's $t$ transformation is obtained,

$$t_k^{(n)}(\gamma, s_r) = \frac{\sum_{j=0}^{k}(-1)^j \binom{k}{j} \left(\frac{\gamma+n+j}{\gamma+n+k}\right)^{k-1} \frac{s_{(n+j)}}{a_{(n+j)}}}{\sum_{j=0}^{k}(-1)^j \binom{k}{j} \left(\frac{\gamma+n+j}{\gamma+n+k}\right)^{k-1} \frac{1}{a_{(n+j)}}}. \tag{2.9}$$

In the case of a sequence of partial sums, $s_r$, satisfying a logarithmic convergence, i.e.,

$$\lim_{r\to\infty} \frac{s_r + 1 - s}{s_r - s} = 1, \tag{2.10}$$

Levin suggested

$$\omega_r = a_r(\gamma + r), \qquad r \in \mathbb{N}, \tag{2.11}$$

which being substituted into equation (2.5) produces the Levin's $u$ transformation

$$u_k^{(n)}(\gamma, s_r) = \frac{\sum_{j=0}^{k}(-1)^j \binom{k}{j} \frac{(\gamma+n+j)^{k-2}}{(\gamma+n+k)^{k-1}} \frac{s_{(n+j)}}{a_{(n+j)}}}{\sum_{j=0}^{k}(-1)^j \binom{k}{j} \frac{(\gamma+n+j)^{k-2}}{(\gamma+n+k)^{k-1}} \frac{1}{a_{(n+j)}}}. \tag{2.12}$$

Finally, Levin also suggested

$$\omega_r = \frac{a_r a_{r+1}}{a_r - a_{r+1}}, \qquad r \in \mathbb{N}, \tag{2.13}$$

which corresponds to the Levin's $v$ transformation

$$v_k^{(n)}(\gamma, s_r) = \frac{\sum_{j=0}^{k}(-1)^j \binom{k}{j} \left(\frac{\gamma+n+j}{\gamma+n+k}\right)^{k-1} \frac{a_{(n+j)}-a_{(n+j+1)}}{a_{(n+j)}a_{(n+j+1)}} s_{n+j}}{\sum_{j=0}^{k}(-1)^j \binom{k}{j} \left(\frac{\gamma+n+j}{\gamma+n+k}\right)^{k-1} \frac{a_{(n+j)}-a_{(n+j+1)}}{a_{(n+j)}a_{(n+j+1)}}}. \tag{2.14}$$

Other Levin-type transformations are reported in the literature (see [25]), but only those originally proposed by Levin are listed above, and are used in this work.

## 3. Methodology

A Levin's transformation can be applied to a well-defined sequence to obtain a new sequence that presents a better convergence than the original one. Still, in this work, the sequence is not completely defined, and the supposition that some Levin's transformation can change the values of lower-order terms to the values of higher-order terms of the same sequence is tested for the virial series. Indeed, the virial series defined by equation (1.1) is a sequence of partial sums in accordance with equation (2.7), i.e.,

$$s_r = \sum_{i=1}^{r} a_i, \quad r \in \mathbb{N}, \quad \text{and} \quad Z = \lim_{r\to\infty} s_r, \tag{3.1}$$

where $a_i = B_i\eta^{i-1}$ and $B_1 = 1$. In this case, $s_r$ is the value of the virial series truncated at the term proportional to $\eta^{r-1}$, whose coefficient is $B_r$. Then, it is supposed that the application of the Levin's transformation $\mathfrak{L}_k^{(n)}$ to $\{s_r\}_{r=1}^{\infty}$ produces, according to equation (2.4),

$$\mathfrak{L}_k^{(n)}(\gamma, s_r, \omega_r) = s_{kn} = s_{n+k}, \tag{3.2}$$

i.e., the $n$th element of the sequence $\{s_{kn}\}_{n=N}^{\infty}$ is equal to the $(n+k)$th element of the sequence $\{s_r\}_{r=1}^{\infty}$, for all $n \geqslant N$. Thus,

$$\mathfrak{L}_k^{(n)}(\gamma, s_r, \omega_r) = \sum_{i=1}^{n+k} B_i\eta^{i-1}. \tag{3.3}$$

As already mentioned, the values of coefficients are precisely known up to $B_{10}$. But, for equations (2.9) and (2.12), suppose that the first unknown term is

$$a_{n+k} = B_{n+k}\eta^{n+k-1} \quad \text{for all} \quad 2 \leqslant n+k \leqslant 10. \tag{3.4}$$

Then, using (3.3) in equations (2.9) or (2.12), $a_{n+k} = B_{n+k}\eta^{n+k-1}$ can be found for all $n+k$ in $2 \leqslant n+k \leqslant 10$. Analogously, for equation (2.14), suppose that the first unknown term is

$$a_{n+k+1} = B_{n+k+1}\eta^{n+k} \quad \text{for all} \quad 2 \leqslant n+k \leqslant 9. \tag{3.5}$$

Using the equation (3.3) in (2.14), $a_{n+k+1} = B_{n+k+1}\eta^{n+k}$ can be found for all $n+k$ in $2 \leqslant n+k \leqslant 9$. In any case, for a given value of $\eta$, the corresponding virial coefficient can be calculated and compared with the values reported in the literature (table 1). It is worthwhile noting that the choice of the values of coefficients reported on table 1 has been made arbitrarily. Indeed, more recent references could be used, such as [26].

In general, the estimation of a virial coefficient can be obtained from several representations of the same Levin's transformation, as shown in table 2. The representations whose virial coefficients values do not deviate more than 1% from the corresponding values in table 1 are used to estimate higher-order coefficients. The representations do not provide good estimates for the coefficients of the order less than $B_5$, while for higher orders, acceptable values are found. This behavior stems from the lack in information supplied to the representations by the virial series truncated on the terms of the order smaller than the fourth, so that a minimum number of the known terms of the series is required.

The methodology is based on determining simple functions $\eta = f(i)$ ($i$ is the index of $B_i$) by using the optimal $\eta$ values which correspond to the best estimates of coefficients from $B_5$ to $B_{10}$. These functions

**Table 1.** The virial coefficients $B_i$.

| $i$ | [9] | [10] |
|---|---|---|
| 1 | 1 | 1 |
| 2 | 4 | 4 |
| 3 | 10 | 10 |
| 4 | 18.3647684 | 18.364768 |
| 5 | 28.2245 ± 0.00010 | 28.2245 ± 0.0003 |
| 6 | 39.81550 ± 0.00036 | 39.81507 ± 0.00092 |
| 7 | 53.3413 ± 0.0016 | 53.34426± 0.00368 |
| 8 | 68.540± 0.010 | 68.538± 0.018 |
| 9 | 85.80± 0.080 | 85.813± 0.085 |
| 10 | ... | 105.77± 0.39 |

**Table 2.** The representations which estimate the terms $a_{n+k} = B_{n+k}\eta^{n+k-1}$, $2 \leqslant n+k \leqslant 10$, by equations (2.9) (transformation $t$) or (2.12) (transformation $u$), and $a_{n+k+1} = B_{n+k+1}\eta^{n+k}$, $2 \leqslant n+k \leqslant 9$, by equation (2.14) (transformation $v$, disregarding the last line in the table).

| $i$ | Levin's representations, $\mathfrak{L}_k^n(\gamma, s_r, \omega_r)$ |
|---|---|
| 2 | $\mathfrak{L}_1^1$ |
| 3 | $\mathfrak{L}_2^1, \mathfrak{L}_1^2$ |
| 4 | $\mathfrak{L}_3^1, \mathfrak{L}_2^2, \mathfrak{L}_1^3$ |
| 5 | $\mathfrak{L}_4^1, \mathfrak{L}_3^2, \mathfrak{L}_2^3, \mathfrak{L}_1^4$ |
| 6 | $\mathfrak{L}_5^1, \mathfrak{L}_4^2, \mathfrak{L}_3^3, \mathfrak{L}_2^4, \mathfrak{L}_1^5$ |
| 7 | $\mathfrak{L}_6^1, \mathfrak{L}_5^2, \mathfrak{L}_4^3, \mathfrak{L}_3^4, \mathfrak{L}_2^5, \mathfrak{L}_1^6$ |
| 8 | $\mathfrak{L}_7^1, \mathfrak{L}_6^2, \mathfrak{L}_5^3, \mathfrak{L}_4^4, \mathfrak{L}_3^5, \mathfrak{L}_2^6, \mathfrak{L}_1^7$ |
| 9 | $\mathfrak{L}_8^1, \mathfrak{L}_7^2, \mathfrak{L}_6^3, \mathfrak{L}_5^4, \mathfrak{L}_4^5, \mathfrak{L}_3^6, \mathfrak{L}_2^7, \mathfrak{L}_1^{(8)}$ |
| 10 | $\mathfrak{L}_9^1, \mathfrak{L}_8^2, \mathfrak{L}_7^3, \mathfrak{L}_6^4, \mathfrak{L}_5^5, \mathfrak{L}_4^6, \mathfrak{L}_3^7, \mathfrak{L}_2^8, \mathfrak{L}_1^9$ |

are obtained both by interpolating the five or six optimal $\eta$ values themselves, and by interpolating their variations (optimal $\eta$ value for $B_6$ less optimal $\eta$ value for $B_5$, and so forth). The mathematical structures of such functions are determined by using the Mathematica computer program, version 8.0. Once the functions are known, they are used to estimate $B_{11}$ and $B_{12}$.

## 4. Estimates of the 11th, 12th and 13th virial coefficients

In this section, the best representations and the corresponding estimates of coefficients are presented. For the $t$ and $u$ transformations, the $t_3^n$ and $u_3^n$ representations, respectively, provide good estimates of coefficients, while for the $v$ transformation, the best estimates are obtained through the $v_2^n$ representations. For the $t_3^n$ representations, the optimal $\eta$ for estimating the coefficients from $B_5$ to $B_{10}$ approximately lie between 0.20 and 0.28, while they are approximately in the interval from 0.01 to 0.08 for the $u_3^n$ representations.

Using the $v_2^n$ representations, the optimal $\eta$ values are, approximately, in the interval from 0.40 to 0.78. This range is about five times broader than the other two ranges, yet a large $\eta$ variation for low $i$ values is not important, while the $\eta$ tendency to reduce its variation as the index $i$ increases is fundamental. Moreover, this range presents an upper bound about 5% greater than the physical one (the geometric maximum packing factor for rigid spheres is about 0.74). Nonetheless, the $v_2^n$ representations are retained for this work, because this physical restriction is irrelevant for the present mathematical purpose. Moreover, packing factors above its physical upper bound, and even above 1.0, are frequently considered in the literature.

### 4.1. T-type representations

Using the $t_3^{i-3}$ representations for $i = 5,6,\ldots,12$, the virial coefficients $B_i$ are estimated. Thus, to predict the coefficients $B_{11}$ and $B_{12}$, the $t_3^8$ and $t_3^9$ representations are respectively used. The optimal $\eta$ values for the $t_3^8$ and $t_3^9$ representations are established by using four trial functions, which are obtained by interpolating the optimal $\eta$ values corresponding to the coefficients from $B_5$ to $B_{10}$. These optimal $\eta$ values are $\eta = 0.2493$ for $B_5$, $\eta = 0.2591$ for $B_6$, $\eta = 0.2703$ for $B_7$, $\eta = 0.2425$ for $B_8$, $\eta = 0.2172$ for $B_9$, and $\eta = 0.2053$ for $B_{10}$. Table 3 shows the values of the obtained virial coefficients, and their percentage deviations from the values reported in the literature.

According to table 3, all $B_{12}$ estimates obtained from the interpolation of the optimal $\eta$ values themselves deviate more than 6% from the reported values. Thus, the estimates using the $\eta$ values from the functions of the optimal $\eta$ variations are the only ones providing virial coefficients close to those reported in the literature. Among the four functions, only the logarithmic and the straight-line functions provide deviations less than 3% for both $B_{11}$ and $B_{12}$. Comparing these values with those obtained by Padé approximants ($B_{11} = 128.6$ and $B_{12} = 155$) [27] one concludes that both methods lead to similar estimates. Therefore, this comparison, as well as the values reported in [9] and [10], lead to the values of $B_{11}$ and $B_{12}$ obtained from the logarithmic and the straight-line functions of the optimal $\eta$ variations.

**Table 3.** Values of the virial coefficients $B_{11}$ and $B_{12}$ estimated by the $t_3^8$ and $t_3^9$ representations, respectively. Percentage deviations from the values reported in the literature are also presented. $(\%)^a$ Percentage deviation from the $B_{11} = 129\pm2$ and $B_{12} = 155\pm10$ values reported by [9]. $(\%)^b$ Percentage deviation from the $B_{11} = 127.9$ and $B_{12} = 152.7$ values reported by [10].

| Functions | $\eta$-variation | | | | | | $\eta$-absolute | | | | | |
|---|---|---|---|---|---|---|---|---|---|---|---|---|
| | $B_{11}$ | $(\%)^a$ | $(\%)^b$ | $B_{12}$ | $(\%)^a$ | $(\%)^b$ | $B_{11}$ | $(\%)^a$ | $(\%)^b$ | $B_{12}$ | $(\%)^a$ | $(\%)^b$ |
| Logarithmic | 127.1 | 1.47 | 0.63 | 156.5 | 0.97 | 2.49 | 133.8 | 3.72 | 4.61 | 170.2 | 9.81 | 11.5 |
| Exponential | 131.3 | 1.78 | 2.66 | 164.9 | 6.39 | 7.99 | 131.1 | 1.63 | 2.50 | 165.5 | 6.77 | 8.38 |
| Straight-line | 126.2 | 2.17 | 1.33 | 153.5 | 0.97 | 0.52 | 130.9 | 1.47 | 2.35 | 165.1 | 6.52 | 8.12 |
| Potency | 126.8 | 1.71 | 0.86 | 147.4 | 4.90 | 3.47 | 133.6 | 3.57 | 4.46 | 169.9 | 9.61 | 11.3 |

Thus, the logarithmic and straight line functions of the optimal $\eta$ variations are also considered to estimate the value of $B_{13}$. Using the $t_3^{10}$ representation, the values 183.68 and 175.45 are respectively found. The value from the logarithmic function is very close to those estimated in [10] (181.19) and [11] (180.82), whereas the value obtained from the straight-line function is between those estimated in [28] (177.40) and [29] (171.28). Therefore, this comparison confirms that the logarithmic and straight-line functions can be used to find estimates of the optimal $\eta$ for high order coefficients.

### 4.2. U-type representations

Using the $u_3^{i-3}$ representations for $i = 5,6,\ldots,12$, the virial coefficients $B_i$ are also estimated. Thus, to predict the coefficients $B_{11}$ and $B_{12}$ the $u_3^8$ and $u_3^9$ representations are respectively used. The optimal $\eta$ values are 0.01714 for $B_6$, 0.06435 for $B_7$, 0.07235 for $B_8$, 0.07209 for $B_9$, and 0.07819 for $B_{10}$ (the 1% minimal deviation is not attained for $B_5$). Table 4 presents the values of the obtained virial coefficients.

Table 4 shows that the $\eta$ values from the logarithmic, exponential and potency functions obtained by interpolating the optimal $\eta$ variations provide good estimates. In the case of a straight-line function, one can also note a good estimate, but only for $B_{11}$. Considering the $\eta$ values obtained from functions of the optimal $\eta$ values themselves, only the logarithmic function provides a $B_{12}$ value which deviates less than 6% from a reported value. This function also provides the best estimate for $B_{11}$. However, imposing the smaller percentage deviations as a criterion, the logarithmic, exponential and potency functions obtained by interpolating the optimal $\eta$ variations are selected to estimate high-order coefficients.

The $B_{13}$ values estimated by using the exponential and potency functions of the optimal $\eta$ variations are 190.03 and 190.80, respectively. It is impossible to obtain an estimate of $B_{13}$ from the logarithmic function, because the $u_3^{10}$ representation does not support the supplied optimal $\eta$ value. The estimated

**Table 4.** Values of the virial coefficients $B_{11}$ and $B_{12}$ estimated by the $u_3^8$ and $u_3^9$ representations, respectively. Percentage deviations from the values reported in the literature are also presented.

| Functions | $\eta$-variation | | | | | | $\eta$-absolute | | | | | |
|---|---|---|---|---|---|---|---|---|---|---|---|---|
| | $B_{11}$ | $(\%)^a$ | $(\%)^b$ | $B_{12}$ | $(\%)^a$ | $(\%)^b$ | $B_{11}$ | $(\%)^a$ | $(\%)^b$ | $B_{12}$ | $(\%)^a$ | $(\%)^b$ |
| Logarithmic | 127.3 | 1.32 | 0.47 | 150.3 | 3.03 | 1.57 | 131.0 | 1.55 | 2.42 | 162.8 | 5.03 | 6.61 |
| Exponential | 128.9 | 0.08 | 0.78 | 156.6 | 1.03 | 2.55 | 132.9 | 3.02 | 3.91 | 170.9 | 10.3 | 11.9 |
| Straight-line | 126.2 | 2.17 | 1.33 | 221.5 | 42.9 | 45.1 | 132.4 | 2.64 | 3.52 | 167.9 | 8.32 | 9.95 |
| Potency | 129.1 | 0.08 | 0.94 | 156.9 | 1.23 | 2.75 | 131.5 | 1.94 | 2.82 | 164.5 | 6.13 | 7.73 |

values are above those obtained by [10, 11, 28, 29]. However, they are close to the values estimated in [30] (190.82) and [31] (185±10).

### 4.3. V-type representations

In the last test, the $v_2^8$ and $v_2^9$ representations are used to estimate the coefficients $B_{11}$ and $B_{12}$. The optimal $\eta$ values are 0.4010 for $B_5$, 0.5445 for $B_6$, 0.6507 for $B_7$, 0.7089 for $B_8$, 0.7464 for $B_9$, and 0.7783 for $B_{10}$. Table 5 presents the estimated coefficients $B_{11}$ and $B_{12}$, as well as their deviations from previously reported values. This table clearly indicates that the $\eta$ values corresponding to the potency and exponential functions of the optimal $\eta$ variations, and the logarithmic interpolation of the optimal $\eta$ values themselves, provide the best estimates for the coefficients $B_{11}$ and $B_{12}$. Using the $v_2^{10}$ representation and the potency function of the variations, which shows the lowest deviations, the coefficient $B_{13}$ is estimated. The value obtained is 172.65. This estimated value is lower than those presented in [10, 11, 30, 31], but it is placed between those in [28] and [29].

**Table 5.** Values of the virial coefficients $B_{11}$ and $B_{12}$ estimated by the $v_2^8$ and $v_2^9$ representations, respectively. Percentage deviations from the values reported in the literature are also presented.

| Functions | $\eta$-variation | | | | | | $\eta$-absolute | | | | | |
|---|---|---|---|---|---|---|---|---|---|---|---|---|
| | $B_{11}$ | $(\%)^a$ | $(\%)^b$ | $B_{12}$ | $(\%)^a$ | $(\%)^b$ | $B_{11}$ | $(\%)^a$ | $(\%)^b$ | $B_{12}$ | $(\%)^a$ | $(\%)^b$ |
| Logarithmic | 131.5 | 1.94 | 2.81 | 166.5 | 7.42 | 9.04 | 127.8 | 0.93 | 0.08 | 149.1 | 3.81 | 2.36 |
| Exponential | 130.8 | 1.39 | 2.27 | 162.2 | 4.52 | 6.09 | 116.6 | 10.1 | 8.84 | 106.8 | 31.1 | 30.1 |
| Straight-line | 135.1 | 4.73 | 5.63 | 187.6 | 21.1 | 22.9 | 119.2 | 7.60 | 6.80 | 117.1 | 24.5 | 23.3 |
| Potency | 128.9 | 0.08 | 0.78 | 152.5 | 1.61 | 0.13 | 124.9 | 3.18 | 2.35 | 137.7 | 11.2 | 9.82 |

## 5. Conclusions

Only representations with $k = 2$ for the Levin's $v$ transformation, and $k = 3$ for the $t$ and $u$ Levin's transformations, are acceptable. This interesting result guided the choice of the representations used to estimate the high order virial coefficients. Moreover, the estimates also depend on the dimensionless $\eta$ value. This is an expected dependence, because the Levin's convergence accelerators modify the terms from the series, not just the coefficients included within these terms.

The $B_{11}$ and $B_{12}$ values have been confirmed in the literature, by using distinctive methodologies, which imply different assumptions on the mathematical behavior of the virial series. Thus, such values are reliable. Meanwhile, the values reported in the literature are in accordance with some representations of the Levin's transformations, highlighting these transformations usefulness in the prediction of virial coefficients.

Should a Levin's transformation be able to change the values of lower order terms to the values of higher-order terms of the considered virial series, then it is expected that: (i) such ability is enhanced for high order terms of the series, which are favored by high $\eta$ values, and (ii) the $\eta$ value variation caused by substituting $i+1$ for $i$ decreases as $i$ increases ($\eta$ tends to some unique value for high-order coefficients). Accordingly, note that, in table 3 to 5, the functions are not selected to achieve the best fit to the $B_5$ to $B_{10}$ values in table 1 (for instance, functions with more than two parameters are not used), but to test the asymptotic behavior of the functions. An interesting result is that all the functions selected in section 4 by comparing the obtained values to previously reported ones, except the straight line function, are asymptotic to the $i$ axis, that is, they satisfy the above condition (ii).

The $\eta$ values corresponding to the $u$ transformation refer to the gaseous state, whose description is accurate enough by using only the low order terms of the series, and the $\eta$ values corresponding to the $t$ transformation concern the liquid state, whose description is accurate enough by using the low and medium order terms of the series. However, the $\eta$ values corresponding to the $v$ transformation refer to the overcooled liquid and vitreous states, whose accurate descriptions also demand the high order terms of the series. Note that the $v$ transformation is the only one producing good results not exclusively from the interpolation of the $\eta$ values variations, but also from the interpolation of the optimal $\eta$ values themselves, confirming the above condition (i). Thus, the $v$ transformation, which is the least specific one among those originally presented by Levin, is preferable. As a consequence of this choice, the $B_{13}$ value near 173 is proposed in this work. Note that the 13 terms long, virial series for rigid spheres developed in this work should be useful in describing the repulsive pressure of overcooled liquids and vitreous transitions. However, this series will not reproduce crystallization, which involves drastic changes in entropy and volume.

## Acknowledgement

Financial support by the Brazilian federal government agency — CAPES — and the University of Campinas — UNICAMP — are acknowledged.

## References

1. van der Waals J.D., Proc. K. Ned. Akad. Wet., 1899, **1**, 138.
2. Boltzmann L., Versl. Gewone Vergad. Afd. Natuurkd., K. Ned. Akad. Wet., 1899, 7, 484.
3. van Laar J.J., Proc. K. Ned. Akad. Wet., 1899, **1**, 273.
4. Mayer J.E., Mayer M.G., Statistical Mechanics, John Wiley, New York, 1940.
5. Rosenbluth M.N., Rosenbluth A.W., J. Chem. Phys., 1955, **23**, 356; doi:10.1063/1.1741967.
6. Ree F.H., Hoover W.G., J. Chem. Phys., 1967, **46**, 4181; doi:10.1063/1.1840521
7. van Rensburg E.J., J. Phys. A: Math. Gen., 1993, **26**, 4805; doi:10.1088/0305-4470/26/19/014.
8. Vlasov A.Y., You X.M., Masters A.J., Mol. Phys., 2002, **100**, 3313; doi:10.1080/00268970210153754.
9. Labík S., Kolafa J., Malijevský A., Phys. Rev. E, 2005, **71**, 021105; doi:10.1103/PhysRevE.71.021105.
10. Clisby N., McCoy B.M., J. Stat. Phys., 2006, **122**, 15; doi:10.1007/s10955-005-8080-0.
11. Tian J.X., Jiang H., Gui Y.X., Mulerlo A., Phys. Chem. Chem. Phys., 2009, **11**, 11213; doi:10.1039/b915002a.
12. Roy D., Bhattacharya R., Bhowmick S., Comp. Phys. Commun., 1979, **93**, 159; doi:/10.1016/0010-4655(95)00106-9.
13. Brezinski C., Zaglia M.R., Extrapolation Methods: Theory and Practice (Studies in Computational Mathematics), North Holland, Amsterdam, 1991.
14. Levin D., Int. J. Comput. Math., 1973, **3**, 371; doi:10.1080/00207167308803075.
15. Smith D.A., Ford W.F., Math. Comput., 1982, **38**, 481; doi:10.2307/2007284.
16. Baram A., Luban M., J. Phys. C: Solid State Phys., 1979, **12**, L659; doi:10.1088/0022-3719/12/17/005.
17. Erpenbeck J.F., Luban M., Phys. Rev. A, 1985, **32**, 2920; doi:10.1103/PhysRevA.32.2920.
18. Bhagat V., Bhattacharya R., Roy D., Comput. Phys. Commun., 2003, **155**, 7; doi:10.1016/S0010-4655(03)00294-7.
19. Bouferguene A., Fares M., Phys. Rev. E, 1994, **49**, 3462; doi:10.1103/PhysRevE.49.3462.
20. De Prunelé E., Int. J. Quantum Chem., 1997, **63**, 1079; doi:10.1002/(SICI)1097-461X(1997)63:6<1079::AID-QUA2>3.0.CO;2-U.
21. Jetzke S., Broad J.T., Int. J. Mod. Phys. C, 1991, **2**, 377; doi:10.1142/S0129183191000524.
22. King F.W., Int. J. Quantum Chem., 1999, **72**, 93; doi: 10.1002/(SICI)1097-461X(1999)72:2<93::AID-QUA2>3.0.CO;2-#.
23. Scott T.C., Aubert-Frécon M., Andrae D., Appl. Algebr. Eng. Commun. Comput., 2002, **13**, 233; doi:10.1007/s002000200100.
24. Weniger E.J., Comput. Phys. Rep., 1989, **10**, 189; doi:10.1016/0167-7977(89)90011-7.
25. Homeier H.H.H., J. Comput. Appl. Math., 2000, **122**, 81; doi:10.1016/S0377-0427(00)00359-9.
26. Wheatley R.J., Phys. Rev. Lett., 2013, **110**, 200601; doi:10.1103/PhysRevLett.110.200601.

27. Guerrero A.O., Bassi A.B.M.S., J. Chem. Phys., 2008, **129**, 044509; doi:10.1063/1.2958914.
28. Clisby N., McCoy B. M., Pramana-J. Phys., 2005, **64**, 775; doi:10.1007/BF02704582.
29. Santos A., López de Haro M., J. Chem. Phys., 2009, **130**, 214104; doi:10.1063/1.3147723.
30. Kolafa J., Labík S., Malijevsky A., Phys. Chem. Chem. Phys., 2004, **6**, 2335; doi:10.1039/b402792b.
31. Ončák M., Malijevský A., Kolafa J., Labík S., Condens. Matter Phys., 2012, **15**, 23004; doi:10.5488/CMP.15.23004.

# 22

# A simple model potential for hollow nanospheres

K. Köksal[1], M. Öncan[1,2], Bulent Gönül[2], Besire Gönül[2]

[1] Physics Department, Bitlis Eren University, 13000, Bitlis, Turkey

[2] Department of Engineering Physics, Gaziantep University, 27100, Gaziantep, Turkey

A new model potential is introduced to describe the hollow nanospheres such as fullerene and molecular structures and to obtain their electronic properties. A closed analytical solution of the corresponding treatment is given within the framework of supersymmetric perturbation theory.

**Key words:** *electronic structure, model potential, analytical solution*

## 1. Introduction

Nanostructures have received great interest nowadays because of their importance in solid state technology and for medical purposes [1–5]. Specifically, quantum dots have been studied to strongly confine and control the electrons in a few nanometers of the volume [6, 7]. Nevertheless, semiconductor quantum dots can be produced using high technological growth devices while the production of metallic quantum dots or clusters also requires advanced technology [8]. However, molecular nanospheres can be obtained more easily (e.g., by chemical synthesis) compared with those of metal and semiconductor quantum dots [9].

Molecular nanoparticles as in the case of metal clusters [10, 11] can be produced in the shape of hollow spheres. In these structures, there is a limited radial motion of the electrons. As an example of hollow nanospheres, we can consider the fullerene molecule which is stable and consists of 60 carbon atoms. The derivatives of fullerene are $C_{70}$,$C_{84}$, $C_{540}$ [12, 13]. Furthermore, $B_{80}$ is a spherical molecule in which one uses the boron atoms instead of carbon [14].

For complex molecular systems which include more than ten atoms, the best way of electronic structure analysis is to use some ab-initio [15, 16] or semi-empirical techniques [17, 18]. Although these are known as realistic calculations, they have time-consuming procedures and give only numerical results. However, since the molecules like fullerene have spherical symmetry, if the motion of an electron can be described in only one dimension (which is radial direction), the whole motion will be described due to the spherical symmetry. Appropriate model potentials can describe the motion of the electrons in radial direction [19–23]. The angular motion of an electron is described by well known spherical harmonics. In other words, the related spherical symmetric one-dimensional radial Schrödinger equation analytically yields the spectra of interest, from which one can readily observe the behaviours and the physics behind the system considered, unlike its corresponding numerical results.

Within the context, we remind that the attractive Gaussian potential

$$V(\lambda; r) = -\gamma e^{-\lambda r^2} \tag{1.1}$$

is very important in modelling the nucleon-nucleon scattering [24] and quantum dots with single or more electrons [25–27], impurity [28, 29] and excitons [30]. In equation (1.1), $\gamma$ shows the height of the radial central potential, $\lambda$ is related to the dot radius ($R$) in such a way that $\lambda = 1/R^2$.

We have recently obtained a closed form of the eigenvalues of this potential [31] and shown that the obtained results are very accurate, particularly for low principle quantum numbers. For completeness, in this article we aim to suggest that such interaction profile, apparently with a plausible modification, should be also suitable for the investigation of hollow nanospheres.

The paper is organized as follows. The next section briefly discusses the model potential. Theory section includes the calculation procedure and the discussion on the results obtained. Some concluding remarks and outlook are given in the final section.

## 2. Model potential

The electronic properties of a spherical symmetric molecular cluster can be obtained by modelling this structure. So far, many of the studies have been performed in the framework of numerical calculations. In previous studies regarding fullerene and endohedral structures, the spherical potential which models the structures is presented in different forms [19–23]. The square potential having the following form

$$V(r) = \begin{cases} -V_0 & \text{if } r_0 \leqslant r \leqslant r_i + d, \\ 0, & \text{if } r < r_0, \end{cases} \tag{2.1}$$

$d$: the thickness of the wall,

$r_0$: inner radius

has been introduced by [20, 21] to describe the fullerene. To describe an atom inside $C_{60}$, Dolmatov et. al. has introduced the Woods-Saxon potential [19] as follows:

$$V(r) = \left[\frac{2V_0}{1+\exp\left(\frac{r_0-r}{v}\right)}\right]_{r \leqslant r_0+\Delta/2} + \left[\frac{2V_0}{1+\exp\left(\frac{r-r_0-r}{v}\right)}\right]_{r > r_0+\Delta/2}, \tag{2.2}$$

where $v$ is a parameter related to the additional atom, $r_0$, $V_0$ and $\Delta$ are potential parameters. Lin et. al. have used a power-exponential potential which has the following form [23]

$$V(\lambda; r) = -\gamma \exp\left[-\lambda (r - r_0)^p / w^p\right], \tag{2.3}$$

where $r_0$ and $w$ indicate the radius and and thickness of the hollow cage. Nascimento et. al. have introduced the gaussian-type potential for modelling the fullerene-type structures as follows [22]

$$V(\lambda; r) = -\gamma \exp\left[-\lambda (r - r_0)^2\right]. \tag{2.4}$$

Using the mentioned potentials, it is possible to obtain electronic structures of hollow spheres. As an example, the last potential has the shape as seen in figure 1. Since the parameter $r_0$ breaks the symmetry and makes the potential discontinuous, the solution of the potential can be performed only by using the numerical techniques. An analytical solution for a potential, $r_0$, should be zero. Therefore, a spherical hollow nanostructure may be described by the combination of an attractive Gaussian potential with an additional term as follows

$$V(\lambda; r) = -\gamma e^{-\lambda r^2} + \frac{\beta}{r^2}, \tag{2.5}$$

where the second term including $\beta$ (this should be positive) is very similar to the barrier term due to the spherical symmetry which is $\ell(\ell+1)/r^2$ [32]. In the case of $\beta = 0$, the potential in equation (2.5) can describe a solid nanoparticle such as metal clusters, while for $\beta \neq 0$, a hollow nanosphere can be described.

Figure 2 shows the potential introduced in equation (2.5). Due to the additional term ($\beta/r^2$), the electrons in the system will be pushed to make a motion in the regions far from the center of the sphere. The increase in the $\beta$ will cause a decrease of the motion area of the electrons. Therefore, this potential can describe a group of spherical symmetric molecular structures and different types of hollow nanospheres.

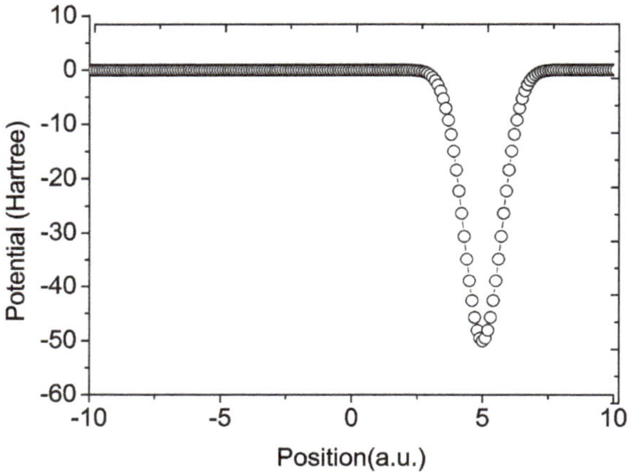


**Figure 1.** Attractive gaussian potential which has a formula given by equation (2.4). As can be understood from the shape, the potential is not symmetric and unsuitable for analytical solution. Here, the $a.u.$ indicates atomic unit and should be considered as Bohr radius.

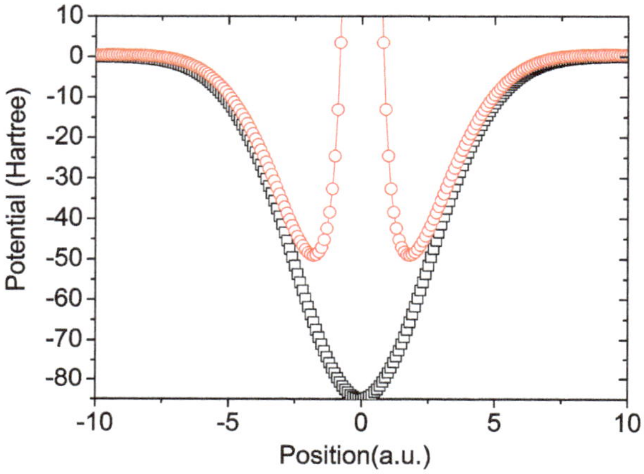


**Figure 2.** (Color online) The model potential to describe a hollow nanosphere. The red circles indicate the exponential part of equation (2.5) and the black squares show the second part in equation (2.5).

## 3. Theory

As seen from equation (2.5), the potential is constructed with two parts. $\beta/r^2$ can be introduced into the barrier term $-\ell(\ell+1)/r^2$ following the previous study by Aygun et al. [32]. In this case, we can describe a new parameter $\ell'(\ell'+1) = -\ell(\ell+1)+\beta$ which results in:

$$\ell' = \frac{1}{2}\left(-1+\sqrt{1+4\ell+4\ell^2+4\beta}\right). \tag{3.1}$$

The attractive Gaussian potential part in equation (2.5) can be expanded as follows

$$V(\lambda;r) = -\gamma+\gamma\lambda r^2 - \frac{\gamma\lambda^2}{2}r^4+\ldots, \tag{3.2}$$

where $\gamma\lambda r^2$ indicates the well known harmonic oscillator potential where $\gamma\lambda = m\omega^2/2$. The role of the terms other than $-\gamma+\gamma\lambda r^2$ is to modify the potential and they may be referred to as modification potentials. The normalized wavefunction of a harmonic oscillator is

$$\chi_{n\ell'}(r) = \sqrt{\frac{2^{n+2\ell'+3+1/2}}{\sqrt{\pi}}\frac{n!}{(2n+2\ell'+1)!!}}\left(\frac{\lambda\gamma m}{2\hbar^2}\right)^{\frac{2\ell'+3}{8}} \mathrm{e}^{-\sqrt{\frac{\lambda\gamma m}{2\hbar^2}}r^2} L_n^{\ell'+1/2}\left(\sqrt{2\lambda\gamma m}r^2/\hbar\right)r^{\ell'+1}, \tag{3.3}$$

**Table 1.** The eigenvalues of the potential for $\ell = 0$ and for different parameters in atomic units. The value of the $\gamma$ is 4 Hartree. One should note that $n$ is a radial quantum number.

| $n$ | $\lambda = 0.0125$ | | | $\lambda = 0.025$ | | | $\lambda = 0.0125$ | | |
|---|---|---|---|---|---|---|---|---|---|
| | $\beta = 0$ | $\beta = 2.5$ | $\beta = 5$ | $\beta = 0$ | $\beta = 2.5$ | $\beta = 5$ | $\beta = 0$ | $\beta = 2.5$ | $\beta = 5$ |
| 0 | -2.9404 | -1.79436 | -1.20055 | -3.37972 | -2.69002 | -2.32524 | -3.55892 | -3.06275 | -2.79796 |
| 1 | -1.94631 | -1.30392 | -0.83482 | -2.78255 | -2.3891 | -2.09817 | -3.12967 | -2.84443 | -2.63231 |
| 2 | -1.00933 | -0.5972 | -0.24876 | -2.20674 | -1.94968 | -1.73062 | -2.71159 | -2.52366 | -2.36291 |
| 3 | -0.12217 | | 0.4336 | -1.65066 | -1.46551 | -1.2973 | -2.30409 | -2.16762 | -2.0433 |
| 4 | | | | -1.11281 | -0.9707 | -0.8373 | -1.90659 | -1.80102 | -1.70174 |
| 5 | | | | -0.59181 | -0.47792 | -0.36901 | -1.51854 | -1.43333 | -1.35173 |
| 6 | | | | -0.08637 | | | -1.13943 | -1.06854 | -0.99989 |
| 7 | | | | | | | -0.76877 | -0.70845 | -0.6496 |
| 8 | | | | | | | -0.40612 | -0.35388 | -0.30266 |
| 9 | | | | | | | -0.05102 | -0.00515 | |

where $m$ and $\hbar$ are electron mass and Planck's constant, respectively, $L_n^{\ell'+1/2}\left(\sqrt{2\lambda\gamma m}r^2/\hbar\right)$ is associated Laguerre polynomial.

Here, the corresponding eigenvalue and superpotential terms of the harmonic oscillator term can be written from the literature [33]

$$\begin{aligned} W_{n=0}(r) &= -\frac{\hbar}{\sqrt{2m}}\frac{\chi'_{n\ell'}(r)}{\chi_{n\ell'}(r)} = -\frac{\hbar}{\sqrt{2m}}\frac{\ell'+1}{r} + \sqrt{\gamma\lambda}\, r, \\ E_{n,\ell'} &= \hbar\sqrt{\frac{2\gamma\lambda}{m}}(2n+\ell'+3/2). \end{aligned} \tag{3.4}$$

Perturbative wavefunctions, energies and superpotentials corresponding to the modification potentials are as follows:

$$\begin{aligned} \Delta V(r;\epsilon) &= \sum_{k=1}^{\infty} \epsilon^k \Delta V^{\{k\}}(r), \\ \Delta W_{n\ell'}(r;\epsilon) &= \sum_{k=1}^{\infty} \epsilon^k \Delta W_{n\ell'}^{\{k\}}(r), \\ \Delta E_{n\ell'}(\epsilon) &= \sum_{k=1}^{\infty} \epsilon^k \Delta E_{n\ell'}^{\{k\}}, \end{aligned} \tag{3.5}$$

where $k$ indicates the perturbation order. If the unknown perturbed wavefunction is $R_P(r)$, the Schrödinger equation can be written as follows:

$$-\frac{\hbar^2}{2m}\left[\frac{\chi''_{n\ell'}}{\chi_{n\ell'}} + \frac{R''_P(r)}{R_P(r)} + 2\frac{\chi'_{n\ell'}}{\chi_{n\ell'}}\frac{R'_P(r)}{R_P(r)}\right] = V_H + V_P + E_{n\ell'} + \Delta E_{n\ell'}. \tag{3.6}$$

Skipping the details of the calculation procedure of the theory to our previous work [31], the total energy eigenvalues can be written as follows:

$$\mathrm{Im}_{n,\ell} = \left(2n + \frac{-1+\sqrt{1+4\ell+4\ell^2+4\beta}}{2} + \frac{3}{2}\right)\sqrt{\frac{\gamma\lambda\hbar^2}{2m}} - \gamma e^{-\left[2n+\frac{1}{2}\left(-1+\sqrt{1+4\ell+4\ell^2+4\beta}\right)+3/2\right]\sqrt{\frac{\lambda\hbar^2}{2\gamma m}}}, \tag{3.7}$$

where $\mathrm{Im}_{n,\ell} = E_{n\ell'} + \Delta E_{n\ell'}$ $\left(\ell' \to \left[-1+\sqrt{1+4\ell+4\ell^2+4\beta}\right]/2\right)$. Therefore, we obtain a simple analytical form of the energy eigenvalues.

## 4. Results and discussion

The change of the parameters in the potential of equation (2.5) refers to the change of the kind of the molecular structure. As can be seen from figure 3, varying the value of $\lambda$ parameter it is possible to control

**Table 2.** The eigenvalues of the potential for $\ell = 1$ and for different parameters in atomic units. The value of the $\gamma$ is 4 Hartree.

| $n$ | | $\lambda = 0.0125$ | | | $\lambda = 0.025$ | | | $\lambda = 0.0125$ | |
|---|---|---|---|---|---|---|---|---|---|
| | $\beta = 0$ | $\beta = 2.5$ | $\beta = 5$ | $\beta = 0$ | $\beta = 2.5$ | $\beta = 5$ | ; $\beta = 0$ | $\beta = 2.5$ | $\beta = 5$ |
| 0 | –1.00933 | | | –2.20674 | –1.56434 | –1.22363 | –2.71159 | –2.24051 | –1.98875 |
| 1 | –0.12217 | | | –1.65066 | –1.28332 | –1.01121 | –2.30409 | –2.03295 | –1.83114 |
| 2 | | | | –1.11281 | –0.87216 | –0.66683 | –1.90659 | –1.7277 | –1.57458 |
| 3 | | | | –0.59181 | –0.41799 | –0.25992 | –1.51854 | –1.38844 | –1.26988 |
| 4 | | | | –0.08637 | | | –1.13943 | –1.03865 | –0.94383 |
| 5 | | | | | | | –0.76877 | –0.68731 | –0.60927 |
| 6 | | | | | | | –0.40612 | –0.33824 | –0.27249 |

the size of the potential well. Furthermore, using the $\beta$ parameter different from zero, it is possible to split the gaussian potential into two parts in the positive and the negative regions. The increase in $\beta$ causes a shrink in the potential.

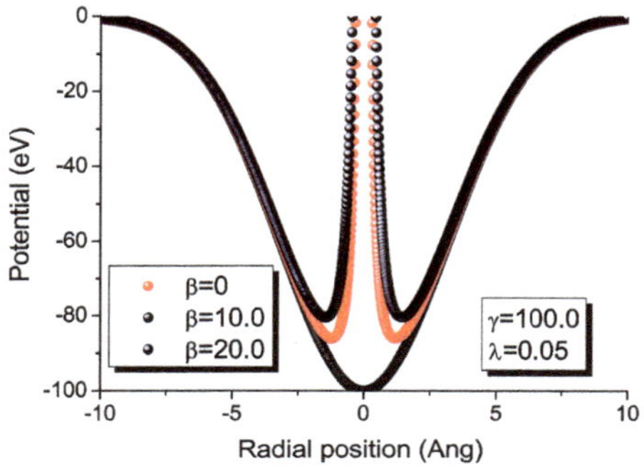


$\lambda = 0.05$

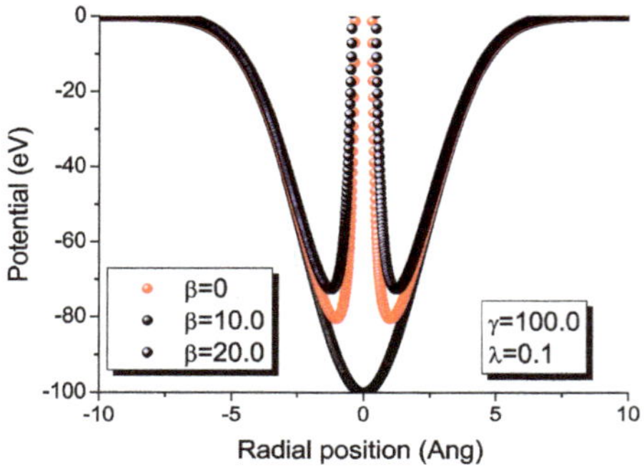


$\lambda = 0.1$

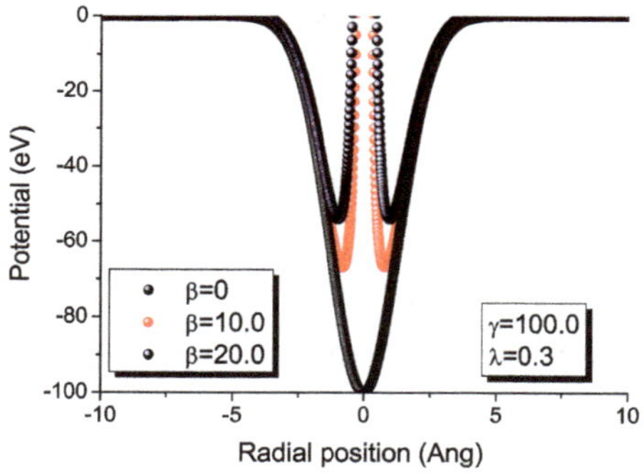


$\lambda = 0.3$

**Figure 3.** (Color online) The change of the model potential with the change of the parameters. Each potential can be considered as the one describing hollow nanoclusters. Here, it should be noted that the units are taken as eV and Angstrom to provide a possibility to compare the model parameters with the structure parameters from literature.

Therefore, the parameters make it possible to produce different model potentials corresponding to different hollow nanospheres. Tables 1 and 2 show the eigenvalues for different parameters. From the tables, it can be concluded that bound state energy values can be obtained from equation (1.1).

## 5. Conclusions

We have obtained an analytical form for the energy eigenvalues of fullerene-type hollow nanospheres by using the technique from our previous study. A simple analytical expression (but not an exact one) is a closed form for the bound states of a modified attractive Gaussian potential by using supersymmetric perturbation theory. The analytical expression can be used to see the treatment of the energy values. Furthermore, we have shown that this potential can be a good candidate to model semiconductor quantum dots and to reveal the electronic and optical properties of the system.

## References

1. Köksal K., Pavlyukh Ya., Berakdar J., J. Sci. Technol. Bitlis Eren University, 2011, **1**, No. 1, 4.
2. Martz L., SciBX: Science-Business eXchange, 2013, **6**, 11; doi:10.1038/scibx.2013.256.
3. Li C., Li L., Keates A.C., Oncotarget, 2012, **3**, No. 4, 365.
4. Ahn S., Jung S.Y., Lee S.J., In: Proceedings of the Conference "APS March Meeting" (Baltimore, Maryland, 2013), B. Am. Phys. Soc., 2013, **58**.
5. Salata O.V., J. Nanobiotechnology, 2004, **2**, No. 1, 3; doi:10.1186/1477-3155-2-3.
6. Akgül S., Şahin M., Köksal K., J. Lumin., 2012, **132**, No. 7, 1705; doi:10.1016/j.jlumin.2012.02.012.
7. Şahin M., Phys. Rev. B, 2008, **77**, 045317; doi:10.1103/PhysRevB.77.045317.
8. Harrison P., Quantum Wells, Wires and Dots: Theoretical and Computational Physics of Semiconductor Nanostructures, Wiley, The University of Leeds, 2011.
9. Guldi D., Martin N., Fullerenes: From Synthesis to Optoelectronic Properties, Developments in Fullerene Science, Springer, Dordlecht, 2002.
10. Ma H., Gao F., Liang W., J. Phys. Chem. C, 2012, **116**, No. 2, 1755; doi:10.1021/jp2094092.
11. Podsiadlo P., Kwon S.G., Koo B., Lee B., Prakapenka V.B., Dera P., Zhuravlev K.K., Krylova G., Shevchenko E.V., J. Am. Chem. Soc., 2013, **135**, No. 7, 2435; doi:10.1021/ja311926r.
12. Pavlyukh Y., Berakdar J., Chem. Phys. Lett., 2009, **468**, No. 4, 313; doi:10.1016/j.cplett.2008.12.051.
13. Pavlyukh Y., Berakdar J., Phys. Rev. A, 2010, **81**, 042515; doi:10.1103/PhysRevA.81.042515.
14. Zhao J., Wang L., Li F., Chen Z., J. Phys. Chem. A, 2010, **114**, No. 37, 9969; doi:10.1021/jp1018873.
15. Scuseria G.E., Science, 1996, **271**, No. 5251, 942; doi:10.1126/science.271.5251.942.
16. Gonzalez Szwacki N., Sadrzadeh A., Yakobson B.I., Phys. Rev. Lett., 2007, **98**, No. 16, 166804; doi:10.1103/PhysRevLett.98.166804.
17. Halac E., Burgos E., Bonadeo H., Chem. Phys. Lett., 1999, **299**, No. 1, 64; doi:10.1016/S0009-2614(98)01226-3.
18. Chen Z., Thiel W., Chem. Phys. Lett., 2003, **367**, No. 1, 15; doi:10.1016/S0009-2614(02)01660-3.
19. Dolmatov V.K., King J.L., Oglesby J.C., J. Phys. B: At. Mol. Opt. Phys., 2012, **45**, No. 10, 105102; doi:10.1088/0953-4075/45/10/105102.
20. George J., Varma H.R., Deshmukh P.C., Manson S.T., J. Phys. B: At. Mol. Opt. Phys., 2012, **45**, No. 18, 185001; doi:10.1088/0953-4075/45/18/185001.
21. Grum-Grzhimailo A.N., Gryzlova E.V., Strakhova S.I., J. Phys. B: At. Mol. Opt. Phys., 2011, **44**, No. 23, 235005; doi:10.1088/0953-4075/44/23/235005.
22. Nascimento E.M., Prudente F.V., Guimares M.N., Maniero A.M., J. Phys. B: At. Mol. Opt. Phys., 2011, **44**, No. 1, 015003; doi:10.1088/0953-4075/44/1/015003.
23. Lin C.Y., Ho Y.K., J. Phys. B: At. Mol. Opt. Phys., 2012, **45**, No. 14, 145001; doi:10.1088/0953-4075/45/14/145001.
24. Buck B., Friedrich H., Wheatley C., Nucl. Phys. A, 1977, **275**, No. 1, 246; doi:10.1016/0375-9474(77)90287-1.
25. Gomez S., Romero R., Cent. Eur. J. Phys., 2009, **7**, 12; doi:10.2478/s11534-008-0132-z.
26. Adamowski J., Sobkowicz M., Szafran B., Bednarek S., Phys. Rev. B, 2000, **62**, 4234; doi:10.1103/PhysRevB.62.4234.
27. Coden D.S.A., Gomez S.S., Romero R.H., J. Phys. B: At. Mol. Opt. Phys., 2011, **44**, No. 3, 035003; doi:10.1088/0953-4075/44/3/035003.
28. Xie W., Superlattices and Microstruct., 2010, **48**, No. 2, 239; doi:10.1016/j.spmi.2010.04.015.
29. Lu L., Xie W., Hassanabadi H., Physica B, 2011, **406**, No. 21, 4129; doi:10.1016/j.physb.2011.07.063.
30. Hours J., Senellart P., Peter E., Cavanna A., Bloch J., Phys. Rev. B, 2005, **71**, 161306; doi:10.1103/PhysRevB.71.161306.
31. Köksal K., Phys. Scripta, 2012, **86**, No. 3, 035006; doi:10.1088/0031-8949/86/03/035006.
32. Aygun M., Bayrak O., Boztosun I., J. Phys. B: At. Mol. Opt. Phys., 2007, **40**, No. 3, 537; doi:10.1088/0953-4075/40/3/009.
33. Dabrowska J.W., Khare A., Sukhatme U.P., J. Phys. A: Math. Gen., 1988, **21**, No. 4, L195; doi:10.1088/0305-4470/21/4/002.

# Permissions

All chapters in this book were first published in CMP, by Institute for Condensed Matter Physics; hereby published with permission under the Creative Commons Attribution License or equivalent. Every chapter published in this book has been scrutinized by our experts. Their significance has been extensively debated. The topics covered herein carry significant findings which will fuel the growth of the discipline. They may even be implemented as practical applications or may be referred to as a beginning point for another development.

The contributors of this book come from diverse backgrounds, making this book a truly international effort. This book will bring forth new frontiers with its revolutionizing research information and detailed analysis of the nascent developments around the world.

We would like to thank all the contributing authors for lending their expertise to make the book truly unique. They have played a crucial role in the development of this book. Without their invaluable contributions this book wouldn't have been possible. They have made vital efforts to compile up to date information on the varied aspects of this subject to make this book a valuable addition to the collection of many professionals and students.

This book was conceptualized with the vision of imparting up-to-date information and advanced data in this field. To ensure the same, a matchless editorial board was set up. Every individual on the board went through rigorous rounds of assessment to prove their worth. After which they invested a large part of their time researching and compiling the most relevant data for our readers.

The editorial board has been involved in producing this book since its inception. They have spent rigorous hours researching and exploring the diverse topics which have resulted in the successful publishing of this book. They have passed on their knowledge of decades through this book. To expedite this challenging task, the publisher supported the team at every step. A small team of assistant editors was also appointed to further simplify the editing procedure and attain best results for the readers.

Apart from the editorial board, the designing team has also invested a significant amount of their time in understanding the subject and creating the most relevant covers. They scrutinized every image to scout for the most suitable representation of the subject and create an appropriate cover for the book.

The publishing team has been an ardent support to the editorial, designing and production team. Their endless efforts to recruit the best for this project, has resulted in the accomplishment of this book. They are a veteran in the field of academics and their pool of knowledge is as vast as their experience in printing. Their expertise and guidance has proved useful at every step. Their uncompromising quality standards have made this book an exceptional effort. Their encouragement from time to time has been an inspiration for everyone.

The publisher and the editorial board hope that this book will prove to be a valuable piece of knowledge for researchers, students, practitioners and scholars across the globe.

# List of Contributors


**L. Gálisová**
Department of Applied Mathematics and Informatics, Faculty of Mechanical Engineering, Technical University in Košice, Letná 9, 042 00 Košice, Slovak Republic

**I.N. Adamenko and E.K. Nemchenko**
V.N. Karazin Kharkiv National University, 4 Svobody Sqr., 61022 Kharkiv, Ukraine

**V. Holovatsky, O. Voitsekhivska and I. Bernik**
Chernivtsi National University, 2 Kotsiubynsky St., 58012 Chernivtsi, Ukraine

**B.D. Tinh, L.M. Thu and L.V. Hoa**
Department of Physics, Hanoi National University of Education, 136 Xuanthuy, Caugiay, Hanoi, Vietnam

**P.K. Mishra**
Department of Physics, DSB Campus, Kumaun University, Nainital-263 002 (Uttarakhand), India

**K. Haydukivska and V. Blavatska**
Institute for Condensed Matter Physics of the National Academy of Sciences of Ukraine, 1 Svientsitski St., 79011 Lviv, Ukraine

**M. Krasnytska**
Institute for Condensed Matter Physics of the National Academy of Sciences of Ukraine, 1 Svientsitskii St., 79011 Lviv, Ukraine
Institut Jean Lamour, CNRS/UMR 7198, Groupe de Physique Statistique, Universite de Lorraine, BP 70239, F-54506 Vandoeuvre-lés-Nancy Cedex, France

**O. Pizio**
Instituto de Química, Universidad Nacional Autonoma de México, Circuito Exterior, Ciudad Universitaria, 04510 México, D.F., México

**S. Sokołowski**
Department for the Modelling of Physico-Chemical Processes, Maria Curie-Sklodowska University, Gliniana 33, Lublin, Poland

**X.-G. Han and Y.-H. Ma**
Inner Mongolia Key Laboratory for Utilization of Bayan Obo Multi-Metallic Resources: Elected State Key Laboratory, Inmongolia Science and Technology University, Baotou 014010, China
School of Mathematics, Physics and Biology, Inmongolia Science and Technology University, Baotou 014010, China

**A. Patrykiejew and S. Sokołowski**
Department for the Modelling of Physico-Chemical Processes, Faculty of Chemistry, Maria Curie-Sklodowska University, 20031 Lublin, Poland

**P.A. Hlushak and M.V. Tokarchuk**
Institute for condensed Matter Physics of the National Academy of Sciences of Ukraine 1 Svientsitskii St., 79011 Lviv, Ukraine

**S.A. Bercha and V.M. Rizak**
Uzhgorod National University, 54 Voloshyn St., 88000 Uzhgorod, Ukraine

**V.I. Boichuk and R.Ya. Leshko**
Department of Theoretical Physics, Ivan Franko Drohobych State Pedagogical University, 3 Stryiska St., 82100 Drohobych, Ukraine

**H.O. Bingol**
Department of Computer Engineering, Bogazici University, 34342 Bebek, Istanbul, Turkey

**U. Çetin**
Department of Computer Engineering, Bogazici University, 34342 Bebek, Istanbul, Turkey
Department of Computer Engineering, Istanbul Gelisim University, 34310 Avcilar, Istanbul, Turkey

**L.I. Nazarov**
Physics Department, M.V. Lomonosov Moscow State University, 119991 Moscow, Russia

**S.K. Nechaev**
Université Paris-Sud/CNRS, LPTMS, UMR8626, Bât. 100, 91405 Orsay, France
P.N. Lebedev Physical Institute, RAS, 119991 Moscow, Russia

Department of Applied Mathematics, National Research University "Higher School of Economics", 101000 Moscow, Russia

**M.V. Tamm**
Physics Department, M.V. Lomonosov Moscow State University, 119991 Moscow, Russia
Department of Applied Mathematics, National Research University "Higher School of Economics", 101000 Moscow, Russia

**V. Dotsenko**
LPTMC, Universitė Paris VI, 75252 Paris, France
L.D. Landau Institute for Theoretical Physics, 119334 Moscow, Russia

**D.O. Kharchenko, V.O. Kharchenko and S.V. Kokhan**
Institute of Applied Physics, Nat. Acad. Sci. of Ukraine, 58 Petropavlivska St., 40000 Sumy, Ukraine

**P. Kostrobij and R. Tokarchuk**
National University "Lviv Polytechnic", 12 Bandera St., 79013 Lviv, Ukraine

**M. Tokarchuk**
National University "Lviv Polytechnic", 12 Bandera St., 79013 Lviv, Ukraine
Institute for Condensed Matter Physics of the National Academy of Sciences of Ukraine, Svientsitskii St., 79011 Lviv, Ukraine

**B. Markiv**
Institute for Condensed Matter Physics of the National Academy of Sciences of Ukraine, 1 Svientsitskii St., 79011 Lviv, Ukraine

**B.I. Lev**
Bogolyubov Institute for Theoretical Physics of the National Academy of Sciences of Ukraine, 14-b Metrolohichna St., 03680 Kyiv, Ukraine

**M.V. Kozachok**
Bogolyubov Institute for Theoretical Physics of the National Academy of Sciences of Ukraine, 14-b Metrolohichna St., 03680 Kyiv, Ukraine
Open International University of Human Development "Ukraine", 23 Lvivska St., Kyiv, Ukraine

**D.Yu. Mishagli**
Department of Theoretical Physics, Faculty of Physics, Mechnicov National University, 2 Dvoryanska St., 65026 Odessa, Ukraine
Institut für Theoretische Physik, Universität Leipzig, 10/11 Augustusplatz, 04109 Leipzig, Germany

**C.C.F. Florindo and A.B.M.S. Bassi**
Institute of Chemistry, University of Campinas—UNICAMP, 13083-970 Campinas, Brazil

**K. Köksal**
Physics Department, Bitlis Eren University, 13000, Bitlis, Turkey

**M. Öncan**
Physics Department, Bitlis Eren University, 13000, Bitlis, Turkey
Department of Engineering Physics, Gaziantep University, 27100, Gaziantep, Turkey

**Bulent Gönül and Besire Gönül**
Department of Engineering Physics, Gaziantep University, 27100, Gaziantep, Turkey

# Index

www.ingramcontent.com/pod-product-compliance
Lightning Source LLC
Chambersburg PA
CBHW082018190326
41458CB00010B/3228
* 9 7 8 1 6 3 2 3 8 6 6 1 8 *